DEAR INDY

DEAR INDY

A Father's Plea for Climate Action

SIMON WHALLEY

Contents

TWO

ECOLOGICAL COLLAPSE

THREE
HOW WE GOT INTO THIS MESS

FOUR

HOW TO GET OUT OF THIS MESS

Dear Indy: A Father's Plea for Climate Action by Simon Whalley

First Printing, 2022

www.dear-indy.com

Softcover ISBN 978-1-7396351-0-7
Hardback ISBN 978-1-7396351-2-1
E-Book ISBN 978-1-7396351-1-4

Cover by Luke Sandalls.

Dedicated to my son Indy,
and all those unable to act.

Dear Indy

I don't know where to begin with this, so I'll just get going. We don't have much time after all. It's strange to be writing you a book that you won't be able to read for quite some time, but I'm writing this for your future. I'm not sure how old you will be when you read this, but you aren't quite ready for it. It's 2022 and you are eleven. I have a few reasons for writing this book. The first is that I want you to know that I cared deeply about your future. I brought you into this world, and I feel it's my duty to do all I can to protect you in the present, and to ensure that I don't bequeath to you a planet spiralling out of control. I also want you to know that I tried my best to do all I could to avert the catastrophe rapidly unfolding around us. You won't see it happening around you. It's quite possible to not notice it, but you have to really try nowadays. Many people seem to be doing quite a good job of ignoring the situation. Unfortunately, or fortunately, whichever way you look at it, your dad cannot ignore it. The thought of what my generation, and those before us have done to cause this disaster consumes me. It's not possible for me to just carry on as normal and drive us closer to the precipice. You must know that there is something not quite right because you hear your mother and father talking about climate change, ecological disaster, societal breakdown, climate refugees, and the sixth extinction. You don't understand the implications of all this, and even if you did, you wouldn't be able to grasp just how severe our situation is. I'm sure that many adults don't understand how dire our predicament is either. I hope that's the case, because if it isn't, then it means they know, and just don't care. I have too much faith in humanity to allow myself to believe that to be true. That brings me to the second reason for writing this book. I hope that by putting all the facts down on paper; other people like me, might also start to worry about the multiple threats we currently face. More importantly than worrying, I hope that other parents will start to act in a manner appropriate for the disastrous situation our collective action has caused.

I want you to know what the scientists were telling us. They told us that if we did not act immediately, then our children's futures would be damaged irreparably. They warned us that there would be no second chance, no magic bullet, no superhero was coming to save us. They told us we only had a few years left to get moving. The race has begun and by the time you read this, you will know whether we got going or whether we buried our heads in the sand, and sold your future so we could carry on enjoying the present. Let me briefly introduce you to the world that was around us in 2022, when you were eleven years old.

Wildfires raged around the globe, record temperatures were recorded on every continent with tens of thousands of resulting deaths and ruined crops, the sea is washing away Pacific islands, cities are flooded at high tide, super strength storms batter coastlines, shake skyscrapers, and flood farmland and cities. Record cold snaps kill tens of thousands, bats drop from the sky, wildlife is wiped out at up to 10,000 times the normal rate, forest the size of a football field is cut down every second, insects disappear along with earth worms and our soil is only good for another 60 harvests (1). 100 million people are without a home around the world. In the world's richest nation, 650,000 people are homeless (2), but there are 18.9 million vacant houses (3). There are 300,000 people without a place to live in the U.K. (4), yet there are 216,000 empty homes (5). Ecosystems are dying, and yet we are encouraged to shop till we drop to keep the economy going. In the United States, the richest 0.1% earn more than 188 times as much as the bottom 90%, and poverty has remained stable while the income of the top 1% has doubled. The concentration of wealth is as high as it was during the 1929 stock market crash (6). Just eight men now have the same wealth as the poorest 50% of humans on planet Earth (7). Almost 700 million people do not have enough food to eat, while 30% of global food is wasted. The average American has $90,460 worth of debt while the richest American has $200 billion to spend on space travel (8). Only 1% of consumer items remain in use after six months (9), and the rest end up buried under the Earth or discarded as 12.7 million tonnes of plastic at sea (10). More people die from over-eating than under-eating (11). 25,000 humans die from starvation every day, while 70 billion farm animals consume 36% of the world's crops (12). The richest plan on escaping Earth to terraform Mars rather than preserve our beautiful home. This is the world in 2022, and everything I've listed above is happening now. In the following pages, I will explain to you how they are all interrelated.

Many people will say that this is hyperbole, that the situation isn't as bad as the scientists are making out. Some even believe that the climate crisis is a hoax and that climate scientists are all involved in a conspiracy to overthrow capitalism. If you are one of these people, I feel sorry for you. To experience life on a ball of rock floating around a burning star, thinking that everyone around you is a liar must not be very enjoyable. If you are one of these people, go straight to the Bringers of Death chapter and read that before you start the book. If you read this and still aren't convinced, this book probably isn't for you. I will make no effort to prove the legitimacy of the scientific evidence. I am not a scientist after all. I'm an English teacher who happened by chance to watch Al Gore's An Inconvenient Truth in 2007, and slowly over the next fifteen years, as I learned more and more about our perilous situation, I decided I had to focus my attention on this crisis. After you were born Indy, my focus intensified, and I felt I had to work harder on the problem. At times, I feel, like many I assume, that I'm too power-less to affect change. I overcome this simply by asking myself what the alternative of action is, and what the outcome of inaction will be. I also realize that ignoring the problems we face will result in those problems passing over to you. I don't think that's what a strong father would do, and it keeps me focussed. It was hard at first to speak out about the cli-mate crisis. No one wants to talk about it. People just want to be happy, and I understand this. I want to be happy too. But, not at the expense of your future. And hell, I hope you will know that I am happy. I love spending time with you and your mother. I love spending time with my friends. I love playing Fortnite with you, Star Wars not so much. I love throwing a ball at you. I love smacking the inflatable ball at your head in the pool as you laugh out loud. I love watching the cats fight. I love watching Odie (our dog) chase our five cat friends around the house. I love watching our bigger and more jealous dog, Lara, bully Odie. I love being outside. I love camping. I love surfing. I love kayaking. I love reading. All of these things make me happy. Being alive is amazing. I've been lucky enough to have lived in the Philippines, Taiwan, Vietnam, and Japan. I have swum with dolphins and sharks, walked through beautiful forests, swum in crystal clear lakes and rivers. Our planet is paradise. Unfortunately, it won't be paradise for much longer. In fact, it already isn't paradise for hundreds of millions of people, and tens of billions of non-human animals.

In the following pages, I will attempt to get you to understand why I am inextricably worried for your future. I will explain the multiple threats we face from the climate crisis, the rising temperatures, the rising seas, the vanishing fresh water, the threat of mass starvation, the refugee moral dilemma and the future wars. I will inform you of the sixth extinction and what it means for your future. I will show you how we got into this mess, and finally, I will give some solutions to help get us out. I wouldn't be much of a father if I just told you all this and gave you no hope. People must think that I'm a pessimist because I choose to focus on such depressing topics. I'm actually an optimist. That's why I focus on these problems. To ignore them will only ensure that they happen. Focussing on them is the only way out of this situation. And who knows, by focussing on them, we might just be able to make a better world for you, and your children, and the generations who come behind us.

"The world is a fine place, and worth fighting for."

ERNEST HEMINGWAY

ONE

Climate Catastrophe

L et's start with the basic problem. For this, we will have to go back to school: very briefly. The climate emergency is being caused by what is known as the Greenhouse Effect. It is so called because our Earth has an atmosphere of gasses surrounding it, and these gasses help to trap heat, just like the glass in a greenhouse traps heat and keeps the plants inside, warm at night. If we didn't have this gaseous atmosphere, our home would be freezing cold. As you'll see from the following diagram, natural greenhouse gasses like carbon dioxide (CO_2); methane (CH_4), and Nitrous Oxide (N_2O) that are in our atmosphere, trap photons of light as they are reflected out to space and these heat up our atmosphere. Since the agricultural revolution, humans have been burning forests which release CO_2 into the atmosphere, and this de-forested land is then less able to sequester CO_2 than when forested. In the industrial revolution, we started burning ancient rocks of coal, and oil and gas, and these gasses have been added to our atmosphere in massive amounts. The number of CO_2 parts per million (ppm) was around 283 ppm in 1800, and it now sits at 420 ppm. We are also eating more and more animals; especially cows and sheep, and these produce methane, lots of it. The amount of methane in our atmosphere has risen from 751 parts per billion in 1800 to more than 1,892 ppb today. Likewise, nitrous oxide has risen from 273 ppb in 1800 to 335 ppb in 2022. This is what is causing the record temperatures, record rainfall, record storms and record everything: climate related.

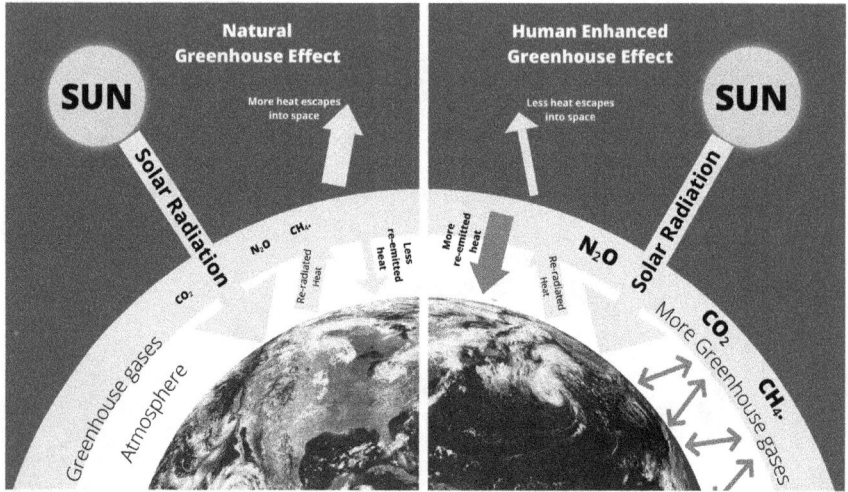

The climate crisis is without doubt the biggest challenge human beings have ever faced. In our 200,000-year history, individual civilizations have risen and fallen, but never have our societies been reliant on each other the way they are today. Be it food security, international space projects or the rare minerals used in the technology human ingenuity has developed; we use resources from all over our incredible planet, and what happens in one area affects other areas and the societies that have arisen there. This was not the case when the Roman Empire, Greenland Norse or Easter Islanders' civilizations collapsed. People in the Americas had no idea the Roman Empire even existed, and they certainly had no idea of its demise. Likewise, of the other civilizations.

Today we face the very real likelihood that a collapse in one area may lead to the collapse in others. Just as the United States worried of the threat of communism knocking down democratically elected governments from Russia, down through China, into Vietnam and the rest of Indo China before toppling Indonesia and finally Australia, we should all be aware that what happens to others will likely happen to us all. We are all on the same living planet and our planet has strict boundaries that cannot be escaped. As we have seen with COVID-19, we are all interconnected, and climate change is not some distant problem that will affect other people in the future. It is affecting us all now.

While our democratically elected leaders have chosen a distant date in the future to finally reach zero carbon emissions, it isn't the politicians who are in charge, it is the planet. Those who are most aware of the necessity for systemic change are not pushing for 2050 at all (13). They believe that this distant target simply allows us to carry on as normal, glibly hoping the technology exists somewhere down the line. This is an incredibly irresponsible risk, a risk taken by politicians who long ago ran out of ideas. The scientists are our bookmakers here and their odds are clear. If we want a 67% chance of avoiding 1.5°C (2.5°F) of warming, then we could safely afford to burn through 230 billion tonnes of CO_2 (230$GtCO_2$) from 2020. If we are feeling more adventurous with our children's futures and fancy a 50/50 flutter, then we could burn through 440 $GtCO_2$. At present we are emitting around 46 $GtCO_2$ annually so this means at present rates we have until 2027 with the 66% odds and around ten years if we fancy a 50/50 roll of the dice (14) (15). After that, we enter unchartered territory. Even after a calamitous 2020 in which economies shrank all around the world due to COVID-19, the accumulative CO_2 in our atmosphere continued to rise. After a brief respite for the planet, 2021 broke the record for greenhouse gas concentrations, ocean acidification, ocean heat and sea level rise. According to Former U.K. Chief Scientific Advisor and Special Envoy on Climate Change, Professor Sir David King, writing in Climate Dominoes: Tipping Point Risks for Critical Climate Systems:

> "This is a code red situation. No government is taking it seriously enough. We must urgently seek productive collaboration between subnational, national, and international bodies to do more to combat climate issues equitably, with determination and speed."

In the coming pages, an argument will be put forth that we should no longer trust our elected officials and we need to take matters into our own hands if our children are to have a safe and healthy future, free of conflict, famine, and mass starvation. First, we will look at how the climate is already being affected and what we can expect in the future. Then we will see the impact we have had on biodiversity and where we might be heading. Then we will go back in time to see how we got into this mess. Finally, we will look at some of the ways in which we can reverse our emissions and biodiversity loss at the same time as reducing inequality. The next decade is critical so let's start with the countdown to 2030.

Rolling the Dice -- the Cube on Regents' Plaza, c.1969.

Countdown to 2030

The most recent Intergovernmental Panel on Climate Change (IPCC) report stated global emissions must decline 45% by 2030. It described what failure to act will bring to our species, and our ability to maintain our civilizations. The Paris agreement required governments to keep warming to within 2°C (3.3°F) of pre-industrial times, although it was widely accepted that nations would endeavour to keep warming to no more than 1.5°C (2.5°F). The 2018 IPCC report explained the vast difference between just 0.5°C (0.84°F) of warming. Although increasing the temperature to 1.5°C (2.5°F) will exacerbate the storms, heat waves, droughts, flooding, and sea level rise that we are experiencing today, pushing the temperature to 2°C (3.3°F) will have many more negative effects, and possibly trigger a runaway climate catastrophe. It is estimated that ten million more people will be at risk from sea level rise should we move above 1.5°C (2.5°F), several hundred more million people will suffer from poverty, 50% more people will be exposed to water stress, the majority of corals will die, an area three times the size of Texas will experience thawing permafrost, 50% more species of plants and animals will lose half of their habitat, there will be declines in yields of maize, rice, wheat and potentially other cereal crops, particularly in sub-Saharan Africa, Southeast Asia, and Central and South America. At 1.5°C (2.5°F) of warming, extreme hot days in most regions will increase in temperature by 3°C (5°F) while 2°C (3.3°F) of warming will increase these hot days to 4°C (6.7°F). There will also be an increase in risk of insect-carried diseases like malaria and dengue fever (16).

When may we see 1.5°C (2.5°F) and 2°C (3.3°F) of warming? According to analysis from Carbon Brief, between 2030 and 2032 for 1.5°C (2.5°F) and around 2043 for 2°C (3.3°F) (17). In May 2022, the World Meteorological Organization (WMO) released a report that analysed data from ten countries and their findings were even more alarming. Over the past decade, their research found that the chances of any one year going temporarily above the 1.5°C (2.5°F) threshold was 20%, but their new assessment more than doubled the risk to 50% for the period 2022-2027 (18). Our leaders tell us that they wish to limit warming to 1.5°C (2.5°F) by getting to net zero by 2050, but the latest science is saying that we will reach 1.5°C (2.5°F) somewhere between 2022 and 2032. Either our elected leaders are insane, or they are massively incompetent.

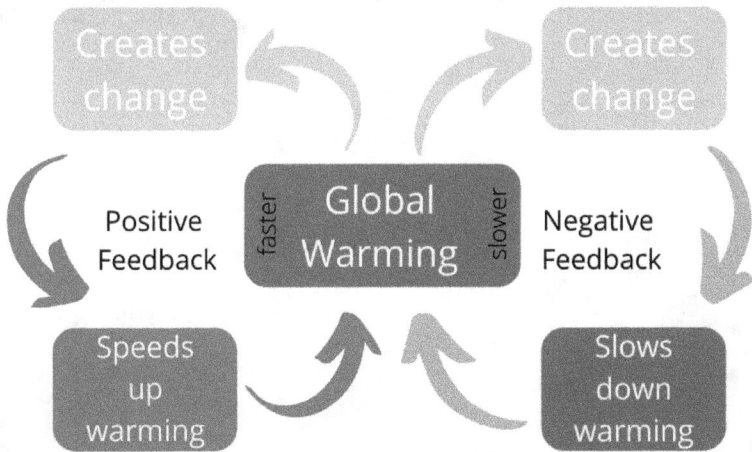

Climate feedbacks can be both negative and positive
U.K. Met Office

Some of the effects of not keeping warming to within 1.5°C (2.5°F) are listed above, however, the impacts listed here are just the tip of the proverbial iceberg. The IPCC report was a watered-down consensus, and the latest climate science warns of the dangers of climatic feedback loops that scientists predict could be triggered if warming reaches 2°C (3.3°F). Feedback loops are commonly known as vicious or virtuous circles, or cycles. They either accelerate or decelerate heating. Positive feedback loops accelerate heating whilst negative feedback loops decelerate heating.

Jamie Henn, co-founder of 350.org said, "*The scariest thing about the IPCC report is the fact that it's the watered down, consensus version. The latest science is much more terrifying.*" The Penn State University climate scientist Michael Mann agreed that the report was "*Overly conservative*" (19).

Durwood Zaelke, founder of the Institute for Governance and Sustainable Development went further, "*The IPCC report fails to focus on the weakest link in the climate chain: the self-reinforcing feedbacks which, if allowed to continue, will accelerate warming and risk cascading climate tipping points and runaway warming*" (19).

Feedback Loops and Tipping Points

The planet's climate systems are more complex than we understand so potential feedback loops and tipping points are plentiful, from the Amazon rainforest becoming a savannah, Greenland and Antarctic melting, and coral systems dying. Once tipping points have been crossed, climate systems can move suddenly from one state to another and the change can be irreversible with the potential to cause cascading events. That said, the positive feedback loop that has climate scientists most concerned about tipping points and runaway warming is the permafrost in the Arctic. As the name suggests, this soil is frozen solid, but the freeze is not, as the name suggests, actually permanent. The permafrost is roughly eighteen million sq. km and covers around a quarter of all exposed land in the northern hemisphere (land not covered in ice) With warming temperatures, this permafrost will thaw. Since the industrial revolution, humans have emitted around 1.5 trillion tonnes ($1500GtCO_2$) of carbon into the atmosphere (20). Under the frozen tundra lies up to four times this amount (21). Even if we stop warming tomorrow, there is a chance that this methane will leak out and very quickly quadruple the amount of CO_2 equivalent in the atmosphere. Even under models where CO_2 emissions are reduced, abrupt thawing of the permafrost, and subsequent methane emissions are still likely to happen. The unknown variables are how much will be emitted, and how fast.

This depends on what humans do in the coming years. The good news is that, although some methane is leaking from the Arctic, large scale emissions have yet to occur, meaning the Arctic is still acting as a carbon sink – it stores more carbon that it emits. The bad news is that the permafrost soils are beginning to thaw, and scientists weren't expecting that for another seventy years (22).

In 2019, scientists in Russia observed a 4.5-meter area of the East Siberian Sea boiling with methane bubbles. Here, the surrounding atmosphere has methane concentrations more than nine times the global average. On land, buildings atop the permafrost are beginning to crumble and rivers are rising and running faster, in turn washing away whole neighbourhoods (23). A year later, frozen methane deposits in the Laptev Sea were observed being released over a large area, off the same East Siberian coast. Although scientists claim the amount is not having a major impact on global warming, they are concerned that the process has been triggered. With the Arctic region warming twice as fast as the global average, the frozen methane deposits have been referred to as the *"sleeping giants of the carbon cycle"* (24).

In turn, the methane release will trigger another feedback loop in that more heat will be trapped inside the Earth's atmosphere, and more soil will thaw in response to warmer temperatures, and the cycle will reinforce itself further and further. Once the permafrost thaws, barring an absolute technological miracle, our climate will have tipped into a complete catastrophic spiral, that of which we can barely imagine. In financial terms, the costs are predicted to add $70 trillion to the next generation's climate invoice (25). Just because the IPCC report excluded the impacts of this feedback loop, it doesn't mean we should exclude it from our concern. In fact, it is this feedback loop that should drive our urgent path to zero carbon as the warming could be very sudden. The IPCC report did not consider the possibility of what is known as "abrupt thaw" and scientists researching the permafrost believe it needs to be built into future modelling if we are to produce realistic projections (26). If we don't act immediately, there will be huge ramifications for each and every one of us on planet Earth.

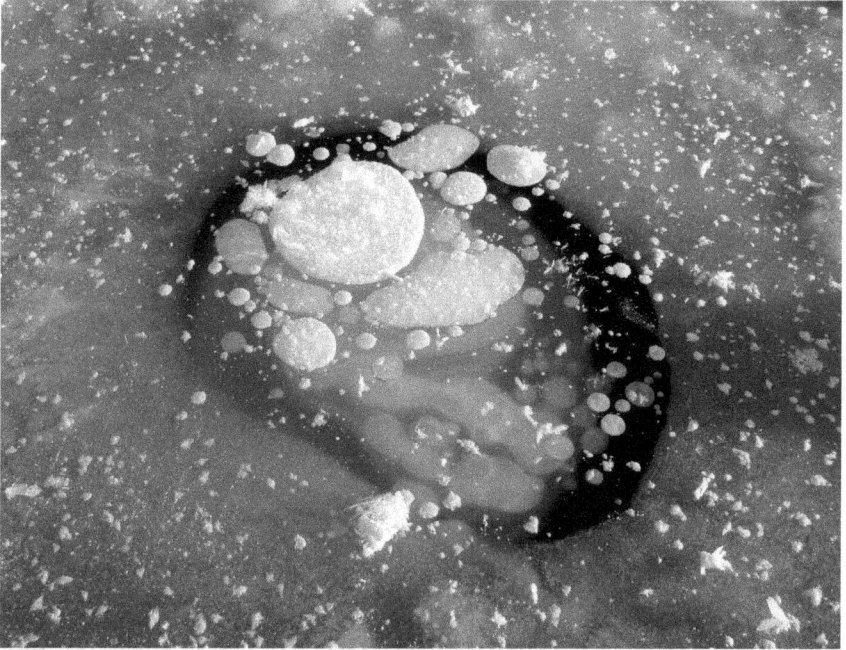

Frozen Methane Bubbles
by U.S. Geological Survey is licensed under CC BY-NC 2.0

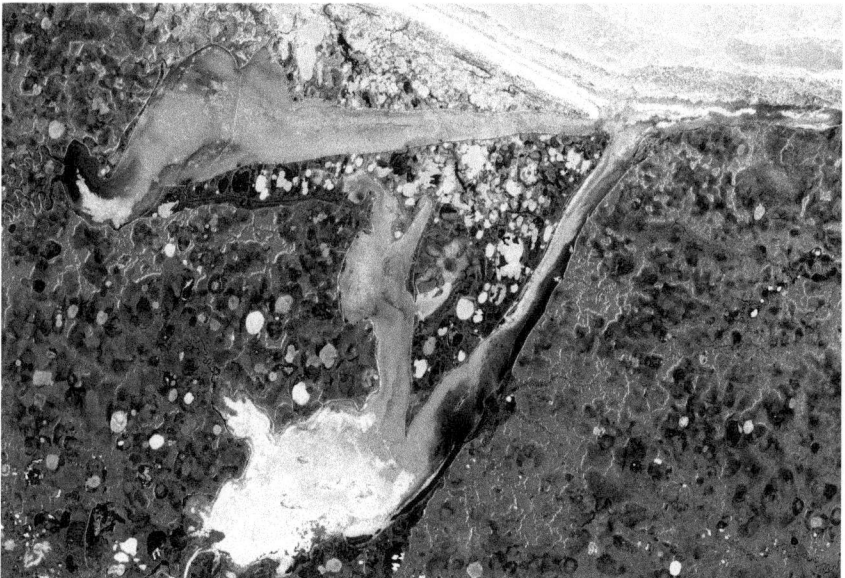

Omulyakhskaya and Khromskaya Bays, Northern Siberia
NASA Earth Observatory

Extreme Heat

As fossil fuels are emitted into our atmosphere, they trap invisible infrared light as it is reflected, and this creates an energy imbalance that causes the planet to warm. Accordingly, the past few years have seen deadly heat waves hit every continent, including Antarctica, which recorded 20.75°C (69.35°F) for the first time on 9th February 2020 (27) (28). Climate scientists are sure that the record warming that occurred here and, in the Arctic, Asia, North America, Africa, Australia and Europe was caused by climate change (29).

Let's take a look at the records that have been broken around the world in the last few years, starting with the usually cold Arctic. As we've already seen, what happens in the Arctic will not remain in the Arctic. Temperature rises at the top of the planet will have massive repercussions for every inch of the globe. Four of the past five years have seen intense temperature surges. Scientists are warning us that this increase in temperature could hasten the melting of sea ice, instigating the albedo effect of darker waters absorbing more heat than lighter ice, and driving temperatures even higher. This could unlock the permafrost and accelerate the methane emissions already underway. In early May 2018, the temperature at the North Pole hit 0° Celsius (32°F). As we learned in school, this is the freezing point for water. This may still seem cold, but this is about 17-19°C (29-32°F) above normal. Most of the entire northern region of the Arctic averaged around 10°C (17°F) above average (30).

The following year started out in similar fashion. Compared to the average from 1958-2002, 2019 was Frankenstein weather. As the new decade got underway, the Siberian town of Verkhoyansk reached 38°C (100°F), setting an all-time recorded high for the northernmost region. Temperatures in the Laptev Sea were between 4-7°C (7-12°F) hotter than normal and the 2020 heatwave lasted for ten days (31).

The Siberian town of Saskylah saw the land surface temperature reach 39°C (102°F) in June 2021 (32). In mid-March 2022, parts of the Arctic were recording temperatures a staggering 30°C (50°F) above normal (33). The effect this is having on the Arctic ice coverage is there for all to see in the graph below.

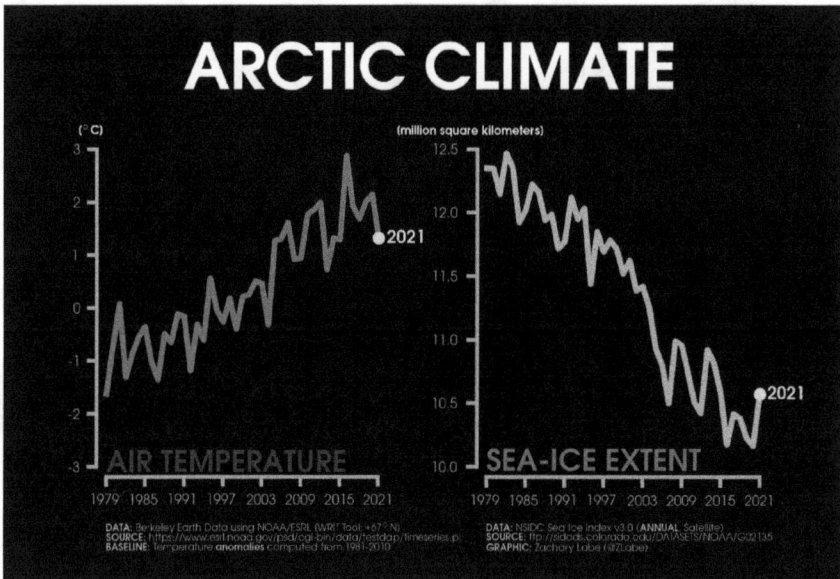

https://sites.uci.edu/zlabe/arctic-temperatures/
Zachary Labe

Whilst the melting ice and permafrost is of grave concern, this doesn't affect sea level rise as this ice is already in the ocean. The ice that covers Greenland, however, is sitting on top of land. If all this were to melt, global sea levels would rise by seven meters. And the ice is melting six times faster than it was in previous decades due to record warming. Since 1972, Greenland's melting has contributed half an inch of global sea level rise (34). The melt season usually runs from June to August, with the most melt occurring in July. Alarmingly, as temperatures over parts of Greenland soared to more than 4.4°C (7.3°F) above normal, some 40% of Greenland experienced melting on 13th June 2019. In a single July day in 2021, enough ice melted to cover Florida in 5cm (two inches) of water. Over a two-day period, 16.9 billion tonnes of ice was lost (35). The air on the west Greenland coast was as much as 7°C (12°F) above the 1981-2010 average (36). Early melting exacerbates the melt later in the year as less ice reflects less sunlight, and more heat is absorbed (37).

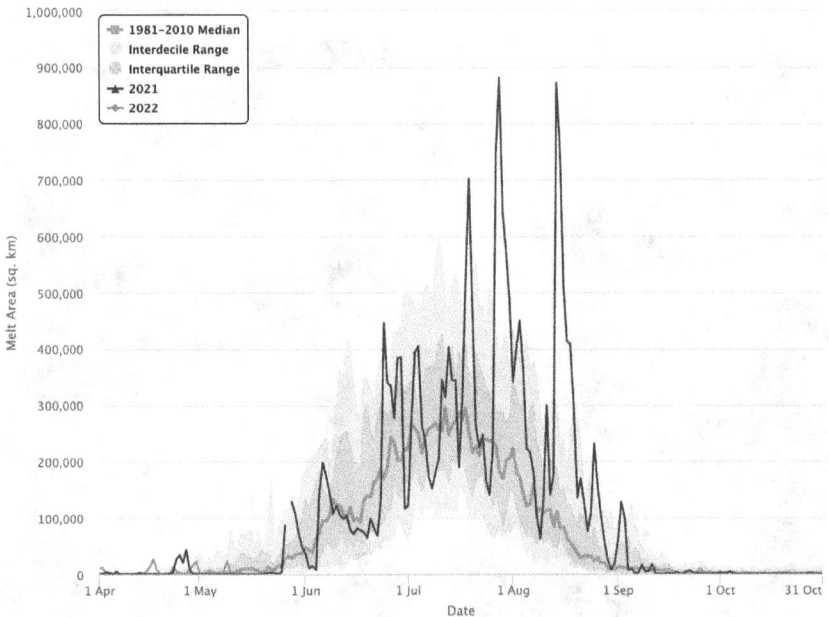

Greenland Surface Melt Extent
National Snow and Ice Data Center

The melting is being exacerbated by an array of tiny life forms who are thriving on the surface of the melting water on the ice sheet. These microbes come in colours from blacks to greys and greens, purples and browns and their dark colour is absorbing more heat and helping to melt the ice even further. Research shows that the algae is responsible for between 13% and 26% of the melting, depending on the area. This is another positive feedback loop with the microbes increasing the melting and then thriving in the additional water produced and warming the water even more, creating even more water to thrive in (38).

The previous graph illustrates just how extreme current melting is, compared with the 1981-2010 average. While the melting has reduced from 2019's record, it is still significantly above the 1981-2010 average, raising the question: Is this the new norm (39)? Scientists tried to answer this question in 2020 and their research suggested that even if the surface melt was somehow reduced, the glaciers would continue in their, *"dynamic state of sustained mass loss"*, meaning a tipping point of no return may already have been reached (40). If this research is proven accurate then the impacts on cities around the world will be profound.

Further south in Alaska, things are also heating up. On March 30th, 2019, temperatures rose almost 22°C (37°F) above normal to 3°C (37.4°F). These heat increases, however cold they seem, are in fact deadly. In separate incidents throughout the state, eight people were killed in April 2019 when their snowmobiles or four-wheelers plunged through the usually thick ice. Others have had lucky escapes in similar accidents (41). 2019 brought Alaska its warmest spring on record and since the 1970s, the temperature has risen by 2.2°C (3.7°F), double what the planet has warmed since the industrial revolution. The all-time record was shattered in the capital, Anchorage, on Independence Day, 2019. The temperature rose to 32°C (89.6°F) on July 4th and the fireworks celebration was cancelled because of the risk of wildfires as the town sweltered in conditions similar to those seen in the sunshine state of Florida. This warming has exacerbated scientists' fears over the melting permafrost and resulting methane leaks (42). Eighty five percent of Alaska is covered in permafrost, and it is beginning to thaw (43). A year later and Fairbanks, Yaktat Airport, Sitka, and the Juneau Airport all saw temperature records fall in May and June (44). Anchorage recorded its hottest ever August 15th with the mercury reaching 25.5°C (77.9°F) (45).

It goes without saying that heatwaves don't have borders. Canada also suffered from a fatal heat wave in July 2018. At least ninety-three people died in Quebec province alone as temperatures topped 35.3°C (95.54°F). This was completely eclipsed in the 2021 heat dome that sat over the Pacific northwest in July when fifty-nine records were shattered and around 500 people died. Canada's hottest ever record was awarded to Lytton in British Columbia where the thermostats rose to 46.6°C (116°F), then 47.9°C (118°F) a day later and finally hit an incredible 49.6°C (121°F). As a prize, Lytton was burned to the ground just a day later as wildfires tore through the area (46).

Southern California saw temperature records shattered as the mercury rose to 48.9°C (120°F) in Chino, near Los Angeles. On June 10th, San Francisco got in on the record-breaking act. Temperatures reached 37.8°C (100°F). Never before had San Francisco experienced that heat in the months of June, July or August. In June 2019, mussels along 140 miles of Pacific coastline were found to be cooked alive in their shells. Mussels are known as a foundation species in that what happens to them impacts all other species in the ecosystem. Kelp and coral are also struggling in warmer waters and starfish are melting (47).

In the summer of 2020, California experienced a ten-day period of extreme heat that caused 300,000 people to lose power and across the country, more than 150 million Americans were exposed to prolonged temperatures over 32°C (89.6°F) with fifty million of those struggling in close to 38°C (100.4°F) heat (46). The following year broke all records in the Pacific Northwest where Oregon recorded its highest ever temperature of 48°C (118°F) in June. Portland topped 47°C (116°F) while Seattle suffered in 42°C (108°F) (47). At least ninety-five people were killed by the extreme heat in Washington State with a further ninety-six perishing in Oregon. A month later, thermostats again got into triple digits in Washington State and Oregon where temperatures are usually below 32°C (89°F).

In Central America, up to a quarter of the population in the sugar growing areas of El Salvador and Nicaragua suffer from chronic kidney disease. The prevalence rises as high as 50.2% of the population in two farming communities. This is thought to be caused by dehydration from working in the fields. This was not a problem just a few decades ago and will only get worse as the global temperature increases (50).

Early 2022 brought exceptionally high temperatures to South America with Argentina, Paraguay, Southern Brazil, and Uruguay experiencing their hottest recorded days of up to 45°C (113°F) (51).

Across the Atlantic, with the Gulfstream moving further north than usual, much of Europe basked in unusually warm conditions in 2019. Norway, Sweden, Denmark, the U.K., and the Baltics all reported crop damage. These increased temperatures led to wildfires in multiple European countries (52). Early June 2019 saw Scandinavia record heat that has never before been felt so early on in the summer. Temperatures above 30°C (86°F) reached inside the Arctic Circle (53). The town where temperatures have risen the most is Longyearbyen in Svalbard. Since 1971, the temperature has risen five times faster than the global average, and today the air is 4°C (6.72°F) warmer. The winters here are an astounding 7°C (11.76°F) warmer, but things are predicted to get even hotter with warming of 10°C (16.8°F) expected by 2100. These temperatures are already proving deadly as an avalanche in 2015 killed two, injured eight others and pushed eleven houses twenty meters off their foundations. Heavy rain fell on the Sukkertoppen (Sugar top) mountain overlooking the town. That rain froze and a mass of snow fell on top, but unfortunately this snow and ice did not stick to the mountain. It collapsed in the middle of the night and slid down into the bedrooms of the sleeping victims. Rain never usually falls in winter, but it does now, and the glaciers that cover 60% of Svalbard are melting along with the permafrost. This is causing buildings to crack, landslides, and damaged foundations (54).

2019 didn't disappoint beach lovers in the south of France as temperatures soared to 45.9°C (114°F) in late June. Never before had France exceeded 45°C (113°F), with the previous record of 44.1°C (111°F) being set in 2003. With warm air drawn from Africa, 4,000 schools were closed as head teachers confirmed they could not guarantee the safety of students. With incredible irony, climate activists, in Paris to raise awareness of the unfolding climate catastrophe, were pepper sprayed by police as temperatures in the French capital hit 32°C (89.6°F) (55). France, Spain, and Italy were all gripped by the intense heatwave. Forty percent of grapes withered on vines in the Languedoc region (56). High in the Alps, near Europe's highest mountain, a new lake was discovered. The lake, formed from melting glaciers at an altitude of 3,400m, was found after a ten-day heat wave and was linked to the warming climate (57). The following summer, ten tonnes of fish were found dead in Enghien-les-Bains Lake in northern France after a heatwave struck the region in August. The heat helped cause a reduction in oxygen levels in the lake and the fish perished (58).

In 2019, Germany saw record temperatures more than 4°C (6.72°F), hotter than the averages for June and the warmest since records began in 1881. The town of Coschen in Brandenburg set a new June record of 38.6°C (101°F), while the Czech Republic was breaking records too, at 38.9°C (102°F). Poland was in on the action with temperatures hitting 38.2°C (101°F) and in Rome, the mercury reached 41°C (105.8°F). It had never previously been necessary for air conditioners in Europe, but the extreme heat led to a spike in sales, and also power demand. Switzerland and Austria also saw June temperature records fall (59).

The 2019 heatwave south of the Pyrenees stretched into its fourth day with temperatures reaching 44°C (111°F) in Zaragoza. Forty out of fifty regions in Spain were placed under weather alert with seven of them categorised as extreme risk and two people died from extreme heat. Southern Spain outdid its northern counterpart in the summer of 2021 as a new heat record of 47.2°C (117°F) was confirmed (60).

The Southeast U.K. saw temperatures rise to 34°C (93°F) in late June, close to the all-time June record of 35.6°C (96°F), set in 1976. London basked in temperatures warmer than Hawaii. Dozens of festivalgoers at the famous Glastonbury music festival were treated for heatstroke, and thirty-one people drowned while trying to cool off in open water (61). Older people are especially vulnerable to heat waves and heat related deaths among the over 65s have more than doubled since the early 2000s. In 2018 alone, 8,500 Brits in this age group died of heat stroke (62).

In unusually frank language about the 2019 heat, Professor *Hannah Cloke* of *the University of Reading* said, *"We knew June was hot in Europe, but [the Copernicus data for June] show that temperature records haven't just been broken – they have been obliterated. It is the hottest June on record in Europe by a country mile".* The record temperatures were made at least five times more likely because of climate change, but it's possible the figure may be as high as one hundred. Heatwaves in Europe are not just becoming more frequent but are also around 4°C hotter (6.72°F) (59). While 2019 broke records across Europe, the following year was even hotter with the European average rising by 0.4°C (0.672°F) (63). Europe saw its hottest ever day recorded in 2021 on the Italian island of Sicily which sizzled in 48.8°C (119.8°F). The continent is warming much quicker that climate models had predicted and researchers at the American Geophysical Union warn that Europe should prepare for more frequent extreme heat events (64).

The Eurasian continent wasn't spared either as wildfires raged in Russia. Smoke from the 2018 wildfires reached as far away as New England. In total, 13,000 sq.km of forest were alight, causing bears to flee their usual habitat and come into contact with humans (65). Georgia broke records on 4th July 2018, as Tbilisi reached 40.5°C (105°F). The following year saw Moscow's average temperature rise to 7.7°C (46°F) which beat the previous hottest average by 0.3°C (0.5°F). The head of Russia's *Gidromedtsentr* weather service, Roman Vilfand, said that *"This year in Russia was the hottest for the entire period of instrumental observations" (66).* There was no let up for Russia in 2020 with the average annual temperature being 3.22°C (5.4°F) higher than the 1961-1990 average. The first half of 2020 saw the country record its warmest temperatures since records began in 1879. In response to the warming climate and melting Arctic ice, Vladimir Putin talked of the benefits a warming world would bring to the nation, including the opening up of the Northern Sea Route which could be used for exporting oil and gas. That our leaders focus solely on the profit-making opportunities that may arise out of the carnage that is unfolding should be a wakeup call to us all (67). Putin failed to mention the impact that thawing permafrost is having on the lives of people and ecosystems in his nation right now.

In South Korea, electricity consumption rose by 6.3% in 2019 as people relied on air conditioners to cool off in temperatures that exceeded 40°C (104°F). Dozens of people died as a result of the heat. Spring and autumn seasons now last only a few weeks each, and summer temperatures are arriving a month earlier (68). June 2020 saw the southern part of the Korean peninsula record its hottest ever June with temperatures reaching 35.2°C (104°F) (69).

To the east of South Korea, Japan was baking. A staggering 1,032 people perished and 22,000 were hospitalised in 2018 as temperatures soared to a new record 41.1°C (106°F). In a 2019 study, researchers declared that the 2018 heatwave in Japan would not have been possible without 'human-induced global warming' (70). 2019 followed the same pattern of intense heat with the usually mild northern island of Hokkaido breaking the nationwide May record by hitting 39.5°C (103°F). Temperatures in May had never before passed 35°C (95°F) in Hokkaido, and the region had never experienced anything warmer than 38°C (100°F) at any time in its recorded history (71). To illustrate that 2018's record was not a one off, Shizuoka Prefecture also recorded 41.1°C (106°F) on August 17th, 2020. The all-time September high was also broken in 2020 when Niigata Prefecture hit 40.4°C (104.72°F) on the 3rd. The 2020 average temperature for Eastern Japan in August was also the highest since records began in 1946. In a three-day period, between August 3rd and 6th, 6,664 people entered hospitals suffering from heat exhaustion (72). The previous 2010 record was broken by 1.7°C (2.8°F). Rainfall was only 30% of its average during the same period (73). The record June temperature was recorded in 2022 in Gunma. The thermostats read 40.2°C (104°F) and 1,337 people were hospitalized with heatstroke.

In Ouargla, Algeria, Africa set a new heat record as thermometers displayed an alarming 51.3°C (124°F) in July 2018. Further east, an Omani town broke the record for the highest low temperature in recorded history. Thermostats in Qurayyat didn't drop below 42.6°C (109°F) for fifty-one straight hours (74). Records continued to fall in the Middle East the following July as many regions experienced +50°C temperatures, with Kuwait and Iraq's Basra recording 53.9°C (129°F) on July 21st (75). The extreme heat is fast becoming the norm in the Middle East with Iraq's capital Baghdad breaking its all-time record of 52°C (126°F) in early August 2020. Things got so hot that the energy grid almost collapsed and many Iraqis, experiencing regular blackouts, had to rely on diesel generators to power their air conditioners. Those without access to generators simply had to endure while pollution levels from the diesel engines rose to unhealthy levels (76). During demonstrations against the power cuts, security forces killed two protestors. Syrians already suffering after a decade of civil war, partly the cause of drought, struggled through August temperatures that rose to 46°C (115°F) (77). Increased heat will only add to the conflict the region has been experiencing the past few decades.

The Indian sub-continent has also been suffering from punishing heat. Temperatures in northern India topped 50°C (122°F) in May 2019, causing water shortages and heatstroke. Rajasthan hit 50.6°C (123°F), and New Delhi saw temperatures pass 48°C (118°F). Many wealthy Indians head to the mountains to cool off in summer with Himachal Pradesh a favourite destination. Unfortunately, it is difficult to cool off in 44.9°C (113°F) heat. Chennai and several other Indian cities are worrying about water shortages as lakes and rivers dry up (78). Out of the fifteen hottest places on Earth in early June 2019, eleven were in India, with the rest being in neighbouring Pakistan (79). Later in the same month, four elderly pilgrims died on their way back from the holy city of Varanasi. They were travelling to the southern district of Kerala in a non-air-conditioned train when they collapsed. The men are among a group of thirty-six to die due to the long-lasting heat wave affecting the country. Nature is also being impacted. Around fifteen monkeys died in Joshi Baba Forest range in Madhya Pradesh.

It is thought they were prevented from reaching the nearest water by a separate troop of monkeys (80). The decade from 2000 to 2019 was the hottest on record and 2019 saw more than 1,500 people lose their lives to heat related causes. The situation had worsened significantly by April 2022 with temperatures in Pakistan reaching 49°C (120°F). People could only work outside in the night-time. Some areas were reporting 50% losses to food crops and Pakistan's minister for climate told reporters that the country was facing an "existential crisis" with glaciers in the north melting at an unprecedented rate and reservoirs drying up. In India, south of Ahmedabad saw the surface-land temperature hit an astonishing 65°C (149°F) (81). Worryingly for food production, soil bacteria cannot survive surface-land temperatures above 60°C (140°F). The extreme heat was causing blackouts at a time when people needed air conditioners the most. Ironically, in many parts of India, domestic coal supplies were at critically low levels and passenger trains had to be cancelled for coal to be transported to power plants so it could be burned and raise the temperature even further. The programme manager at the Gujarat Institute of Disaster Management said, *"the extreme, frequent, and long-lasting spells of heatwaves are no more a future risk. It is already here and is unavoidable (82)."*

Down under, Australia was also wilting in extreme temperatures. 2018 was the third hottest on record in Australia, after 2013 and 2005. Queensland experienced daytime temperatures of 42.6°C (109°F) in Cairns and 44.9°C (113°F) in Proserpine in late November (83). New South Wales and Victoria broke day time highs, and South Australia and the Northern Territory recorded the second highest temperatures ever. The new year didn't bring any respite to Australia with Adelaide sizzling under record temperatures of 46.6°C (116°F) in January 2019. No major Australian city had ever experienced temperatures so hot (84). Humans can use technology to cool off in the intense heat, but animals have no such luxury. Hundreds of flying fox bats fell from the sky as the heat literally fried their brains (85). Koalas are already threatened by deforestation and disease. They now have to contend with climate change as well. They are moving out of their trees, where they spend most of their time, in search of water. As 2020 got underway, amid record fires, New South Wales saw Penrith hit 48.9°C (120°F) (86). The lead up to the summer in 2020/2021 was the hottest ever and maximum, minimum, and mean temperature records were all broken during November 2020 (87). The warmest year ever recorded in Australia and Oceania was 2019 (88). By January 2022, the records were tumbling again as Western Australia recorded temperatures of 50.7°C (123°F) (89).

The southernmost region on Earth saw its temperature record broken twice in quick succession in 2020 with an Argentinian weather station logging 18.3°C (65°F) on the peninsula in early February and then an island off the coast becoming the first area to top 20°C (68°F) a week later (90) (28). Until recently, the sea ice in Antarctica was thought to be growing, but new research shows that growth peaked in 2014, and since then, the extent of sea ice has nosedived. While again, this is sea ice, so it won't raise sea levels, it will expose more dark water to sunlight which will cause further warming. Sea ice at both poles is now in retreat and could have disastrous consequences for the future of our planet. The decline in Antarctic Sea ice has happened three times faster than any melting ever recorded in the northern hemisphere. While the Arctic is an ocean surrounded by continents, the Antarctic is a continent surrounded by oceans, and the westerly winds that circulate around it were thought to protect it from the globally warming world. Now scientists are not so sure and are frantically trying to understand what happened in 2014 to trigger the sudden melting (91). At the same time the Arctic was recording temperatures 30°C (50.4°F) above normal in mid-March 2022, East Antarctica saw even more abnormal temperatures that were 40°C (67.2°F) higher than average. This led scientists to use words like "historic," "unprecedented" and "dramatic" to describe the unfolding scenes at both poles (33).

Globally, the planet experienced its second hottest year ever recorded in 2019, just 0.4°C (0.67°F) cooler than 2016. If 2019 was feeling a little inferior, 2020 arrived on the scene, full of confidence, to claim the prize of warmest year ever recorded on Earth. The average temperature was 1.2°C (2°F) above the 1859-1900 level. This was also a year in which our civilization was brought to its knees because of the ongoing pandemic. The World Meteorological Organization (WMO) says that the warmest six years on record have all occurred since 2015 (92).

With the carbon already in the atmosphere, the extreme heat we are witnessing now will become the new normal. The impacts of a warming planet will be felt by us all. In fact, they already are. Storms, droughts, and heatwaves have increased by a third in just the past ten years (93). Scientists analysing models in Switzerland and the U.K. declared that heat events like that of 2008 were unprecedented prior to 2010 and don't occur in historical simulations (94).

As fossil fuels continue to be pumped into the atmosphere, forests are felled and oceanic ecosystems destroyed, the extreme heat we have seen the past five years is only going to get worse in the next decade. Anthony Arguez at NOAA's National Centers for Environmental Information in Asheville, North Carolina warns that *"After the last five years, we've really separated ourselves from the past,"* and he adds, *"It looks pretty likely that we're going to have a whole lot of top 10 years (95)."*

In total, the natural-disaster loss events have more than tripled in the past forty years with July 2021 alone witnessing two $25 billion flood disasters. It is debatable whether we can continue to call these events 'natural' when they are being fuelled by human activity. The world is moving into uncharted territory, and our planet has only warmed by 1.2°C (2°F) so far. What will the future look like if we continue on our current path? The current trajectory we are on would see temperatures rise by between 1.5°C (2.5°F) between 2030 and 2032 and the next threshold of 2°C (3.6°F) a decade later (96). A possible scenario sees the temperature rising by 5°C (9°F) by 2100 (97). Try to imagine the heatwaves, storms, wildfires, droughts, and floods that will become normal if we allow this to happen.

If things are this bad at just 1.2°C (2°F) of warming, more than 2°C (2.5°F) will likely be dire for the future of our species. Also, take into account that these are global average temperatures. At the poles and over land the temperature could be double. For every 1°C (1.68°F) of warming, areas affected by heat waves like 2018's will increase by 16% and are predicted to happen in two out of every three years if temperatures reach 2.7°C (4.5°F), and every year at 3.6°C (6°F) (53)

Our planet hasn't seen temperatures this hot in fifteen million years, when the sea level was forty meters higher than it is today. The Arctic and Antarctic were home to vast forests. This warming was caused by volcanic activity in North America and the warming happened 1,000 times slower than the human caused warming we see today. If 2100 seems a long way away, children born today will be in their seventies (98). If that still doesn't concern you, we will lose all the world's coral reefs at 2°C (3.3°F) of warming, hundreds of millions of people will be on the move in search of food and water. There will be forty-one more marine heatwaves than now. All the Arctic Sea ice will be gone. The average length of a drought could be ten months. At least 388 million people might be exposed to water scarcity. Maize production may be 9% lower, and wheat 4% lower at 2°C (3.3°F). The population will have risen to around eleven billion. Sea level may have risen by fifty-six cm. Flooding from sea level rise will cost upwards of $11.7 trillion each year at just 2°C (3.3°F). And the most alarming change will be under the ground in the Arctic region discussed earlier. Here, at least 6.6 million km^2 of permafrost might thaw. The result of this thawing could trigger the emissions of billions of tonnes of methane, and what scientists call runaway climate change (99).

It's worth looking at these temperature rises we keep hearing about. 1°C, 2°C, 3°C, 4°C. We know they will melt ice, raise sea levels and displace millions. But what of the dangers of living in this warmer world? The wet-bulb temperature is the temperature that a thermometer reads when covered in a water-soaked cloth (100% humidity) and gives a more accurate indication of the strain on the body at high temperatures than the conventional air (dry-bulb) temperature. For example, according to research in Environmental Research Letters, at 100% humidity, a temperature of only 35°C (95°F) is thought to be the limit of human tolerance. This is because the body cools itself by sweating, and air already saturated with water cannot absorb more moisture from the body surface, effectively shutting down the body's cooling system. If we can't get rid of our waste heat, our organs fail and death results. Such conditions are rare in 2022, but even at a wet bulb temperature of between 29-31°C, tens of thousands have died in recent years. The elderly and young are more at risk, and the temperature and humidity will be deadly at lower relative humidity for these groups. These temperatures will make it extremely difficult to work outdoors.

As cities become hotter, they will also likely become more violent. Temperature-aggression theory has been found to elevate interpersonal violence by increasing discomfort, frustration, impulsivity, and aggression.

"It is worse, much worse than you think", is the opening line to David Wallis-Well's seminal, but terrifying book, The Uninhabitable Earth. He describes life on our planet at varying degrees of warming. At 5°C (8.4°F), large parts of Earth would be unliveable for humans. At 6°C (10°F), New York would be hotter than Bahrain today. At 4°C (6.72°F), wildfires will burn sixteen times more land in the U.S. West. There will be hundreds of drowned cities. Going outside will be a dangerous act across India and the Middle East at just 2°C (3.3°F) of warming. Even if we meet the Paris goals, Wallace-Wells says the deadly heatwave that killed thousands in India and Pakistan in 2015 will have become annual events. At 4°C (6.7°F), the 2003 heat wave that claimed the lives of around 2,000 people each day in Europe will be a standard summer event. That means more than 35,000 people will be killed each and every summer by intense heat. Keeping temperatures under 1.5°C (2.5°F), of warming, as opposed to 3°C (5°F), will save between 110 and 2,720 heat-related deaths in fifteen U.S. cities (53). The Union of Concerned Scientists are predicting that hundreds of U.S. cities will experience a whole month above 37.8°C (100°F) by mid-century. Scientists expect the extreme heat to cause large scale relocation of residents as Boston becomes the new Columbia, South Carolina, and Chicago experiences the heat of Lafayette, Louisiana (100). The pilgrimage to Mecca will become impossible for the two million Muslims currently making the trip (101). Wallace-Wells adds that even if we keep warming under 2°C (3.3°F), half the population will be exposed to deadly heat waves more than twenty days a year. If we don't keep warming under 2°C (3.3°F), that number will rise to three quarters.

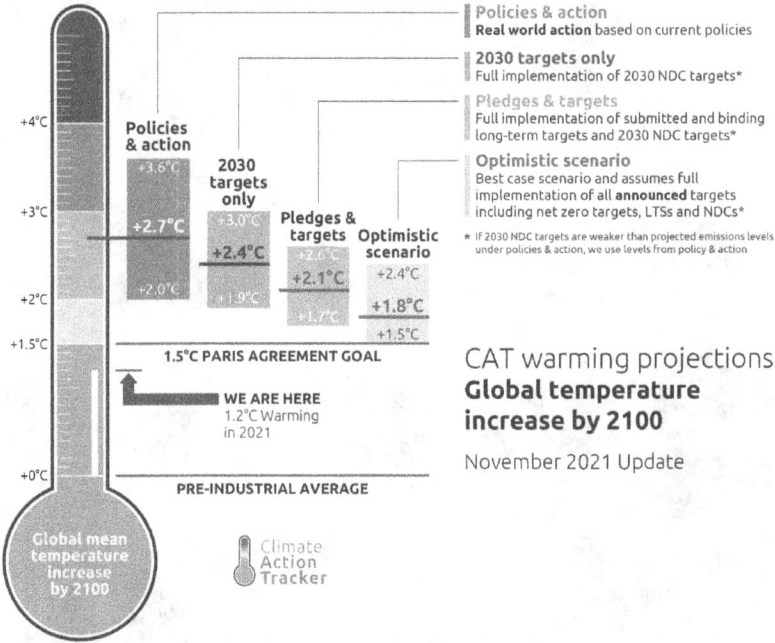

CAT warming projections
Global temperature increase by 2100

November 2021 Update

Policies & action
Real world action based on current policies

2030 targets only
Full implementation of 2030 NDC targets*

Pledges & targets
Full implementation of submitted and binding long-term targets and 2030 NDC targets*

Optimistic scenario
Best case scenario and assumes full implementation of all announced targets including net zero targets, LTSs and NDCs*

* If 2030 NDC targets are weaker than projected emissions levels under policies & action, we use levels from policy & action

1.5°C PARIS AGREEMENT GOAL

WE ARE HERE
1.2°C Warming in 2021

PRE-INDUSTRIAL AVERAGE

Climate Action Tracker Warming Projections
www.climateactiontracker.org

Hans Joachim Schellnhuber, director emeritus of Germany's Potsdam Institute for Climate Impact Research stated that at 4°C (6.7°F) of warming, "Earth's carrying capacity estimates are below 1 billion". Kevin Anderson from the U.K.'s Tyndall Centre for Climate Change concurred that "Only about ten per cent of the planet's population would survive at 4°C." If anyone has seen David Attenborough's documentary on Netflix, Breaking Boundaries, you will be familiar with Johan Rockström, the Director of the Potsdam Institute, and he states "It's difficult to see how we could accommodate a billion people or even half of that.... There will be a rich minority of people who survive with modern lifestyles, no doubt, but it will be a turbulent, conflict-ridden world (102)".

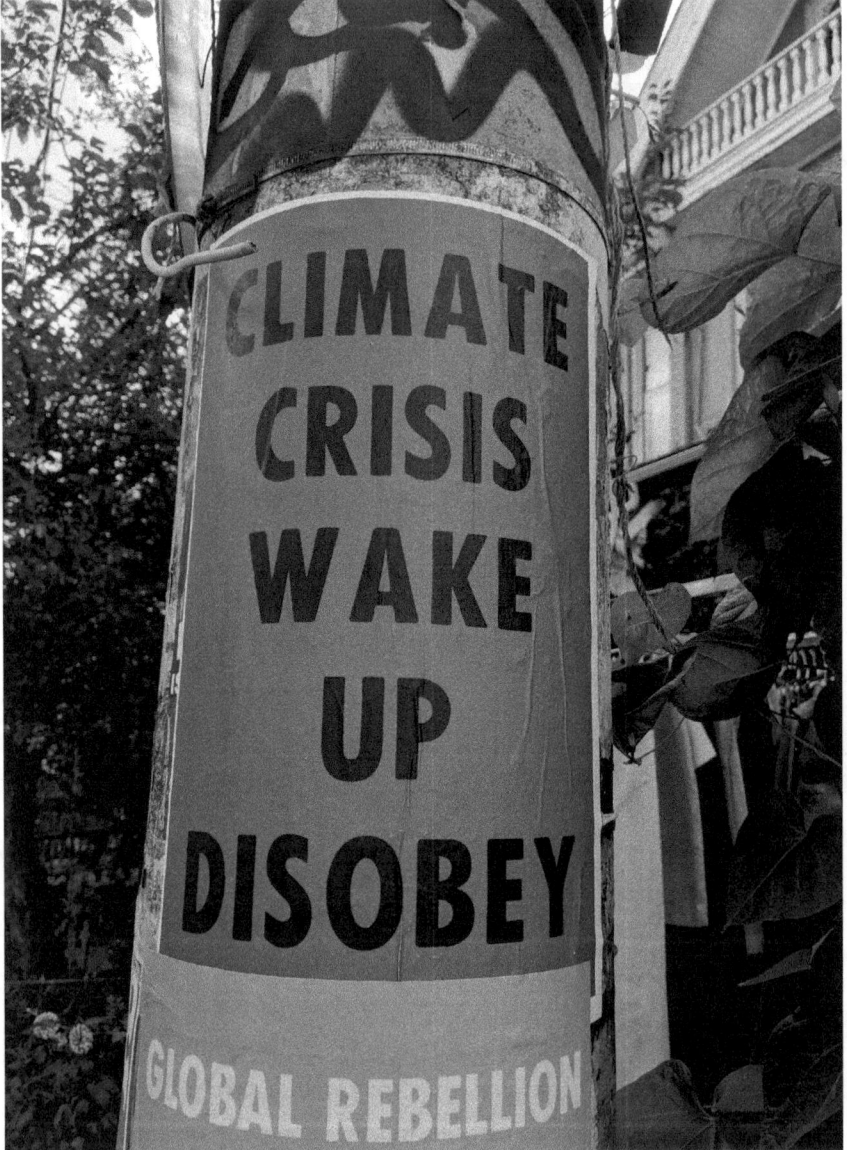

Climate Crisis Wake Up Disobey sign, Kensington Market, Toronto, Ontario, Canada
by gruntzooki is licensed under CC BY-SA 2.0.

Dear Indy,

As someone who grew up in Wales, a little more heat was always something that I craved. With temperatures rarely getting beyond 30°C, and most of the time being below 20°C (68°F), the biggest event of the year was a trip to southern Europe for some beach action that didn't involve frostbite. I think you understand what I'm talking about. On our last visit, we spent the day on the splendid beach at Tenby and we were the only family to have jackets on. While kids frolicked in the frigid waters, we shivered on the beach.

In Japan, the situation is somewhat different. As a family that enjoys camping at the beach, the reality of a warming climate hit home while we were living in Kyushu. For the first time, our visits to the beach were limited to early in the season as it was simply too hot to enjoy yourself without getting burnt very quickly. The temperature was usually in the very high 30s and our house sweltered in the summer months. We opted instead for the cooler temperatures on offer in the mountains where the temperature is 3-4°C (5-6.7°F) cooler.

Fortunately, Hiroshima Prefecture is a little cooler and our island home is a little cooler again, but as the planet warms, we may need to move further north to escape the oppressive heat. While many in the U.K. might be jealous of a warmer climate, they don't really appreciate how lucky they are to live in a country with a mild climate. They may soon understand how lucky they have been as heat waves become increasingly common and food production is affected.

People will certainly be able to enjoy more time at the beach, but the trade-off will be more expensive food, a need to purchase air conditioning units and more and more vulnerable people dying from heat related illnesses.

We have grown up in the perfect conditions for human development but due to our actions, these conditions are about to change, and increased heat will affect us all.

Natural Disasters

With increased temperatures, more water is evaporated, resulting in more moisture in the atmosphere. For every degree Celsius of warming, the atmosphere will hold an additional 7% of moisture. This is known as the Clausius-Clapeyron equation (103). When the figurative dam breaks, all this water comes down in a deluge. In the past decade extreme rainfall events have increased by more than 50%. They are now four times more likely than they were in the 1980s (93). With just one degree of warming, our world has already changed. 2018 brought a cavalcade of disasters, one after the other. In the U.S., Hurricane Florence dumped seventy-six cm (thirty inches) of rainfall, left a million without power, killed forty-eight, and caused $60 billion worth of damage. Hurricane Michael slammed into Florida and killed thirty-four while causing damages of $30 million.

The wettest year on record in the United States occurred in 2019. Weeks of flooding in the Midwest submerged communities and swamped millions of acres of farmland and killed almost forty people. Almost $1 billion of damage to the soybean crop and livestock were reported. Only 22% of the soybean crop in Kansas was harvested, while at the same stage the previous year, the figure was 63%. Corn production was extremely low, with less than half the amount of land planted compared to previous years. Indiana fared even worse at only a fifth of its normal level (104). Hurricane Ida landed in Louisianna in September 2021 and took the lives of twenty-two people. The storm left around one million without electricity as 30,679 electricity poles were toppled. This was more than were tumbled by Katrina, Ike, Delta and Zeta combined (105). The storm brought chaos as far north as New York where fifty-two people died (106). The total annual cost of extreme weather events in the U.S. was estimated at $99 billion in 2020 and is getting worse (107).

North of the border, severe drought and wild weather contributed to a massive decrease in income for farmers in 2019. Farmers in Alberta saw their income decline by 68% as a wet early spring morphed into a long hot and dry spell that withered crops, and then an early snow fall. Cattle ranchers ran out of grass and had to purchase feed at a 20-25% premium or send animals to an early slaughter (108).

Across the Pacific Typhoon Mangkhut struck the Philippines, and after leaving, 127 people were dead, and more than $30 billion worth of damage was left in its wake (109). The storm moved on to Hong Kong and other cities on the Chinese coast, where it uprooted trees, swayed tower blocks, shattered their windows, and turned roads into rivers (110). Further north, Typhoon Jebi careened into Taiwan and Japan, causing six deaths in Taiwan and eleven in Japan. Jebi broke sustained wind records at fifty-three weather stations and damages hit around $15 billion (111). It is clear why climate change has increased the risk of these storms occurring. Warmer waters mean more water evaporates into the atmosphere, and with more moisture in the air, rainfall becomes heavier. A warmer atmosphere is also thought to slow down hurricanes, so they linger over areas longer causing more damage.

In what was a terrible year for Japan, 200 people were killed in landslides and flooding caused by a torrential downpour in July 2018. In just two hours, Hiroshima, Okayama, and Hyogo prefectures received between 300–500 mm of rain. This is 1.5 times the average monthly rainfall for all of July. Hundreds of thousands of people had to leave their homes and take shelter on rooftops as the water rose around them (112). Heavy rainfall is becoming the norm in Japan and landslides are now a regular occurrence every summer. In 2021, the west of the archipelago suffered record deluges that saw rivers burst banks throughout Hiroshima, Fukuoka, Tottori, Saga, Kagoshima, and Kumamoto prefectures. Roads and railways were closed as landslide alerts were broadcast.

Indonesia hosted some extremely wet weather in 2019. At least twenty-nine people died as flooding led to landslides, exacerbated by deforestation. The capital Jakarta, which is sinking due to ground water extraction, suffered two fatalities. The government is planning to move Jakarta's ten million inhabitants as a result of the predicted flooding (113). Approximately 95% of northern Jakarta could be submerged by 2050 (114).

In late May 2021, seventy-one rivers in China were exceeding warning levels with parts of central and southern China recording record rain fall. The country had already seen 10.3% more rain than average with rainfall in some parts of China almost doubling from the year previous. On 23rd May 2021, twenty-one people were killed while taking part in an ultramarathon race in north-western Gansu. The runners were hit by a blitzkrieg of hail, freezing rain and gales (117).

In Bangladesh, ground zero for climate change, almost 200,000 people were displaced by flooding in July 2019. Thirty people died and 66,000 homes were destroyed. Neighbouring Nepal recorded seventy-eight deaths from the monsoon rains, Pakistan reported forty-six deaths and India seventy-eight. The flooding left more than four million people at risk of food insecurity (118). In June 2022, more than 7.2 million people were afected by flash flooding in northeastern Bangladesh. According to the United Nations Office for the Coordination of Humanitarian Affairs, heavy monsoon rains caused 472,856 people to be evacuated to 1,605 shelters, and experts consider the flooding to be worse than the extreme flooding that occured in 1998 and 2004.

India was hit by a left-right combo in late May 2021. Cyclone Tauktae landed from the Arabian Sea and Cyclone Yaas formed in the Bay of Bengal. More than one million were evacuated as record rainfall fell in some parts, like Similipal. The Indian sub-continent has been hit by an increasing number of cyclones the past few decades as the waters warm. Between 1891 and 2000, the Arabian Sea, where the temperature is usually below 28°C (50.4°F), recorded ninety-three cyclones but since 2000, twenty-eight have formed. If we extrapolate that figure out for the same duration, 154 will occur by 2109. That's if temperatures stay the same, but we know they are only going to rise. The warmer waters in the Bay of Bengal, where temperatures are always above 28°C, saw 350 cyclones between 1891 and 2000. With 14% of the Indian population living in coastal areas, and this number expected to grow by three times in the next forty years, these cyclones will cause more and more damage to both infrastructure and food supplies (119). The warming waters in the Indian Ocean are also leading to increased locust swarms on the Arabian Peninsula and in Africa. This will be discussed shortly.

One of the biggest storms to hit Africa slammed into Mozambique in April 2019, killing at least thirty-eight people, and destroying 35,000 homes. Cyclone Kenneth caused huge flooding that endangered 160,000 people. As the cyclone approached land, its wind speeds were some of the fastest recorded, but after moving inland, it slowed down and dumped immense amounts of rain. This storm was only six weeks after cyclone Idai had taken at least 1,000 lives. Mozambique had never before been struck by two strong storms in one year (120).

In April 2022, South Africa bore the brunt of extreme rainfall; 450mm in 48 hours, that washed away roads, bridges, and hillsides, and destroyed homes in KwaZulu-Natal province. The South African President, Cyril Ramaphosa said *"This disaster is part of climate change. We no longer can postpone what we need to do ... to deal with climate change. It is here, and our disaster management capability needs to be at a higher level."* By 13th April, more than 306 people had died in the worst rainfall for sixty years (121).

In mid-June 2019, rain and hailstorms, the like of which had never been seen before, lashed south-east France causing widespread damage to crops in the "orchard of France". The storm, which only lasted for ten minutes, destroyed between 80-100% of some farmers' crops. The hailstones were so big they smashed car windows, damaged homes, schools, and public buildings. The mayor of Romans-sur-Isere called the event 'apocalyptic' (122). Only two weeks later, the area recorded the warmest French temperature ever of 45.9°C (82.6°F).

A freakish hailstorm also descended on Guadalajara in Mexico in early July 2019. A carpet of ice 1.5m deep covered the city, half-burying vehicles. Temperatures had been very high with highs of 37°C in Torreon, just north of Guadalajara. The temperature dropped from 22°C (39.6°F) to 14°C (25.2°F) around 2am and then suddenly the hail fell, damaging 200 homes and sweeping away dozens of vehicles. While the area has seen these hailstorms before, the severity of this was a *"never-before-seen natural phenomena"*, according to the state governor Enrique Alfaro (123).

It's not just extreme rain events, typhoons and hurricanes that are getting stronger and more frequent. Tornadoes are also on the increase. Five hundred twisted through the U.S. in May 2019, and this was 200 hundred above average. Seven people died (124). It's difficult for scientists to be sure if climate change increases the likelihood of tornadoes, but the jet stream that controls the weather has become unpredictable, and the normally west to east stream is behaving wildly. The small island of Taiwan experienced a rare tornado in early July 2019. 7,400 homes were left without power, large vans were throw around and a few people were injured by flying debris. The armed forces dispatched around eighty troops to the southern Pintung County to help with reconstruction (115). On the mainland, China was hit by two tornadoes in May 2021. The twisters killed twelve, injured 280 and destroyed twenty-seven homes in Wuhan and Jiangsu. The state weather forecaster predicts more extreme weather, including heat waves and floods due to the warming atmosphere (116).

Records continued to be broken in 2020 with the Atlantic hurricane season producing the highest number of named storms ever recorded. A staggering thirty storms were recorded with thirteen of them reaching hurricane force and six of them classed as major. This more than doubled the average season that sees twelve named storms with six hurricanes, three of which are usually classed as major. At least 274 people were killed or missing in Central America (125).

The Philippines experienced the strongest typhoon to ever make landfall on Earth on November 1st, 2020. The storm recorded winds of 313 kph (194 mph) and killed thirty-one people, destroyed a quarter of a million homes and caused more than $1 billion in damages (125).

China saw flooding that killed 278 people and recorded $32 billion of damage in June 2020. This was the third most expensive non-U.S. related disaster in history and the country's annual losses between 2006 - 2018 were estimated at $25.3 billion. A year later, more than 300 people lost their lives after deadly flooding hit Henan province. In total, there were forty-four weather disasters that cost at least $1 billion in 2020, with the record, forty-seven, being set in 2013. Twenty-five of these billion-dollar disasters occurred in the United States (125).

With the planet only warming by just over a degree, natural-disaster loss events have more than tripled in the past forty years. Since 2004, the number of events has already doubled (126). What can we expect to happen if we continue on our current path?

One of the first things that has scientists concerned is the Atlantic Meridional Overturning Circulation (AMOC), or *the* Gulf Stream. This is the ocean current that begins in the warm Gulf of Mexico, carries warm water up the northeast coast of the United States, and across the Atlantic to Northern Europe. The Gulf Stream is often credited with stopping countries like the U.K. and Ireland from freezing over in winter. What has scientists concerned is that the current has been slowing down recently, and it is possible that the *Gulf Stream* may 'switch off' completely (126). The slowing down is being caused by the increasing amount of freshwater melt coming off the Greenland ice sheets as a result of manmade climate change. Between 1900 and 1970, 8,000 cubic kilometres flowed into the seas surrounding Greenland, and that figure rose to 13,000 cubic kilometres from 1970 and 2000. During the 20th century, the current slowed down between fifteen and twenty per cent (127). Scientists believe that Europe will still continue to get warmer due to climate change, unless the Gulf Stream turns off completely. They are unsure of this likelihood, although Europe is expected to see increased storms and colder winters, and rainfall in West Africa, India and South America will be severely disrupted. Sea levels in North America will rise and the effects will be felt as far away as the Amazon rainforest and Antarctic ice sheets (128).

Whilst Europe may avoid completely freezing over, the Gulf Stream, along with the jet stream, are expected to drive extreme weather events. In research published in the journal Nature, scientists predict 'climate chaos' as water gushes from Greenland into the oceans. They expect weather to be strongly impacted within decades and temperatures on both sides of the Atlantic to vary more wildly from year to year (129).

The jet stream is like the weather distribution manager for North America. It decides where to send high and low pressure, and how strong to make them. When the jet stream is managing the weather effectively, weather rushes along a torrent from west to east, bringing rain every three to five days. Unfortunately, as the Arctic warms, and the temperature difference between the equator and the North Pole narrows, the jet stream acts unpredictably. The flow is strongest when the temperature variance is highest, but as the Arctic is warming faster than the equator, the flow has slowed down. Recently, the weather is not really rushing, it's more like a child dragging their feet, hanging around in areas for long periods. The June 2019 fires in Alberta, Canada, were caused by a high-pressure ridge hanging out and causing drought and fire. Then further out east, low pressure ridges cause rain and flooding which festers (130). Whatever weather pattern emerges, be it a drought, heavy rainfall, or heat wave; that pattern basically persists for longer and amplifies the associated risks.

It's not only floods and wildfires that are being blamed on the lingering jet stream. The abnormally high number of tornados that hit the U.S. in May 2019, was also thought to be the work of the slower, wavering jet stream. It's particularly difficult for scientists to pin the blame on climate change for this increase in tornados due to their rareness and the difficulty in observing their changes, but scientists believe a warming climate will help to produce the storms that manufacture tornados. *"Extreme warmth in and around Alaska, along with the reduction of Arctic Sea ice, affects the flow of the jet stream. We can't say that the rapid Arctic warming is causing this particularly pattern, but it certainly is consistent with that"*, said Jennifer Francis, senior scientist at the Woods Hole Research Centre (124).

Extreme storms around the world are anticipated to happen more frequently, and be more destructive, as the planet warms. From typhoons in the Pacific, to cyclones in the Indian Ocean, and hurricanes in the Atlantic, the storms we have seen in the past few years will become the new norm, until they too, are overtaken by larger, more destructive storms. Already, storms worldwide drop between 5-10% more rain than previously (131).

In Asia, there is evidence that typhoons are getting stronger. They are beginning to threaten the region's mega cities. Over the past thirty-seven years, typhoons to strike East and Southeast Asia have intensified by 12-15%. The number of the larger category four and five storms has doubled in that period, and in some areas, they have tripled in number. Their destructive power has also increased by 50%. Climate experts are warning cities from Tokyo to Guangzhou, Hong Kong, and Jakarta to prepare for future super strength typhoons. Even smaller storms are thought to pose a risk to these metropolises as infrastructure losses are expected to rise from $3 trillion in 2005 to $35 trillion in 2070 (110).

Asia is home to ninety-nine of the hundred most vulnerable cities on Earth, and the loss to life could be enormous in the coming decades as the tides surge, the winds rush, and the rains lash at both cityscapes and rural areas straddling the coast. The poorest will be hit the hardest, and even at just half a degree of increased warming, 1.5 billion are at high or extreme risk, according to a risk assessment carried out in May 2021 (132).

The world's second largest ocean has also witnessed hurricanes that produce 10% more rain in recent years. If our politicians fail to act, and the temperature is allowed to increase to 3-4°C, then rainfall from hurricanes could increase by 1/3 and wind speeds could be 53 km (33 miles) per hour faster. In a 3°C Warmer world, Hurricane Katrina, which killed almost 2,000 people in 2005, could drop 25% more water (131).

Cyclone Gafilo, which barrelled into Madagascar in 2004, and left 250 people dead and 300,000 homeless, would drop twice as much rain in a 3°C (5.4°F) warmer world, while Cyclone Yasi which struck Australia in 2011, would carry around 1/3 more water. Australia is a rich developed country, and may be able to ride out these storms, but low-lying countries like Bangladesh, and low-lying islands won't be so fortunate. Storm surges will wash away land, and leave millions without homes, food, and water.

The African continent will be hammered by devastating floods and storms leading to considerable disruption to farming. The population on the African continent is the fastest growing with annual rates of 2.5%, more than double that of any other landmass (133). As more and more people are added to the population, intense storms and flooding will damage vital food crops. Western, and central Africa will be worst hit with Niger, Nigeria, and the Democratic Republic of Congo among the most vulnerable to flooding. Scientists predict that the severe rain-fall witnessed today will become seven or eight times more frequent by 2100 (134).

Globally, one in four cities do not have money available to protect them from the impacts of the climate crisis. In a survey of over 800 cities, it was found that 43% of them, home to 400 million people, did not even have a plan in place to adapt to the coming crisis (135).

Wreckage left by Typhoon Haiyan in Tacloban, the Philippines, 21 November 2013.
DFID - UK Department for International Development is licensed under CC BY 2.0.

Hurricane Florence
by Astr o_Alex is licensed with CC BY-SA 2.0.

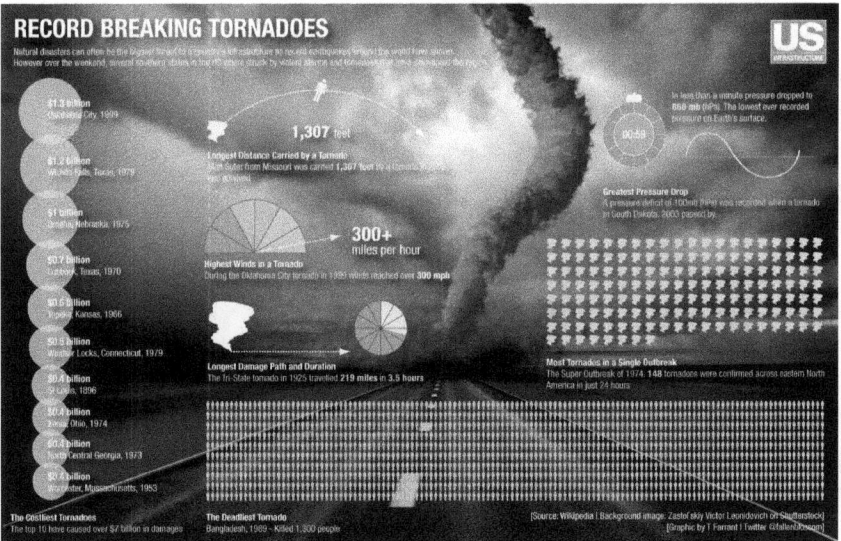

Record Breaking Tornadoes
by GDS Infographics is licensed under CC BY 2.0.

Hiroshima Landslides
Worst of the Disaster Area by GetHiroshima is licensed under CC BY-NC 2.0.

People take refuge on the roofs of buildings after Cyclone Idai in Mozambique
by DFID - UK Department for International Development is licensed under CC BY 2.0.

Dear Indy,

Living in Japan, natural disasters are nothing new for most people. The vast majority are used to earthquakes, flooding, landslides, and typhoons and those older than you, tsunamis.

The climate crisis promises to make things much worse though. Already, in the past ten years we have been close to numerous disasters that will be exacerbated by the rising temperatures. In 2014, some of my students and colleagues were affected by the 2014 landslides in Hiroshima Prefecture that claimed the lives of seventy-four people and caused widespread damage to property and infrastructure.

Four years later, water from the mountain above the village we lived in until you were six years old rushed down on top of the houses and roads. Just a year earlier, we had sold the house and moved to Kumamoto in Kyushu. Large parts of Tenno were covered in soil and debris. Cars were lifted on top of each other. Schools were closed, and parts of the train line were washed away. Our old pre-WWII house survived but those surrounding it were not so lucky. 126 people lost their lives due to flooding and landslides in more than twelve prefectures.

Shortly after moving to the southern island of Kyushu in 2017, torrential rain again caused huge flooding and landslides that caused the deaths of thirty-four people and required the assistance of 40,000 soldiers.

So far, we have been lucky when it comes to typhoons like Hagibus and Jebi. We avoided the worst there and our current home in the Seto Sea is somewhat protected from the super strength monsters that rear out of the Pacific Ocean, by the fourth biggest island of Shikoku. We are, however, in harm's way as our house is halfway up the highest mountain on the island and there have been landslides in the village in the last five years. Japan is at the front line of climate change and things will get far worse unless we cut emissions immediately.

Wildfires

It hasn't been only storms that have shook the world in the past few years. Wildfires also burn their way through 865 million acres (3.5 million sq. km) a year, from California, to Greece, Australia, Scandinavia, and even the Arctic (136).

In California, the deadly Camp Fire scorched more than 150,000 acres and took at least seventy-seven lives in 2018. In all, 10,300 homes were lost to the flames. Further fires torched the state more of a black than golden. Total wildfire damage is estimated to have cost $148.5 billion in 2018 alone (137). The trend is unsettling with six of California's ten most destructive fires occurring since 2015. In total, more than 8.5 million acres were burned in the United States in 2018. Now, whilst fires are a natural, and necessary phenomenon, the fires that incinerated the United States in 2018 were 30% larger than the ten-year average. Since 1972, California's burned area has increased more than five times, and research published in Earth's Future stated this increase was in fact due to the climate crisis (138).The U.S. and the Arctic saw record levels of fire activity in 2020 and according to insurance giant Aon, 2020 was the fifth most expensive wildfire year, behind 2017, 2018, 2015 and 2010 (125). As heat records were obliterated in 2021, 37,803 wildfires scorched the western United States and Canada and by mid-summer more than three million acres had been burned compared to 2.1 million acres from 32,059 fires by the same date in 2020 (139).

Climate change can't be blamed for all these fires, but it certainly makes them more likely, and impacts their size. Warmer and drier forests are just waiting for a spark (140). Two separate reports from

Environment and Climate Change Canada have linked wildfires to climate change. Due to warming, there is a longer fire season, drier fuel for the fire and more lightning. Cases of lightning rise by around ten to twelve percent with each degree of warming (141). Additionally, around four weeks after the winter snow has melted, the ground is considered dry enough to burn. As the planet warms, snowpack melts faster, and the fire season is brought forward as a result. Due to the warmer temperatures, increased evaporation also leaves the ground dry and parched. The warmer winters also allow bugs like the mountain pine beetle to survive and reproduce quickly in the spring. These beetles kill trees and brush, leaving them dry and more likely to ignite (142).

Colorado has also seen an intense escalation in the number of acres burned. In the 1960s, less than a hundred thousand acres succumbed to fire, but by the 1980s, this figure had leapt to more than 200,000. As the West warmed by more than 2°C (3.6°F) from 1970 onwards, the figure for the 2000s was just under 900,000 acres. In the Pacific Northwest, large fires have increased by 1,000%, the Northern Rockies by 889%, and in the Southwest by 462% since the 1960s. The fire season has lengthened by seventy-eight days, and more people are living in fire prone areas than ever before (143). The fire season in the Coastal Plains of the South-eastern United States has extended by as much as 156 days over the past 120 years and rainfall has decreased, meaning the likelihood of lightning ignited wildfires has increased significantly (144).

Europe also suffered with intense fires in 2018. Greece saw its worst fire in a decade. More than 700 people escaped into the sea to be rescued by the coastguard. Four people were not so lucky as their bodies were pulled from the ocean. Twenty-six adults and children were found hugging each other just meters from the sea, as flames stopped their escape. In total, 103 people perished (145).

In late June 2019, as temperatures reached 44°C, 250 firefighters in Spain battled against fires that destroyed more than 81 sq. km (20,000 acres). Two planes and five helicopters were trying to contain the blaze and around 400 people were evacuated from the village of Entrepinos (146). North of the border, where the mercury hit a staggering

45.9°C (82.6°F), 700 French firefighters were working to put out sixty different fires in the Gard (56).

As thermostats in Europe reached record temperatures in 2021, wildfires followed. In Greece, the flames surrounded the historical city of Athens with residents warned to stay indoors to avoid smoke pollution. In neighbouring Turkey, one-hundred blazes broke out across the scorched country and tourists in Mediterranean resorts were evacuated with fires raging around them as they escaped on boats (147).

In response to the fires raging in southern Europe in June 2022, on the back of record temperatures and prolongued drought, Victor Resco de Dios, professor of forest engineering at Lleida University said *"The concept of a fire season is losing its meaning right now. We have the fire season all year-round."*

Even the usually wet British Isles was on fire in late April 2022. A fire swept across 170,000 sq. m (42 acres) of Canford Heath nature reserve in Dorset, home to rare species of bird and reptile. Families from twenty homes were evacuated from a nearby housing estate. An official from the local fire and rescue services said an unseasonably dry April had led to the fire spreading quickly (148).

Australian wildfires are pretty common, but the fire season has lengthened, in some cases by months, and they've become more severe. In what has become known as the Black Summer fires which burned between 2019 and 2020, more than 1.8 million hectares were lost to the flames. The fire was so large that it accounts for 44% of all land burned since 1988. Around 3,000 homes were consumed by the flames, people were urged to stay indoors as the smoke engulfed Sydney, but the worst impacts were felt by wildlife with approximately three billion animals either perishing or being displaced (149).

Across the Indian Ocean, 25, 210, 000 square meters of forests went up in smoke in June 2019, as homes and livestock were lost. This destruction led to conflict between humans and animals. Indian fire incidents have risen by an astonishing 49.3% in the past three years. These fires have been linked to the rising temperatures and lower winter precipitation caused by climate change. As elsewhere, fires are not only larger now, but the season lasts longer (150).

South America saw some of its worst fires in fifty years in 2017. Three firefighters died battling 321,000 acres in central and southern Chile. Forest fires are common in Chile, but the 2017 outbreaks were caused by a prolonged drought that was attributed to climate change (151). Troublingly, fires in the world's largest rainforest, the Amazon, are getting larger and more frequent. This is mainly due to dry conditions, the use of slash and burn techniques to clear land for grazing cows, and more people living in the forest. The forest is also susceptible to globally changing climate patterns (136).

Perhaps, the most astonishing fires of 2018, were recorded in northern Europe. Sweden faced fifty wildfires at the same time and was forced to call on the international community for help. That help arrived from Italy, France, Germany, and Norway. While wildfires happen every year in Sweden, these fires occurred during a long heat wave where temperatures hit 32°C (57.6°F). "We have forest fires every year, but never so many big ones in such a short time," said the acting chief of an affected fire station, Gunnar Lundström. "It's an extraordinary summer. We've hardly had rain in two months, and it's been very hot. We never used to get temperatures above 30°C (54°F)." It is clear that as temperatures continue to rise, the number and size of wildfires will rise too (152). This is an example of a positive feedback loop in that as more trees burn, more carbon is sent into the atmosphere to mix with oxygen and produce CO_2 which in turn raises the temperature and causes more wildfires.

The Arctic was ablaze in 2019 with multiple fires, one the size of 300,000 soccer pitches, burning in Alaska, Siberia and Alberta, Canada. In just the month of June, these fires emitted fifty mega tonnes of CO_2 into the atmosphere. This is equal to the entire annual carbon emissions of Sweden. Whilst fires in the Arctic are normal, the fires in 2019 were unprecedented in their latitude and intensity. New fires were igniting every day as the fire season went on (153). By 2021, Siberia was witnessing fires that were larger than all other fires on the planet combined.

Prior to 2017, the region could expect around two large fires a year but in 2021, there were between thirty-forty large fires being fought and a senior pilot-observer with Russia's federal Aerial Forest Protection Service said *"Now is crazy. There are too many fires and pretty much all of them are major"* (154).

An even more worrying impact of wildfires is being felt under the ground high up in the Arctic Circle. Here, so-called zombie fires are burning through Canada, Alaska, Greenland, and Russia. They get their name because they are capable of coming back to life after being extinguished. They achieve the seemingly impossible by burning through underground methane deposits in the winter and when the snow melts and the soil dries out, they ignite once again. These were once considered a rare phenomenon, but they are now causing up to 38% of fires in the far north (155). This is a major concern in our fight to contain even more dangerous warming as methane deposits sitting under the permafrost are at danger of rapidly escaping into the atmosphere.

Wildfires are an interesting example of a natural disaster that have both positive and negative feedbacks. As the forests burn, carbon stored in them is emitted into the atmosphere. The carbon then mixes with oxygen to become CO_2, and more heat is trapped in the atmosphere leading to more fires. In each of the past twenty years, eight billion tonnes of CO_2 has been added to the atmosphere due to wildfires. After allowing for the regrowth of forests, scientists estimate that between 5-10% of all CO_2 emissions are caused by wildfires. As we attempt to decarbonise the economy, the contribution of wildfires will likely increase. In 2017, blazes in Northern California's wine country emitted as much carbon as all the cars and trucks for an entire year. Wildfires also produce aerosols, one of which, black carbon, absorbs heat while it hangs in the air, and heats the atmosphere roughly two thirds as much as CO_2 (156).

Conversely, some of the aerosols originating in wildfires help to reflect sunlight, and in turn they cool the climate when injected into the upper atmosphere. After fires in Canada in 2017, the level of aerosols in the atmosphere over Europe was higher than after Mt. Pinatubo blew her top in 1991. The volcanic eruption in the Philippines cooled the planet by around 0.4°C (0.72°F) from 1992-1993 (157). Unfortunately, this cooling effect only lasts at most for a matter of months. And once the skies do clear, the particles will fall to the ground when it rains. This creates another albedo affect as the dark ash from wildfires lands on the ice and speeds up heating. This was the case in 2012 when wildfire pollution was a major cause of the record surface melting of the Greenland Ice sheet (156). This will become a growing problem as wildfires move farther north, as they have been in recent years.

The Amazon rainforest is expected to experience more frequent but abnormally intense fires in the coming decades. The negative feedback loop seen here is that once forests have burned once, they tend to burn again. After a fire, the forest is thinned out and allows more light to reach the ground which leads to drought, and more flammable grasses appearing (136).

The effect a warming climate has on wildfires cannot be overestimated. By 2050, burned areas are expected to increase by 60%, and fire season will last twenty-three days more. Australia, Europe, the boreal forests in Russia, and Canada as well as North America will be especially affected. Burned areas are vulnerable to erosion, flooding, mudslides and landslides, and the debris and sediment from fires also pollutes water supplies. By 2050, the annual global cost of climate change caused wildfires might be more than $100 billion (158). By 2090, Europe could see a 200% increase in the amount of land burned. Russia is expected to experience a steep increase in the number of fires by the end of the century (136). Overall, the Union of Concerned Scientists (UCS) predicts that by the end of the century, wildfires will become four, five or even six times more damaging (159). Even more worryingly is that in the U.S. since 2013, areas of forest that are burned are not growing back. One third of areas burned in the Rocky Mountains since 2000 have no trees regrowing at all. It is estimated that around 15% of future burned areas in the Rockies may not regrow and north of the border, in Canada, about 50% of forests are predicted to vanish by 2100 (160).

AUSTRALIA FIRES EXPLAINER 2
by brucedetorres@gmail.com is marked with Public Domain Mark 1.0.

Nuked: Middletown, California, After the Wildfires
by Bob Dass is licensed under CC BY 2.0.

Drought

The warmer climate also affects rainfall as more water is evaporated causing droughts. A warmer atmosphere can also hold more water meaning that when it rains, it often arrives as a deluge of water that floods areas and washes away homes and soil. Europe saw record temperatures in 2018, and the continent basked in a heatwave that lasted most of June and July. Precipitation was also very low. Fields were burning up, and many farmers couldn't grow enough grass for animals. Many farmers chose to send animals to early slaughter and feed meant for winter was used in the summer months. Production of British staple crops such as potatoes, carrots and onions dropped by around 20%, wheat by 25%, and the average weekly shopping bill rose by 5%. Milk yields were also depressed, and the fertility of pigs was affected (161). Whilst it is not possible to blame these droughts on climate change alone, scientists are studying whether the change in ocean surface temperatures possibly blocked the high-pressure weather system and kept the hot dry conditions in place for such a lengthy period. *"It's (drought) certainly been intensified by climate change and these events will unfortunately become more frequent,"* said Jean-Pascal van Ypersele, one of the world's leading experts on climate change (162).

The United States is also suffering from drought. California's seven-year drought ended in 2019, the state went 376 consecutive weeks with drought of one form or another stretching back to 2011. Arizona, New Mexico, Oklahoma, and Texas have all been experiencing exceptional drought which is the highest category on the drought scale. Unfortunately, when it does finally rain, the soil is so hard that the water just

runs off (163). Incredibly, by using the Palmer Drought Index, an unbelievable 100% of the U.S. West was in drought in 2021 and this broke a
122-year record. Even using the standard U.S. Drought Monitor, a staggering 90% of the West was in some sort of drought. A climatologist
at Desert Research Institute and the Western Regional Climate Centre
referred to the situation as 'truly historic' and that the word 'crisis' was
no exaggeration (164). By 2022, the western U.S. was facing its worst
"megadrought" in 1,200 years and scientists writing in Nature Climate
Change stated the drought was made around 40% worse due to climate
change. They expressed a strong probability that the drought would last
until 2030 (165).

Another country that is buckling under multiple climate threats is
Australia. The driest continent experiences its fair share of droughts,
floods, heatwaves, and wildfires, but the drought in 2018 was estimated
to be the worst in 800 years. The eastern states experienced their lowest
September rainfall on record and the second lowest for any month.
Precipitation was 11% lower than average for Australia as a whole and
the lowest since 2005. The government estimated that crop production
would drop by 23%. Many farmers were culling their animals in an
effort to reduce water use. Record heat in 2019-2020 took New South
Wales into its third consecutive year of severe drought. This period saw
the lowest amount of rainfall in one hundred years. Lakes dried up,
millions of fish were killed, and farmland was parched (166). Australia
has warmed by more than a degree Celsius since 1910, but most of
that added heat has come since 1979. Andrew King from the University
of Melbourne believes climate change is making droughts worse. As is
always the case, it is difficult to blame weather events solely on climate
change, but with increasing temperatures, evaporation will be higher
across Australia and this will likely worsen drought conditions, not
improve them (167).

The civil war raging in Syria has killed hundreds of thousands of
people and produced millions of refugees, sparking angry conversations
in Europe about the responsibility to accept people in need. The war
was in part caused by a drought from 2006 to 2009 that left many

farmers unable to grow crops with many migrating to Damascus in search of jobs. Scientists say this is part of a century-long trend toward warmer and drier conditions in the Eastern Mediterranean (168). More than twelve million people in Syria and Iraq are threatened by drought with the Norwegian Refugee Council predicting that *"The total collapse of water and food production for millions of Syrians and Iraqis is imminent"* as the Euphrates and Tigris Rivers run dry. According to the Danish Refugee Council's Gerry Garvey, the water crisis is only going to get worse and *"It is likely to increase conflict. There is no time to waste. We must find sustainable solutions that would guarantee water and food today and for future generations"* (169).

A debate also raged across the U.S. about responsibility to accept immigrants from Central America. Then president Trump declared he would build a wall on the U.S.'s southern border to repel would be migrants who had made the arduous journey through the desert. A caravan of people from Guatemala made their way north in late 2018 in desperate need of security. Drought was again largely to blame for this exodus of people. The weather has become completely unpredictable in recent years and crops are failing. As a last resort, people are leaving the country and heading to the United States in search of work (170).

As temperatures rise in southern Europe, tensions are also heating up between Spain and Portugal. Rivers are drying up, reservoirs are shrinking rapidly, some by 60% in just five years, crops are ruined, and wildfires are becoming more frequent. Hydroelectric power production is waning due to a lack of water which is pushing up electric prices (171).

France saw record heat that led to prolonged drought throughout the summer of 2019. Water was restricted in sixty-one regions with nuclear power generation being impacted due to low flow on the Rhone River which supplies coolant water for EDF's nuclear plants (172). Things didn't improve in 2020 with fifty-five departments in France imposing water restrictions after the third driest July since 1994. The harvest was badly affected, and farmers had to choose which plants to water and which to leave die (173).

The island nation of Taiwan is also waking up to the possibility of a future without water. In May 2021, following its worst drought in fifty-six years, some of its reservoirs were at 5% capacity and its famous Sun Moon Lake was showing off its dry cracked lakebed. Mass fish die offs were being reported as the island lamented the loss of usually very destructive, but essential typhoons. Not one of the many storms that usually batter Taiwan each year made landfall in 2020 and the rains from these typhoons are required to fill up the reservoirs. The island's leaders were clouding seeds in an attempt to make it rain while one of the island's most important industries was assuring its customers that supplies would not suffer. The semiconductors used in cell phones and cars make up around 4% of GDP and use huge amounts of water to produce (174). The future for Taiwan may entail a choice of providing water for industry or agriculture.

Central Asia, which stretches from the Caspian Sea in the west to China in the east and from Iran and Afghanistan in the south to Kazakhstan in the north could be about to become a hyper arid desert. The area is already dry, but research has identified it is under threat from climate change and land degradation which is causing seas of sand to grow at unprecedented rates. The area is home to half a billion people who are now seeing their crops ravaged by drought and deserts reaching towards the cities (175) .

In northern Africa, Morocco is facing up to life in a water stressed environment. Their second largest reservoir, Al-Massira, has shrunk by 63% in just three years. As Moroccan water supply declines, demand is expected to rise between 60-100% in most large cities by 2050. A further problem for Morocco is that both industry and domestic users are given priority over water for agriculture while farming provides employment for a third of the population. It is further susceptible to conflict as it is in the middle of the Fragile States Index (171).

Somalia has been facing its worst drought in thirty-five years, with more than two million people at risk of starvation by the end of 2021. Eighty percent of the country was affected and a further 3.2 million people were living hand to mouth, unsure where their next meal would come from. There has been significantly less rain than usual, and crops and livestock have been devastated. The failure of the rainy season is partly blamed on Cyclone Idai that hit Mozambique in March 2019. The cyclones stopped rain heading further north. A two-year long drought ended in Somalia in 2017 (176). The dry season has become longer in the past few years and with two thirds of the population living in rural areas, Somalians are in a very precarious situation, and many have already been displaced. Half the population is reliant on foreign food aid, and with other international events such as the crises in Yemen and Syria, the response from the international community has not been sufficient thus far (177).

On the northwestern border of Somalia, lies Djibouti, a small country with a population of around a million. The country is one of the hottest and driest on Earth. Most food is imported from Ethiopia, as is electricity, and water is extremely scarce. Djibouti has warmed by around half a degree Celsius in the past thirty years, and with just that half degree, the effect has been catastrophic. The warmer temperatures have dried out wells and rising sea levels have polluted coastal water sources with salt. The country used to produce an abundance of fruits and vegetables, but now, nothing grows. Natural pools used for swimming have long since dried up, and the once flowing Ambouli river can now be crossed by motorbike. Many villages have been deserted, and livestock numbers have plummeted due to the disappearance of vegetation. This has resulted in malnutrition for many people, and most now rely on water being delivered in blue barrels by the government. As the country straddles the strategically important Horn of Africa, only twenty-nine km (18 miles) away from Yemen, the government leases out land for military bases to France, the US, Italy, Japan, and China, who want to protect their energy supplies. 4.8 billion barrels of crude pass through the Bab el-Mandeb Strait every day to provide energy for these rich nations. The crude oil is then burned and causes the world to warm even more. That Djiboutians are reliant on $300 million rent to pay for water due to warming temperatures caused by burning the very crude these visiting nations are here to protect, is ironic to say the least (178).

In Southern Africa, Namibia declared a national disaster as the country was gripped by drought in 2019. Around 500,000 people were thought to be affected, out of a total population of just 2.4 million. The government took the unusual step of auctioning off 1,000 wild animals, including buffalos, springbok, oryx, giraffes and elephants. The auction was in response to the deaths of 63,700 wild animals thought to have died due to poor grazing conditions brought on by dry weather in the national parks. The government planned to raise $1.1 million to help fund conservation (179). As wild animals were being sold to the highest bidder, the number of livestock being slaughtered in the country dropped by 66% in 2020 due to the drought. As 63,700 wild animals died from thirst, 116,000 cows were fed adequate amounts of water to still make it to slaughter in 2020 (180). Could the two facts be related? We will see in a later chapter.

The situation in Africa makes for grim reading. The capital of Zimbabwe, home to 4.5 million people, saw half of its inhabitants lose access to water in July 2019. More than two million people in Harare were without running water due to drought with people depending on water merchants to meet their dire needs (181). Kenya also experienced a 50% drop in rainfall in May 2019, with the number of affected people rising from 2.1 million to 2.4 million. In September 2021, the President declared a national disaster (182).

More than a quarter of Lesotho's population of 2.2 million suffered from food insecurity due to the double Cyclone disasters to hit the country in 2018, Mozambique had 1.85 million people in need of aid. South Sudan faced one of the worst situations of all with more than seven million people facing a catastrophe of food insecurity and conflict. Around 50,000 people were experiencing "famine-like" conditions and 860,000 children were severely malnourished. In Ethiopia, there are ten million people in danger because of consecutive years of drought. The 2019 Gu long rains arrived eventually after a delay, but they were too late for the growing season and the situation regarding food security is expected to worsen. In the maize growing south, Zambia experienced their worst drought since 1981, and production dropped significantly enough for exports to be banned. Zimbabwe also saw a 70% drop in Maize production, while cereal output was nowhere near the required level of 1.8 million tonnes. The harvest of 852,000 tonnes left many people in severe crisis (183). Madagascar was facing a massive drought in March 2020. Around 730,000 people became food insecure as the rice crop dropped 75% with maize and cassava also affected (184).

The obvious paradox here is that the poorest continent on Earth, with more that forty-five million people in fourteen countries struggling to find food, is reliant on food aid from rich nations who acquired their wealth by burning the very fossil fuels that are causing the dire situation facing Africa today.

In South Asia, a disastrous drought has been gripping both Pakistan and India the past few years. Hundreds of villages have been abandoned to the parched elements. The situation is worse than the 1972 drought that affected twenty-five million people. Millions of people on either side of the border are struggling to survive in record temperatures that have topped 50.8°C (91.4°F). In May 2019, 43% of India was suffering from drought conditions. In the Indian states of Karnataka, 80% of districts were hit by drought and crop failure, while the number in neighbouring Maharashtra was 72% (185). 20,000 villages had severe drinking water crises, and eight million farmers struggled to survive. More than 5,000 farmers have committed suicide in the past five years. Crops withered in the scorched sun, and livestock starved to death. A total of thirty-five dams had no water left in them. People who could afford it, paid for private water to be delivered, but this was usually the dredged muddy remains of the dams, and the salty liquid made people ill. Even thirsty cows refused to drink the water. There was a 50% increase in patients suffering from diarrhoea, and gastritis. Each day began afresh with people searching for water in bore wells, but the groundwater had been depleted unsustainably, and was running dry. This ground water provides 40% of Indian water needs.

Already, fights were breaking out in the sixth largest Indian city of Chennai as all four of its reservoirs ran dry. The city is home to 4.4 million and most of these people were reliant on government tankers to distribute their water. Some small restaurants closed, and many people were being told to work from home to save their company's water supply. The city's metro system turned off the air conditioning in stations and hotels were rationing water for their guests (186). Special trains were used to transport water from Vellore to help alleviate the drought. In mid-July 2019, four daily trains were used to deliver water from 125 km away. Each train carried 2.5 million litres of water (187).

The situation in Pakistan is similar with 2019 bringing the harshest and longest drought in recent memory. Many families were selling their livestock or seed stock to survive and migrating 200 kilometres to Hyderabad. 71% of people were suffering from food security, and a third severely. Aid groups called for urgent food supplies, medical assistance, clean water, and assistance with agricultural recovery (188).

Climate change is also affecting lakes around the world. The lakes are warming faster than the air or seas. The increased evaporation is drying out entire areas and having disastrous effects on the people who rely on them for food and water. Lake Poopó used to be Bolivia's second largest lake, but it has completely disappeared. People had been living on the lake for thousands of years, but the villages have all been abandoned. Lake Chad in Africa is only a twentieth of the size it was just thirty-five years ago. Poyang Lake in China has been reduced from 4,500 square kilometres to just 200 square kilometres today. Poor water policies are partly to blame for the disappearance of once thriving lakes, but according to Catherine O'Reilly, an aquatic ecologist at Illinois State University, *"The fingerprints of climate change are everywhere, they don't look the same in every lake"* (189).

Going forward into the brave new world our actions are creating, the biggest concern globally, and regionally, will be a shortage of water (190). H_2O is possibly the most common molecule found on Earth, but while almost 70% of the world is covered by water, only 2.5% is fresh, with the remainder containing salt. Of that small percentage, only 1% is easy to access with most frozen in glaciers and snow. Only 0.007% of Earth's water is easily available for the 7.9 billion humans, and all other animals on the planet (191). If you were to put all that water inside a ball, it would be just 35 miles (56km) in diameter (192). Even without the shadow of climate change, with a rising population expected to reach ten billion by mid-century, water would be a major concern. By 2020, a staggering 30-40% of the world was affected by water scarcity with many areas having no access to clean drinking water (193). If temperatures increase by 2°C (3.6°F) by 2042, it's estimated that a quarter of the planet will be in permanent drought (194).

Few people living in developed countries ever have to think about water. They just turn on the tap and fill their cup. For people unfortunate enough to be born in developing countries, there is no such luxury. Obtaining clean drinking water can be a laborious process, and success is never guaranteed. Currently, one in every three people lack access to freshwater, either because it is unsafe, unavailable, or unaffordable. Most of these people live in developing nations, but with the climate crisis, people in some developed nations may find out first-hand, just how lucky they have been. By 2025, water researchers warn that 4.5 billion people will either be running out of water or be unable to afford the precious blue gold (192).

When droughts do occur, hosepipe bans are usually enforced on individuals, but the majority of Earth's freshwater is actually not consumed by individuals, but by agriculture, which accounts for 92% (195). Only a fraction is used in our homes for showers, cooking, and drinking. As people become more affluent, as they are in China and India, they will want to use dishwashers, washing machines, and consume more meat, and this will only drive-up demand for water. Globally, seventy billion farmed animals are killed for food each year. This industry uses 29% of all our freshwater resources at a time when billions of humans struggle to access water. Meat consumption is expected to double between 2000-2050 (196). At the same time, the population is increasing by 220,000 people every day, or 150 more people every single minute. As eighty million humans are added to our population every year, the population is heading to eleven billion by 2100. All these extra mouths to feed are going to put enormous stress on our water supplies just as the world begins to run out of fresh water. If our global population hits the projected ten billion mark by mid-century, then our demand for water is expected to grow by as much as 50% (192). By 2030, our demand for water is expected to outstrip supply by 40% and half of humans will face severe water stress (197).

Our warming climate is already affecting precipitation around the world, causing droughts and migration. As the temperature continues on its upward trajectory and the climate crisis grows, we can expect the impacts to grow too.

Many areas are predicted to see drying soils and an increased drought risk. The Mediterranean, Africa, Central America, southwest U.S., and the subtropical areas of the southern hemisphere are predicted to be most affected by declines in rainfall and increased evaporation.

When the rain does fall, it is forecast to be in shorter and more intense spells. This will help to raise water levels in rivers, lakes, and reservoirs, but as the soils will be parched and dry, much of this water will runoff the land and not increase soil moisture or raise groundwater levels. Droughts will become an important part of our future unless steps are taken to reduce carbon emissions immediately (198).

Across Africa, at least $66 billion needs to be spent on providing universal access to water and sanitation, but there is a huge short-fall (199).

The situation in southern Europe is expected to deteriorate as droughts are expected to happen more often and the severity will increase. By 2040, many parts of Spain are projected to be water stressed (171).

The largest rainforest on the planet should not be vulnerable to droughts, but due to extensive deforestation, mainly to satiate the world's demand for cheap meat, the Amazon is undergoing huge changes. Recent studies have shown that deforestation leads to both warming and less rain in deforested areas. This then leads to less mois-ture in the soils and has the potential to cause recurrent droughts, like those seen in 2010, 2012-2013, and then in 2016 (200). Central America is likely to experience less rainfall and increased droughts which will affect food security, especially of the poorest (201).

Whilst, undeniably, it will be necessary to use desalination plants to provide our growing appetite for water, these plants carry significant financial, environmental, and social costs. For every gallon of fresh-water produced, around two gallons must be taken from the ocean. This isn't just sea water that is taken from the ocean. It is an ecosystem containing phytoplankton, fish, and invertebrates. The damage caused by desalination plants hasn't been fully studied, but it is estimated to be akin to losing the biological productivity of thousands of acres of habitat. In addition to the problem of removing seawater, there is also the added problem of discharging the salt brine and chemicals after removal.

Removing water from the oceans, separating the salt, and then dumping the salt back into the depleted ocean will increase salinity. This is denser than the ocean water which has half the salinity of the water being dumped and will sink to the bottom of the ocean near the plant. There will be a layer of brine on the sea floor with elevated salt concentration and this increases the problems for marine organisms along the seafloor. Lower levels of oxygen have also been recorded on the seafloor near desalination plants. It is estimated that around 3.4 billion fish and other marine organisms are killed annually by desalination plants. Imagine the number of deaths if desalination plants provide for the majority of freshwater supplies. If you aren't worried by the deaths of billions of fish, consider that fishermen lose at least 165 million pounds of fish a year and their future potential catch declines by 717.1 million pounds (202).

In addition to the salty brine discharged, the desalination process involves the use of many chemicals. Ferrous chloride, aluminium chloride, biocides, anti-foaming additives, and detergents are some of those used. Heavy metals such as copper, zinc, and nickel can also be released into the waste stream from equipment that corrodes (203). Seawater also contains boron, of which only 50-70% is removed when desalinated. Boron has been found to cause reproductive and developmental problems in animals. It also causes irritation of the human digestive track (202).

Desalination plants are between two and four times more expensive than traditional options. The Carlsbad plant in California cost $1billion. This plant provides around 10% of the region's water, but accounts for a quarter of the cost (192). In rich developed nations, these costs can be afforded, but in poorer countries it will be more difficult, if not impossible.

As climate change affects our water supply, desalination plants contribute to climate change. Removing salt from water uses nine times as much energy as surface water treatment and fourteen times more energy than groundwater protection. In the U.S., the energy needed to move, clean and use water already accounts for around 5% of annual emissions. Desalination plants on average use about 15,000 kilowatts for every million gallons of fresh water produced (204). Like air conditioners adding heat to the world outside, as they cool the inside, removing salt from seawater contributes to the climate crisis that necessitates the need for desalination in the first place.

The final problem of desalination is social. Corporations like Coca-Cola and Nestle will make huge profits as water is completely commodified, and unless people have enough money, they may not be able to afford something that once fell from the sky for free.

"WHEN THE WELLS DRY, WE KNOW THE WORTH OF WATER."

BENJAMIN FRANKLIN

Starvation

According to the United Nations Food and Agriculture Organization (FAO), there are 815 million people who suffer from chronic hunger. That's approximately a tenth of all humans on planet Earth (205). This is not a problem of scarcity. We already grow enough food for ten billion people. It is a problem of economics and efficiency. Millions of people around the world, especially in Africa and Asia simply do not have the money to buy food, while globally, 1.3 billion tonnes of food is wasted annually, and this is enough to feed over three billion (206).

The other problem involves what we choose to eat. The United States alone could feed 800 million more people if they just gave the food they feed to livestock to the people going hungry. This is not to single out the US, as many developed countries also follow an unsustainable diet. It is simply to highlight the problem of food inefficiency that exists. For example, raising chickens to edible weight requires 4.5kg of edible food to produce 1kg of edible chicken. That number rises to an astonishing 25kg when raising cows for food (207). The graph on the next page highlights the enormous inefficiency of raising animals to eat.

In the coming decades, the warming temperatures are expected to devastate food supplies. For every degree of warming, cereal crops decline by 10%. Our increased population is expected to top ten billion by the middle of the century and could be more than eleven billion by 2100. A 3°C warmer world would see crop yields fall by 30%. Corn yields in the world's biggest producer, the U.S., are expected to drop by half at 4°C of warming. China, Brazil, and Argentina could see yields fall by at least a fifth. At 2°C of warming, the Mediterranean and most of India is projected to be riven by droughts, and food security will be hit hard (208). Our species could require more food than it's possible to produce at just 2.5°C of warming. At 3°C, droughts will affect Central America, Pakistan, the western United States, and Australia. Central America is estimated to see significant drops in food production, with beans dropping by 19%, and maize by up to 21% by 2050 (209). The U.N. is predicting that cereal prices could rise by 29% due to the unfolding crisis (210).

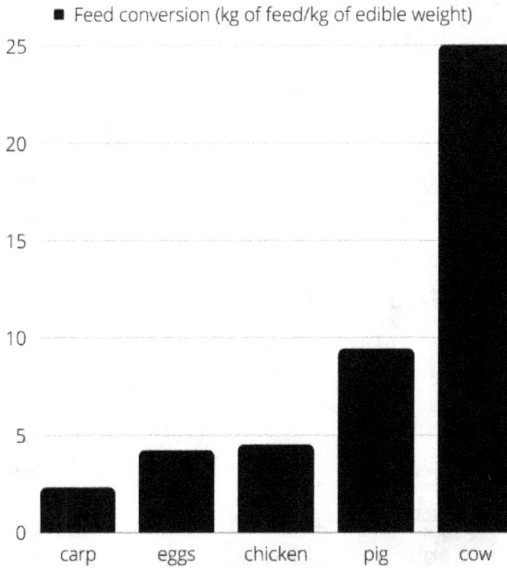

Feed conversion (kg of feed/kg of edible weight)
Vaclav Smil

In Africa, 65% of people are employed in agriculture which accounts for 32% of GDP. Food production has increased since 2000, but the rise has not been enough to satiate demand. This growth has mostly come from using more land for agriculture which has had a devastating impact on wildlife. This will be discussed in the next chapter. The growth has now plateaued, and by 2030, Egypt predicts that wheat production will fall by 15% if temperatures increase by 2°C. Morocco also expects its wheat production to rapidly fall after 2030. In sub-Saharan Africa, 95% of food grown is reliant on rainwater, which makes it especially susceptible to reduced rainfall and increased temperatures (211). The warming climate is causing Biblical plagues in Africa with Kenya, Ethiopia, Somalia, Djibouti, Eritrea, and Uganda all suffering from a locust invasion in 2019. Over 1,000 sq. km of crops were devastated and tens of millions were impacted. Heavy rains from two Cyclones fell on Yemen and the warm moist conditions were the perfect breeding ground for the locusts who then spread out across the Arabian Peninsula and across the Horn of Africa. With warming seas, these storms will grow in number and climate scientists at the Food & Agriculture Organisation (FAO) warn that this will lead to larger and more frequent locust outbreaks in the future (212). The poorest continent, which has contributed the least to the climate crisis, is paradoxically one of the most vulnerable places on Earth, and food security will be a major problem over the next 80 years (134).

A further problem is that the food we are producing is less nutritious than it was. Since 1950, protein, calcium, iron, and vitamin C content has declined by a third. As more and more carbon is added to the atmosphere, our food becomes less nutrient dense. This could lead to the deaths of 150 million people in developing countries by 2050 (208).

People in rich countries may think they will be buffeted from the threat of starvation, but today, the globally integrated food system means that what happens in one country will affect all the others. In the U.K., a staggering 88% of fresh fruit is imported and only 58% of fresh vegetables are grown locally (213). Japan is another rich country that relies on imports to feed its population. Japan's self-sufficiency has decreased from 79% in 1960 to only 39% in 2015 (214).

At present, the world is able to cope with location specific droughts or crop failures by sourcing from unaffected areas. The climate crisis may throw a spanner in the works. The four largest corn exporters on the planet are the USA, Brazil, Argentina, and Ukraine. All areas are expected to see significant drops in yield, between 8% and 18% at 2°C

(3.6°F), and 19% and 47% at 4°C (7.2°F). These losses are by themselves not insignificant, but the chances of them all suffering yield losses at the same time is what has scientists worried. This situation is known as simultaneous crop failure. At 2°C (3.6°F), the risk of all four crop areas failing is 7%. If the temperature rises to 4°C (7.2°F), this risk soars to a staggering 86% (215).

While storms, wildfires and sea level rise receive most attention, the real worry revolves around the food supply. As the population increases and extreme weather events proliferate, the U.N. is warning that over half a billion, in India alone, will be vulnerable to food shortages should the temperature rise by 2°C (3.6°F) (210). As we saw, we will likely hit that figure in the next two decades and without a massive, concerted effort, the temperature will continue on its upward trajectory, meaning less stability in the global food system. Many will point out that northern areas will be able to grow more food and while this might be true, the added amount will be nowhere near replacing what we lose further south. If proof is needed to support the reality of our situation, it arrived in early 2022 when the price of pasta in the U.K. rose from 55p to 70p as a direct consequence of the extreme heat and flooding that hit Canada in 2021. Canada grows two-thirds of the world's durum wheat – a key ingredient in the staple (216).

As we will see in a later section, climate change is likely to increase the likelihood of conflict and while the Ukraine war was not a result of climate change, the fighting there can give us a glimpse of our future. Combined, Russia and Ukraine supply around 30% of global wheat exports. North Africa and the Middle East are particularly impacted by the tension as they consume the most wheat per capita (128 kg) and were reliant on the two countries for 59% of their supply in 2020. Due to the crisis, cooking fuel in Sudan has jumped 56% and oil by 67%. This helped to push wheat prices up by 61% between August 2021 and February 2022 (217). Globally, the price of wheat rose by 37% by April 2022 and corn by 21%. Ukraine supplies around 16% of world corn exports (218). To compound matters even further, an early record-breaking heat wave in India cut their wheat harvest and their plan to make up some of the export shortfalls from the Ukraine war has now been shattered. Nine cities in India saw temperatures pass 45C (113F) in April, but March saw the hottest temperatures since records began. Wheat is particularly sensitive to heat and this heat wave stunted crops (219). India can now expect similar heat waves every four years and coupled with drought in Paraguay and Brazil which reduced soybean production, typhoon related flooding in Malaysia which impacted palm oil supply and we start to see a pattern emerge with food prices continuing to rise as production is limited due to increased climate related disturbances to crops (220).

With increased temperatures and proliferating natural disasters, more and more crops will be affected and those who can ill afford the increased prices will be hit hard. Throughout history, food shortages have been the cause of civilization collapse, and as we saw in Syria, the stress that follows food insecurity often leads to violence and conflict. As warming destroys fertile lands, it will also melt ice and raise seal levels, which in turn will reduce our available farming land.

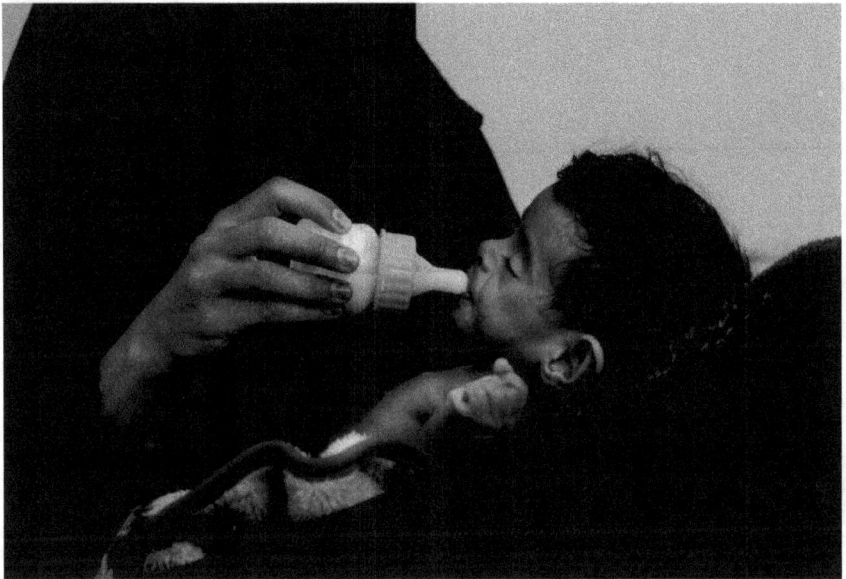

502 Starvation in Yemen
Felton Davis is licensed under CC BY 2.0.

"THE GREATEST THREAT TO OUR PLANET IS THE BELIEF THAT SOMEONE ELSE WILL SAVE IT."

Robert Swan

Dear Indy,

It's hard for most of us in the developed world to understand the problem of going hungry. We all use the phrase "I'm starving" when we are hungry, but few of us understand what real hunger feels like. Most of us don't even know what the term kwashiorkor means . This is a form of malnutrition caused by a lack of protein. It only affects people who don't have enough food to eat. It doesn't affect anyone in developed nations.

We are finally seeing what the future could hold in Spring 2022 with the combined impacts of the Russian invasion of Ukraine and global warming hitting wheat supplies hard. With Russia and Ukraine supplying around a third of wheat exports, market prices went sky high with the war delaying harvest and preventing transport. India was hoping to make up the shortfall but their own crops were ruined by record heat in early 2022 when temperatures hit 50°C.

This was a year after global durham wheat prices rose sharply as the world's largest exporter, Canada, experienced wildfires first, and then floods which destroyed crops.

While many people in developed countries might have been able to afford the rise in prices thus far, many in the global south have not been so lucky. Things will not stay like this forever though. As temperatures continue to rise, and water supplies diminish, many people will face the threat of starvation. This will ultimately lead to conflict and will lead to further food supply problems and further conflict.

The answer to this problem is obviously to prevent further warming by transitioning to renewable sources of energy like solar, wind and tidal, but it is also paramount that we use the food we have as efficiently as possible. That is going to mean eating further down the food chain. We can not longer sit idly by as 800 million of our brothers and sisters suffer from malnutrition and we continue to feed 70 billion farmed animals. If we are to live on this planet sustainably, then we need to start eating the plants we feed to animals directly.

Sea Level Rise

The most obvious consequence of a warming world is melting ice, and the accompanying sea level rise that will bring. Water also expands as it gets warmer which further increases sea level. Between 1901 and 2010, global sea levels rose by about 19cm, or 1.7mm annually. But between 1993 and 2010, the annual rise of the seas almost doubled to 3.2mm a year (221). Like everything associated with climate change, it's difficult to estimate how much the waters will rise. All we know is that rise they will. How much, and how fast depends on what humans do, or don't do, in the next decade.

The scientific community had for years believed that sea level rise by 2100 would be less than a meter. In 2016, two reports came out shortly after one another stating that in fact, sea levels will rise by one meter, even if countries meet the pledges they made in Paris in 2015, and keep warming lower than 2°C (3.6°F). Even at 2°C of warming, around 20% of the world population will need to migrate away from the coast. Let's, for arguments sake, imagine that humans do carry on burning fossil fuels, meat and dairy consumption continues to increase unabated, and deforestation continues apace, how much sea level rise can we expect in this scenario? The two reports claim that sea levels could rise to more than two meters by the end of the century (222).

The World Resources Institute predicts that the number of people affected by flooding will double by 2030. More than 147 million will be impacted and the costs will stretch to $712 billion a year. By mid-century, the researchers found that costs would increase to $1.7 trillion a year and impact 221 million. Most of this flooding will happen in south and south-east Asia, including Bangladesh, Vietnam, India, Indonesia, and China. A different report stated that once-in-a-lifetime floods today could happen every day along much of the U.S. coastline by 2050, and with just 5-10 cm of rise, flooding chances will double. Flooding is expected to cost the U.S. $38 billion by mid-century, a huge increase from the $1.8 billion in damages in 2010 (223) (224).

We have already seen that temperatures are expected to rise to 3.1–3.5°C (5.6–6.3°F) by 2100. Two-hundred-seventy-five-million people currently live in areas that will be flooded at 3°C (5.4°F). Asia will be the worst affected with the greater Osaka, Hong Kong, and Shanghai areas alone home to 31.1 million people. Cities around the world, including Miami, New Orleans, Houston, Atlantic City, Charleston, Boston, Virginia Beach, New York, London, Rio De Janeiro, Mumbai, and Alexandria will also be submerged. The good news is that sea level rise will not happen suddenly when we hit 3°C of warming, but the bad news is that it will be irreversible (225).

With just a couple of meters of sea level rise, which is almost certainly locked in for 2100 (should we not cut emissions drastically), our world will be markedly different. World maps will have to be redrawn, and vast numbers of the human population will have to migrate further inland to get out of the water's way. Large deltas that currently provide food for many millions of humans are actually sinking as young sediments compact rapidly. So, as the sea rises, the ground also sinks in many vulnerable places, which more than doubles the potential sea-level rise (226).

With the albedo effect, the worry is that the positive feedback loop of darker water absorbing more heat might lead to runaway climate change. Greenland and Antarctica contain 99% of the planet's fresh water. It would almost certainly take centuries to occur, but if all Greenland and Antarctica, were to melt into the world's oceans, the seas would rise by 66 meters (227). Both are melting and the pace quickens every decade. At 4°C (7.2°F), a third of the Antarctic ice shelves are expected to collapse. Even at 2°C (3.6°F), a quarter of a million sq. km of ice will be at risk (228). The situation in Greenland is just as dire with scientists now claiming that even if emissions were to stop today, the ice sheet would continue to melt into the Atlantic Ocean and raise sea levels (229).

While the complete collapse may take centuries, according to former NASA scientist and grandfather of modern climate science, James Hanson, we may see several meters of sea level rise within 50-150 years. Three to four meters of sea level rise by 2070 would give us very little time to mitigate against the rising waters (230).

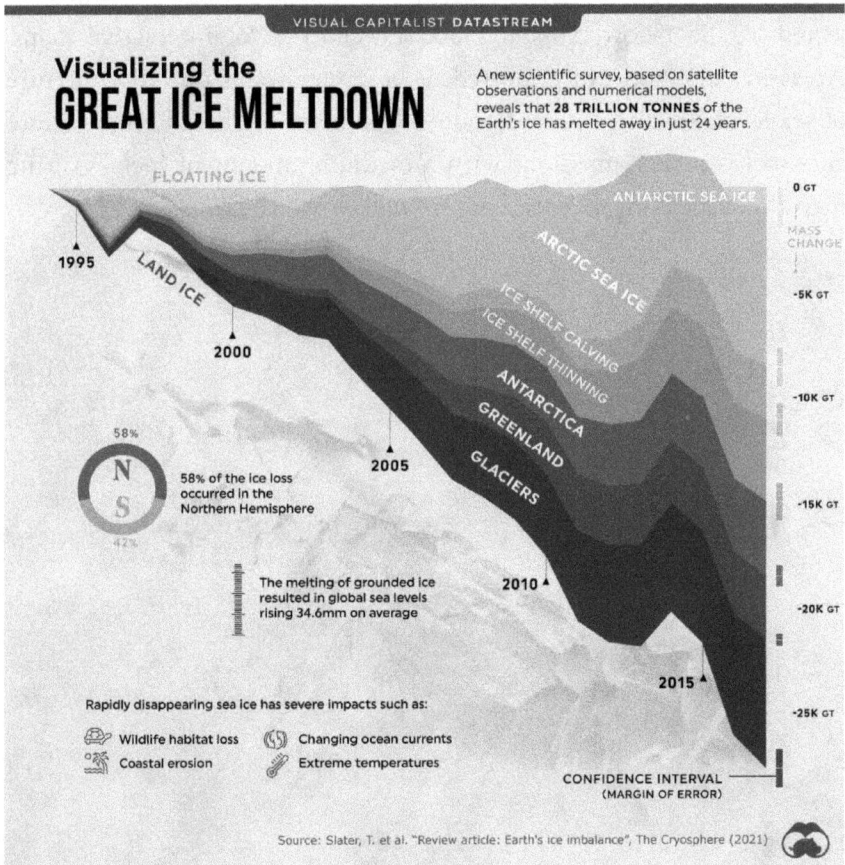

Visualizing Earth's Global Ice Loss Between 1994-2017
https://www.visualcapitalist.com/

A visit to Climate Central's Coastal Risk Screening Tool gives a good visual idea of what we can expect in the coming decades. Large swathes of red (gray in the print edition) hug the coasts of many countries with entire cities appearing at risk, according to their mapping tool. The South-East Asian breadbasket of the Mekong Delta is particularly concerning, as is China's east coast, a vast area of Bangladesh, and many coastal cities in Japan (231). The rising seas will exacerbate the already multiplying problems we face as a growing population will be migrating away from the coasts, our most fertile land will be lost to the oceans and as we will see later, nuclear power stations will be in harm's way.

The following maps are taken from Climate Central's "Land Projected To Be Below Annual Flood Level In 2050" interactive maps. Areas in dark gray are projected to be under water due to a mixture of sea level rise and annual flooding. The maps are based on moderate cuts to fossil fuel emissions with a medium amount of luck. Visiting https://coastal.climatecentral.org/ is well worth the time.

LAND PROJECTED TO BE BELOW ANNUAL FLOOD LEVEL IN 2050 (Southeast Asia)
https://coastal.climatecentral.org/

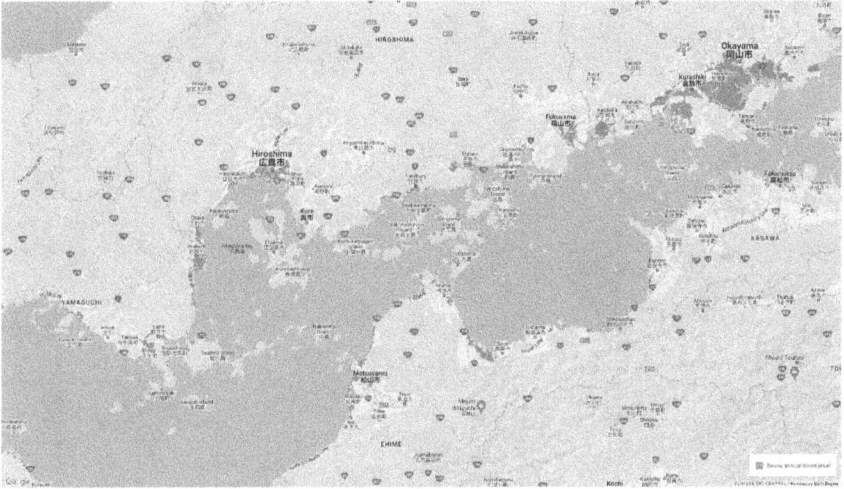

Hiroshima, Japan
https://coastal.climatecentral.org/

Shanghai, China
https://coastal.climatecentral.org/

Netherlands and U.K.
https://coastal.climatecentral.org/

Mozambique, Africa
https://coastal.climatecentral.org/

Dear Indy,

Sea level change is something that is going to affect billions of people on our planet, ourselves included. Many have made the coasts their home, attracted by beautiful views and access to the oceans.

As the ice caps melt and the waters rise, hundreds of millions will be displaced, and all our maps will need to be redrawn. The impacts will be enormous.

Our current home will be greatly affected. While our house is high enough up the mountain to keep dry, the bulk of infrastructure on Osakikamijima will be at threat of rising seas by the time you are my age. The area around your school where the supermarket, residential housing, three schools and town office are located is all built on reclaimed land. Unless massive amounts of money are spent on shoring up flood defences and changing the sewage systems, this entire area may be flooded annually within twenty years.

Most of the coast around Osakikamijima will be at risk of flooding and my school is in a perilous location, itself sitting on reclaimed land. How is this island, already struggling with a declining population, going to cope with the huge financial demands that will be necessary to protect buildings, homes, and infrastructure?

Our previous home in Tenno is also built on reclaimed land and sits in the path of our rising oceans. Hiroshima will feel the brunt as much of the city is predicted to be annually under water in the coming decades.

With sea level rise all but locked in, regardless of what we do to eliminate further emissions, the future for coastal populations is extremely dire. As people decide to move away from coasts, more pressure will be exerted on forests for timber, and remaining farmland. The knock-on effects to nature could be calamitous for future generations.

Air pollution

B reathing is one of the few things we do without needing to be conscious. We do it around 20,000 times every single day. We do it so often that we take it for granted. It's difficult for many people to notice any change in the air they breathe, but at Mauna Loa in Hawaii, scientists have been monitoring the air since 1958. What this has told us is that humans have affected the composition of the air drastically. Prior to the industrial revolution, there were approximately 280 carbon dioxide molecules in the air per million molecules of dry air (280 ppm). Through burning fossil fuels, we have pushed that number up to more than 420 ppm. Carbon dioxide isn't the only air pollutant. Generally, any chemical, particulate matter or biological agent that changes the makeup of the atmosphere is considered a pollutant.

Already, poor air quality is believed to cause eight million deaths per year. 4.2 million people die from breathing dirty air outside, and a further 3.8 million die from smoke exposure in homes. A staggering 99% of the population live in areas that exceed the safe World Health Organization (WHO) limits (232). When we think of air pollution, most people think of cities in China and India as the culprits for deaths, but although these cities have some of the dirtiest air on the planet, many people die from air pollution in countries where air pollution isn't considered a problem.

In many nations, the number of deaths attributed to air pollution is falling. In the U.K., the number of deaths from small particle pollution have decreased from 42,000 to 25,000 since 1990, but one in twenty deaths are still attributed to small particle pollution (233). The United States has witnessed a decline from 112,000 to 105,000 over the same period. Similarly, deaths in Germany have fallen from 58,000 to 39,000. As other countries have industrialized, the opposite has happened with deaths in China rising from 970,000 to 1.85 million and India increasing from 740,000 to 1.67 million (234). To put this into context, around 1.35 million people die in traffic accidents each year around the globe while there are around half a million homicides. This is perhaps not surprising as Delhi, home to almost twenty million people, has recorded 1,000 pieces of fine particulate matter (PM2.5) per cubic meter. According to the World Health Organization, anything over 25 is dangerous. Breathing the air when PM2.5 content is between 950-1,000 is claimed to be equivalent to smoking forty-four cigarettes a day (235).

The problem of air pollution isn't limited to deaths. Academics at University College London (UCL) found that increased carbon dioxide in the atmosphere might be making us dumber. Just at the time when we need to use all our brainpower to solve the enormous problems we've created, our ability to concentrate and make decisions could be diminished. Research from Yale University stated that polluted air can reduce people's level of education by one year (236).

CO_2 levels are expected to top 660 ppm by 2100 even if we adhere to the Paris agreement's strictest targets. Our brainpower is forecast to decrease by around 15% at this level. While outdoors in large cities, CO_2 levels often exceed 500 ppm, the levels in poorly ventilated buildings can regularly exceed 1,000 ppm. A study of bedrooms in Denmark found that after a night's rest, CO_2 levels could exceed 2,000 ppm. This affected student's performance the following day. Unfortunately, the insulation being recommended in homes and schools to mitigate the effects of climate change could be making the problem of indoor CO_2 levels worse, as more CO_2 is trapped inside (237).

A more worrying aspect of our atmospheric makeup regards oxygen content. Oxygen is necessary to keep body cells, organs and the immune system ticking over. In samples taken from dinosaurs, oxygen levels were between 28-35% of current levels. Burning coal and deforestation have drastically reduced oxygen content to around 21%, but in large cities it is only around 12-17%.

These levels are not sufficient for the body to function at full efficiency and can lead to cancers and other diseases, but if oxygen levels drop to 6 or 7%, life is not possible (238).

Air pollution, Beijing
◈ ◈ is licensed under CC BY 2.0.

Diseases

One of the less obvious impacts of the climate emergency is disease. Disease relates to any problem that causes pain, dysfunction, distress, social problems, or death in humans. Diseases are spread in many ways, and they are the number one killer on our planet. One such way that disease spreads is from animals.

On average sharks are responsible for six human deaths annually, wolves another ten, lions around twenty-two, elephants five hundred, hippos five-hundred, crocodiles around a thousand. Even man's best friend kills around 35,000 of us, while humans themselves are responsible for knocking off half-a-million homo sapiens. The prize though, for the most efficient killer of humans, goes to one of the smallest creatures on the planet, the humble, but very annoying, mosquito (239).

Mosquitos kill one million humans every single year. They transport diseases like dengue fever, malaria, yellow fever, chikungunya, West Nile virus and Zika. Until now, deaths from mosquitos have been limited to the tropics on either side of the equator. But, with a warming planet, deaths from mosquito bites are about to become more ubiquitous closer to the poles. One of the most common diseases spread by mosquitos is malaria. According to the WHO, almost half of the humans on planet Earth are at risk of malaria. Out of 241 million cases in 2021, 627,000 people died (240). Malaria is most likely to spread at 25°C so areas that are already warm, like sub-Saharan Africa are unlikely to be negatively affected as transmission slows once a certain temperature has been reached, but areas with cooler climates will see a rise in vector-born disease (241).

The Zika virus will also become less common in the tropics with a warming world as its favoured temperature is 29°C, but Australia, southern Iran, the Arabian Peninsula, and more parts of North America will play a welcome host to the Zika carrying Aedes aegypti mosquito (242).

Another concern is ticks who enlarge their territory in a warmer world. The tick-borne disease, Babesiosis, infected 1,100 people in fifteen states across the US in 2011. 82% of these cases occurred between June and August when temperatures are at their highest. Longer and warmer summers will increase the number of these infections which cause fevers, nausea, and headaches (85). While dogs are usually the most affected in our families, it has been found that the little diggers change their appetite when temperature rises. Ticks carrying the deadly Rocky Mountain Spotted Fever (RMSF) were more than twice as likely to choose a human over our four-legged companions, when given the choice. The study placed a dog in one box and a human in the other and a tube with ticks in, joined the boxes. Ticks use their sense of smell to find food and while the ticks moved towards the dog at 23.3°C, when the temperature rose to 37.8°C, they were 2.5 times more likely to move towards the human (243).

Lyme disease is another winner on our warming globe. By the 2080s, the ticks which carry the disease will move out of their southern U.S. home and venture further north into Canada, where their habitat will increase by 213% (242).

The increased temperatures and flooding will also increase the propensity of cholera. Cholera is often spread through dirty water, and places with poor sanitation will be hardest hit. *"Cholera likes warm weather, so the warmer the Earth gets and the warmer the water gets, the more it's going to like it. Climate change will likely make cholera much worse"*, according to David Morens, senior scientific adviser at the National Institute of Allergy and Infectious Disease (242).

Even more alarming than our current diseases spreading out of their traditional homes, is the re-emergence of old viruses like the bubonic plague and smallpox. These are just two of the diseases awaiting us if the temperature keeps rising. As we are currently finding, we are in a constant game with viruses continuously evolving in step with human resistance but buried under the permafrost are diseases that will take humanity completely by surprise if they escape along with the copious amounts of methane. Already, there have been victims. A twelve-year-old boy died and more than twenty were hospitalized in 2016 when the melting permafrost in Siberia exposed a reindeer carcass infected with anthrax. The reindeer had died around seventy-five years previously, but the summer heatwave thawed the frozen soil and the anthrax got into water and soil close by with more than 2,000 reindeer also becoming infected. The virus then got into the food supply which led to the human infections (244).

It is thought that bacteria can live for as long as a million years in the frozen permafrost as there is no oxygen or light. Some of these bacteria could be from victims of the 1918 Spanish flu outbreak that claimed the lives of as many as 100 million. People who died from the bubonic plague and a smallpox outbreak in the 1890s were also buried under the permafrost near the banks of the Kolyma River in Siberia (244).

It is even possible that the vanquished Neanderthals could come back to haunt Homo sapiens as their disease ridden 30,000 - 40,000-year-old remains have been found in the Russian permafrost.

Migration

Migration has been a fact of life since the dawn of time. It's as natural to all flora and fauna as breathing and drinking. When situations arise, we escape our potential demise. Survival mode kicks in and away we go. Fight or flight.

In his intriguing book, Sapiens: A Brief History of Humankind, Yuval Noah Harari explains that since humans first arose in East Africa 2.5 million years ago, we have been on the move, and we haven't stopped. Two million years ago, some of our ancestors got itchy feet and left for North Africa, Europe, and Asia. The humans living in these vastly different areas evolved different characteristics to suit their terrain, but all were part of the human (homo) family. In East Africa, Homo rudolfensis arose, while in Europe and West Asia, Homo neanderthalensis was chief and in Eastern Asia, Homo erectus became dominant for almost two million years. There were many other human siblings early in our development, but our ancestors eventually arose in East Africa and called themselves Homo sapiens, or 'Wise Man'. Ultimately, around 70,000 years ago, our species became the domineering force, and both bred with and somehow displaced the many other human bands, leaving us as the only human family on Earth. We still have proof of this fascinating past in our DNA. People living in Europe and the Middle East today have around 1-4% Neanderthal DNA while Melanesians and Aboriginal Australian DNA is up to 6% Denisovan (245).

Sapiens wandered within their territorial range looking for food, and sometimes they went outside of their natural terrain. According to Harari, natural disasters may have instigated exploration, or it may have been conflicts between groups, the emergence of too many people for an area to sustain or possibly at the behest of a leader. Around 45,000 years ago, Sapiens crossed into Australia and likely went on to wipe out all the megafauna that existed there.

Not content with the settling of Australia, around 16,000 years ago, we headed off across the land bridge that connected Siberia and north-western Alaska to settle the American continent. Within 2,000 years, humans had made it all the way to Tierra del Fuego, at the southern tip of South America. Unfortunately, we left a trail of absolute destruction in our wake. Rodents the size of bears, herds of horses and camels, oversized lions, mammoths and mastodons, saber tooth tigers and giant sloths were just some of the incredible fauna that made way for the arrival of man (245).

Man has been marching in different directions ever since, whether it is the Mongols marching from east to west and raping and pillaging everything in their path, or Europeans sailing west to east and conquering all of Africa, India, and South-East Asia, stealing resources and enslaving people as they went. In 1521, the Spanish defeated the Aztec empire and followed that up by destroying the Incans in 1532. The first European landed in Australia in 1606 and in 1620, the pilgrims set sail from Plymouth to escape religious persecution in England. The British colonialization of Australia began in 1788 bringing all of the Americas, Africa, most of Asia and Oceania into the European sphere (245).

Migrations haven't ended there. The Great Famine hit Ireland between 1846 and 1852. At least a million Irish starved to death and more than two million people emigrated between 1845-1855, as the British government withheld vital crops that could have mitigated the crisis (246). Further migrations into the U.S. have occurred from Italy, Puerto Rico, and Cuba, while many African people were forcibly enslaved and shipped to Europe and the Americas by Europeans and their descendants. The same fate was experienced by many Indians. The Japanese migrated into the Ainu land of Hokkaido and destroyed their culture while the Chinese crossed into Taiwan to escape communist rule.

The reasons for migration are often different but these migrations continue apace. With the advent of globalisation and free market economics, free flow of capital should have resulted in the free flow of human resources, but this has only been true of the European Union. Today, roughly 244 million people migrate globally. The majority of these people are known as economic migrants as they are on the move looking for work. People displaced internally were at a record high of 55 million with an additional 26.4 million refugees (247). The Russian invasion of Ukraine has led to the displacement of an additional twelve million Ukrainians with five million leaving the country to find refuge.

A number of crises including the civil war in Syria, ongoing violence in Iraq and Afghanistan saw Europe struggle to deal with an influx of more than a 1.3 million people in 2015. More than 476,000 applied for asylum in Germany but German authorities estimated that more than a million had been counted. Europe approved 292,540 asylum applications in 2015 with Germany accepting 140,910, Sweden 32,215, Italy 29,615, France 20,630, Holland 16,450, and the U.K. 13,905. These are the lucky ones. 3,770 perished on the long dangerous journey north in the same year. More than 1,200 people drowned in the Mediterranean in April of that year alone.

While the sudden migration shocked Europe and caused widespread debate over the responsibility to accept refugees, the majority of refugees do not head for Europe at all. Turkey, Lebanon, Jordan, Iraq, and Egypt took in 4.8 million Syrian refugees. Turkey took the greatest number with 1.9 million, Lebanon around 1.2 million, Jordan more than 650,000, Egypt 132,375, and incredibly, Iraq, where three million Iraqi civilians were internally displaced is temporary home to 249,463 Syrians (248). As previously mentioned, the Syrian crisis was exacerbated by climate change as a mismanaged drought helped to displace more than two million Syrians. The displacement increased the level of social unrest that led to the civil war, and scientists expect the changing climate to affect regions elsewhere in the coming years (249).

Migration at the Southern border of the United States averages around half a million people per year, while the number up to June 2019, had risen to 676,315 (250). It's hard to say how many of these people are economic migrants moving for the chance of a better life, or how many are escaping the impacts of a warming world. What we do know is that in 2019, they were being kept in squalid cages with many so cramped that it was impossible for people to lie down, sometimes for a week at a time. Officials from the government's Homeland Security department visited five detention centres along the border and reported that the conditions there should shame any country. Children and babies who were removed from their parents were kept on concrete floors in cages with no access to water or soap. Although the maximum children can be kept in these conditions is seventy-two hours, some of these kids had been kept like this for nearly a month (251). The U.S. celebrated its creation on its founding principle of human dignity on July 4th as children were kept in what the Salt Lake Tribune recently declared as *"concentration camps for children"* (252).

It seems that President Trump was modelling the southern border migrant situation on the system that successive Australian Prime Ministers have been running. In a country that, like the U.S., was stolen by white people from brown people, brown people trying to enter the country by boat to seek asylum are towed by the navy to islands in Papua New Guinea and Nauru where they are kept in tents behind barbed wire fences, in some cases for years. The U.N. and human rights groups have condemned this situation and believe they contravene various human rights charters. Reports of safety concerns, suspicious deaths, assault, and sexual abuse have been made, and in some instances, people even decide to return to their home country where they face huge risks, rather than stay any longer in these camps (253). It emerged in 2022 that the company running Australia's "processing" operation on Nauru was profiting massively from the misfortune of others. It made a $500,000 profit for each refugee kept on the island. In total, the company launched in 2017 with $8 million in assets, made a profit in 2021 of $101 million. The cost to the Australian taxpayer is a bewildering $4 million per refugee per year, or $12,000 a day.

These are just some of the ways in which refugees around the world today are treated when seeking refuge. While countries like those in Europe, the USA and Australia do accept refugees who can handle the initial time spent in concentration camps, many other countries who have become rich by burning fossil fuels admit only a tiny few. Between 2007–2016, South Korea accepted just 6,645 refugees. Portugal only registered 5,610 over the same period and Japan, with a population of 120 million and the world's 3rd largest economy accepted just 24,038. Two countries that have actually gotten rich by selling fossil fuels are Qatar and the United Arab Emirates. They allowed only 872 and 4,832 respectively (254).

Bangladesh, home to 163 million, 28% of whom live along the coast, is at the front line of the climate crisis. Already, rural areas are being abandoned due to rising seas and extreme weather, but by 2050, one in seven Bangladeshis is expected to become a climate refugee. Around twenty-five million people will be looking for a safe haven. The obvious place to head will be to India, who until independence from Britain, Bangladesh was a part of (255). To counter this migration, India has built the longest border wall on the planet. The wall, made from double-fence barbed wire, stretches 4,023 km (2,500 miles), and requires a police force of 70,000 to guard it (256).

Over the coming decades, the number of people on the move, including those heading to Europe and the United States, is going to increase dramatically. The wealthy nations who have benefited from burning fossil fuels and cutting down rainforests are already struggling to cope with people heading north. How are they going to react to many tens of millions knocking on their doors?

The World Bank estimates that by 2050, more than 140 million people from Sub-Saharan Africa, South Asia, and Latin America will need to move because of the climate crisis. The Institute for Economics and Peace (IEP) predicts there could be 1.2 billion people displaced globally by 2050. These climate refugees will be in addition to the millions of economic or political migrants already on the move (257).

Many cities in South and Central America will be at very high risk of climate hazards in the coming decades. Droughts, extreme temperatures, and heavy rains will be the major impacts seen across the region, and as we've seen with the Venezuelan crisis, borders cannot contain people when they travel in numbers (258).

Scientists at the U.S. Department of Defence-funded Strauss Centre state that the warming climate has changed the weather in Africa and the Middle East with more frequent droughts and floods occurring. Land is being desertified and heat waves are killing crops and farm animals. As the planet warms, farmers, fishermen and herders will need to move away from affected areas. Already, North Africans and people from the Sahel region just south of the Sahara are beginning to move due to the effects of climate change. Due to the killing of Libyan leader Muhamad Gaddafi, Libya has become a hotspot for people smugglers and people are being bought and sold in slave markets throughout the country (249) (259). In early July 2019, a migrant centre in the capital Tripoli was bombed and more than forty men, women and children were killed. At least eighty migrants were injured in the strike which hit the compound that houses around 616 migrants and refugees. Many of these people had been returned to Libya while trying to cross the Mediterranean to Europe. More than 6,000 migrants from Eritrea, Ethiopia, and Somalia, as well as other nations, are being kept in detention centres across Libya, and many of these centres are run by local militias who are accused of rape and torture (260).

Stephen Cheney, retired U.S. military corps brigadier general, cautioned, *"If Europe thinks they have a problem with migration today ... wait 20 years. See what happens when climate change drives people out of Africa – the Sahel [sub-Saharan area] especially – and we're talking not just one or two million, but 10 or 20 [million]. They are not going to South Africa, they are going across the Mediterranean"* (261).

Should emissions continue to rise, the only certainty we will be faced with is that the number of climate refugees will rise too. People in the rich countries, most responsible for the warming, are faced with the following dilemma: carry on as normal and be witness to migrations never witnessed by any human beings in the history of mankind, or make changes now, and give the poorest on our planet a chance.

The death of Aylan Kurdi
by urcameras is marked with Public Domain Mark 1.0.

Dear Indy,

Growing up as what the Japanese call a halfu kid, with one Japanese parent and one British, you understand something of what it is like to be different. Fortunately, your dad happens to be the "right" colour, speaks the "right" language, and has the "right" passport. If I had been born a slightly darker complexion, spoken a different language, and had a different passport, things might not have been as easy as they have.

Since I arrived in Asia in 2000, I have been welcomed with open arms, first in the Philippines, then in Taiwan, Vietnam and Japan. I've been invited into people's houses and supplied with beer in the middle of the mountains of Cebu Island. I've found myself at parties by the river in Maolin, Taiwan, gotten a job at a shipping company in Ho Chi Minh City and run my own bar in the snowy north of Japan.

All of these things have happened because I was lucky enough to be born at 52.1307° N, 3.7837° W and with the right skin colour. If I had been born at 17.6078° N, 8.0817° E and happened to have darker skin, the opportunities I've had in life would most likely have been denied me.

People don't want to admit it, but we live in a kind of apartheid system where those of us born into relative wealth get given opportunities that are routinely denied to those who are simply born at the wrong coordinates.

With the climate crisis, this system is about to get even harsher for those of us born at the wrong latitude and longitude. Already, we are seeing millions internally displaced by climatic change or natural disasters. Currently, these millions are being helped by those within their own arbitrary borders, but as the situation gets worse, where will they go? Who will accept them? As food becomes scarce, as scientists are predicting, how are those of us who look and sound different, going to be treated?

We are already seeing a rise in xenophobia around the world, and we live in a world of abundance at present. How is this going to change as our abundance turns to scarcity?

War

The Republican Party may not believe that climate change is being caused by human activity, but the U.S. military certainly do not share this anti-scientific claim. In 2014, the Secretary of Defence, Chuck Hagel, stated:

> "Climate change is a "threat multiplier"...because it has the potential to exacerbate many of the challenges we already confront today – from infectious disease to armed insurgencies – and to produce new challenges in the future."

He went on to stress the loss of glaciers will strain water supplies, hurricanes will increase instability and drought and crop failure will create mass migration. In addition, 1,930 km² (1,200 mi²) of coastline in the Caribbean Sea will be lost by 2064 according to the former Secretary of Defence. It seems the government in the U.S. is well aware of how climate change will impact national security, even if the Republican Party pretends, or believes, it does not (262).

Seventeen retired military officers also wrote a letter to then Secretary of State Rex Tillerson advising them to remain committed to combating climate change, in part because 128 U.S military bases would be threatened by ninety-one centimeters of sea level rise. These bases are worth $100 billion. They wrote:

> "Climate change poses strategically significant risks to U.S. national

security, directly impacting our critical infrastructure and increasing the likelihood of humanitarian disasters, state failure and conflict (263)."

It is not only the U.S. military that believe climate change to be a threat to national security; the Secretary-General of the United Nations Antonio Guterres warned that military strategists around the world view climate change as a threat to global peace and security (264).

There is evidence that the warming climate has already helped to cause conflict. The civil war in Syria, whilst climate scientists are quick to stress, cannot be attributed solely to climate change, there was a large drought in the Fertile Crescent in Syria, Iraq and Turkey leading up to the conflict. Livestock died, food prices rose, and children were sickened, leading to 1.5 million rural residents moving to the suburbs of the already full Syrian cities. This happened at the same time that many refugees started to arrive after the Iraq war. Scientists stated it was unlikely that the weather conditions experienced would have existed without the warming climate. These conditions exacerbated the problem of high unemployment and poor government which led to the civil unrest (265).

Further back, in 2007 the U.N. declared that the conflict in Darfur was the opening salvo in climate induced wars. The original cause of the conflict which began in 2003 was thought to be a regional rebellion, but the U.N. Environment Programme (UNEP) study suggests the true source was failing rains and creeping desertification. Environmental degradation and the changing climate led to tensions between farmers and herders over decreasing land as the Sahara encroached more than a mile each year. Rainfall was also down by 30% over forty years, and these factors combined to reignite the war between North and South Sudan. Crop yields in the Sahel are expected to drop by as much as 70%. As the desert moves south, rainfall diminishes and crops fail, it's difficult to imagine these shocks will not lead to major disturbances in Sudan and across Africa in the coming years. (266). In fact, research in Nature finds there will be a 54% increase in the possibility of civil war in Africa by 2030 (267).

For the sceptics out there, who do not believe a changing climate will drive wars, research from the past fifty years has found a link between the fall of Rome and rising temperatures or an alteration in rainfall. Research in Nature even contends there is an observed connection between climate change and violent conflict over the past 12,000 years. This takes in all of human history since the agricultural revolution (268).

Unless we all undergo a massive revolution in consciousness and start to show love to our evolutionary brothers and sisters, conflict looks more likely, not less. And, in fact, the research from Nature found that a temperature increase of around 3°C (5.4°F) in any given month increased the chance of violent conflict by 14%. (268).

While it is difficult to highlight climate change as the root cause of any conflict, the mass migrations discussed earlier will definitely lead to a rise in xenophobia. How is it possible to say 'definitely'? Because we are witnessing it now. On both sides of the Atlantic, as migrants make their way north in search of a better life, to escape poverty, war, or climate change, we have seen a rise in nationalism which has led to chants of *"Build that wall"* in the U.S., Brexit in the U.K., a rise in far-right populism across Europe and North America, and a string of attacks on migrants across the two richest continents. That the climate crisis is largely being driven by the richest countries in the northern hemisphere and that many migrants might be homeless down to their actions seems to escape many people.

The author of the book, 'Climate Wars: Why People Will Be Killed in the 21st Century', Harald Welzer, states it is hard to see that climate change induced migration won't lead to social disruption and the chance of violent clashes in the coming decade. As many parts of the planet become inhospitable to humans, people will have no choice but to change how and where they live. This will lead to more nativism, xenophobia and talk of building walls, not bridges (269).

Climate wars will not be limited to Africa, or the destinations African climate refugees are able to escape to. Central Asia, South Asia and

South America will likely play host to outbreaks of climate prompted violence.

The temperature in Central Asia has warmed by twice the global average and is on the front line of the climate crisis. Uzbek and Kyrgyz farmers straddle the Uzbekistan and Kyrgyz Republic border, and both need water to grow their crops. The rising temperatures are leading to increased political instability and violence in the area. This violence is not new and in the past twenty-seven years, hundreds of people have been killed, partly because of land and water. There are daily fights between villages here over water, and according to Altynbek Kadyrov, from the U.S. State Department's Agency for International Development (USAID) in Kyrgyzstan, if they do not fight, they may go hungry in the winter (270).

Water will likely become the blue gold of the future. With supplies dwindling fast and demand set to outstrip supply by 2030, the possibility of water wars will increase over the coming decade. In 2011, the United States Senate Committee on Foreign Relations issued a report called 'Avoiding Water Wars'. In the report they express concern for the aforementioned Central Asian region, including Afghanistan, Pakistan, Tajikistan, Kazakhstan, Kyrgyzstan, and Turkmenistan, and also India. They see the ability of governments to provide water to its citizens as essential for maintaining political, economic, and social stability (271).

In 2012, Chris Arsenault predicted the next water war would be between North and South Yemen (272). Whilst the players might be different, the civil war which began in Yemen in 2014 is believed in part to have been caused by water scarcity. This water scarcity has been exacerbated by climate change as well as water mismanagement (273).

The Middle East is especially at risk due to its lack of water and extreme heat. In fact, eighteen out of thirty nations projected to face water scarcity by 2025 are in the Middle-East and North Africa. Water wars are not unknown in the region with at least two wars in modern history being attributed to water. In his book, Water Wars: Coming Conflicts in the Middle East, Adel Darwish writes that grazing rights on the River Senegal led to a war between Senegal and Mauritania in 1989. He adds that Syria and Iraq have fought small skirmishes over the Euphrates River. In his boldest claim, the writer says that he heard the former Israeli Prime Minister, Ariel Sharon, saying on record, that the 1967 war against the Arabs was over water (272).

Many of the world's major rivers pass through multiple countries on their way to the sea. With diminishing supplies, countries upstream may choose to divert water to areas inland leaving countries downstream with limited supplies. Situations like this could increase tensions, especially in areas suffering from lack of food and water.

With more than 780 million people already suffering through a lack of clean water to drink, the climate crisis we are witnessing is going to increase this number exponentially. By 2030, the UN predicts that almost half (47%) of the world's population will suffer from water stress (272). When people have no water, the first thing they will do is go looking for some. This is a survival instinct we all share. It is likely that wars of the future will be fought over blue gold and not the black gold fought over today.

Dear Indy,

As I'm writing this, we are preparing to migrate again. This will be the third time you have been transplanted into a new school and home environment. So far, we have been what are called economic migrants. As a foreigner who teaches English in Japan, unless you live in Tokyo or Osaka, staying in one city is pretty difficult. This time we are off to a small island near Hiroshima. This will not be our last move. Due to the increasing temperatures and concern over food availability, we are planning on heading back to Hokkaido, where mummy is from. Hopefully, when you read this, it will be from our small organic farm in the frozen north of Japan.

We've been discussing what is best for your future the last few years. We want to do our best to give you every chance to be able to provide food, water, and shelter for you in what are sure to be challenging times. If we were both Japanese then we would go back to Hokkaido without a second thought, but as a multiracial family, we are a little worried about how foreigners are going to be treated in the future. People may think this concern is misplaced, but I hope anyone who has read this far will understand our thinking.

The U.K. has just announced that it will close its doors to non-English speakers and lowly skilled workers. The climate crisis isn't the cause of this so how will Britons react when hundreds of millions of people start showing up looking for a place to live?

Japan is certainly no better with immigration to the country not encouraged. Even with a falling birth rate, the government does not want to open its borders. Any foreigners allowed in to do menial work have to leave after their contract finishes.

How will people around the world treat the 'foreigners' in their midst when food becomes scarce?

Economic Collapse

O f the two differing collapses on the horizon, let's look at the less frightening of the two first: economic. Financial crises are nothing new to people over the age of twenty. The 2008 crash shook the world economy and led to a lost decade of growth in Europe, bruising austerity programs that resulted in the wide-spread use of food banks and the rise of anti-immigration policies which eventually led to the election of Donald J Trump and Brexit.

Unfortunately, the 2008 crash will appear like a blip unless our governments tackle the climate and biodiversity crisis immediately.

At the end of 2018, a group of investors who manage $32 trillion of wealth demanded urgent cuts in carbon emissions with the complete phasing out of coal if the economy is to avoid a crash four times the size of 2008. The investment firm *Schroders* goes as far as to say that there could be $23 trillion of annual losses in the long term unless we take extreme action. Some actions necessary include the removal of fossil fuel subsidies, the cessation of financial backing for new coal plants and the implementation of a substantial carbon tax of over $100 a tonne (274).

Failure to address the challenge could result in more frequent and stronger weather events including flooding, droughts and storms, infrastructure damage, agricultural losses, wildfires, and commodity price spikes. House prices could plummet, and losses could triple over the next thirty years (275).

One industry with particular reason to worry is the insurance industry. As waters rise, storms rage and fires spread, home insurance is becoming unavailable or unaffordable for many. Insurance companies paid out a record $219 billion for the two years 2017-2019. The only option available to them is to raise the insurance premiums to cover the costs, and this pushes the price too high for many people to afford. Insurers in California and Florida are already pulling the plug on insuring buildings in fire and flood prone areas and leaving home-owners to cover damage costs themselves or sell up and move on. If insurance companies continue to write risky policies, with combined emergencies, they may face bankruptcy as they try to cover rising pay-outs. Merced Property & Casualty from California filed for bankruptcy in 2018 after California's Camp Fire left it unable to pay out millions of dollars. The black summer in Australia, coupled with flooding and hail-storms in the first three months of 2020 cost insurance companies $5.3 billion. Some parts of Australia have seen their insurance premiums rise by 178% since 2007 and many are faced with either paying up, selling up or taking a risk. After a poll of people affected by fires in the Snowy and Bega Valleys, 50% said they had no insurance, while 25% were underinsured (276).

The situation is set to get worse before it gets better, and we can expect the number of insurers to claim bankruptcy to rise along with the temperatures and sea level (277). If large insurance companies are left vulnerable, this could lead to a massive financial crash. The incredible irony here is that the very same insurance companies that risk bankruptcy and collapse are the very same companies underwrit-ing coal expansion around the world. Coal is responsible for 44% of all global energy-related carbon dioxide emissions and the number one cause of the climate crisis, yet insurance companies are providing these companies with the insurance they must have in order to operate, re-gardless if that pushes them closer and closer to financial ruin. Logic should dictate that insurers stop funding and insuring the expansion of the almost 800 new coal-fired power plants currently planned or under construction today. Without insurance, these could not be built, and green energy would be financed instead. Some international companies are doing just that, but there are many that are not. In 2017, around $6 billion of premiums were sold to the global coal industry.

It's not just coal companies either. The top forty U.S. insurance companies have more than $450 billion invested in coal, oil, and electric utility stock. By investing in and insuring these fossil fuel companies, they are worsening the climate crisis that will lead to their possible collapse (278).

At present, the fossil fuel companies have around five times more assets in the ground than can be safely burned to stay within 2°C of warming, and they have every intention of burning the lot (176). Unfortunately, the problem we face is that so much money is invested in fossil fuel interests, that if we are to leave them in the ground, investors could face massive losses, and the knock-on effects to the economy at large could be profound. However, failure to leave them in the ground could lead to something far worse, complete breakdown of societies. This led the world's largest fund manager, appropriately called BlackRock, to sell over half a billion dollars in shares in coal companies in 2020 (279).

The climate crisis also risks causing another mortgage led financial collapse like 2008. The U.S. federally backed Fannie Mae and Freddie Mac mortgage lenders currently hold mortgages, many of which are in areas which do not require flood insurance. If and when these areas are flooded due to rising seas and extreme weather events, owners may be unable to pay to fix them and they will need to be abandoned. The federal government will then be left with these assets. Areas prone to fire will also see house insurance become prohibitively expensive and constant rebuilding will become unaffordable. These houses will become stranded assets. Former Freddie Mac executive Ed Golding says that *"At some point, it's going to crystallize and everyone's going to pull out,"* of homes in harm's way. According to a report in Politico, many in the industry can see the crisis approaching but no one is yet acting to prevent it (280).

Regardless of insurance or mortgages, the result of not addressing the climate and biodiversity crisis is going to hit the global economy like a wrecking ball. It is estimated that $33 trillion could be wiped off the stock market, according to investment giant Aviva (281). The World Wildlife Fund is warning that the loss of nature is set to slash global economic growth by £368 billion ($464 billion) by 2050. The U.S. and Japan will be hardest hit with each country facing a bill of $85 billion a year by mid-century. While it is estimated that transitioning to net zero carbon will cost 1% of annual GDP, the cost of ignoring the crises will eventually snowball to 20% of GDP (282). With dwindling food supplies, depleted water sources and an economy in tatters, however, there could be a much bigger problem lurking.

Emergency text on calculator screen on the hundred dollar bills
focusonmore.com is licensed under CC BY 2.0.

Dear Indy,

You are too young to understand about our financial system so let's just say it is prone to break down every now and then. The last crash came in 2008, shortly after I met your mother. We first met when me and a friend were trying to open a bar in Hakodate, Hokkaido in the Summer of 2007. Mummy had just come back from three years in Canada, and we were looking for a bartender who spoke English. We hired your mummy. Long story short, you were born four years later.

After the bar had barely opened, the first murmurings of a problem in the financial system started to be heard. By the end of 2008, the global banks were down on one knee with their arms wide open. They were begging our politicians to bail them out after their recklessness crashed the entire global economy.

As millions became unemployed and families struggled to remain in their homes, the taxpayers paid the banks trillions of dollars. We were told that failing to do so would result in a complete economic meltdown. After receiving their record pay outs, the banks then went on to pay their staff bonuses for a job well done and then started kicking families out of homes that they could not afford to pay for.

You could not have made this up. The banks who had caused the problem were saved by society through their taxes while the needs of the families who made up society were ignored, and their homes were repossessed by the bankers. It was a classic case of socialism for the rich and capitalism for the poor.

The response from our politicians was to cut public services like libraries and community centers in a supposed attempt to balance the books. This has been going on ever since.

What happened in 2008 will be dwarfed by the impacts of the climate and biodiversity crises. Governments and banks will be crushed by their stranded assets and huge annual losses. Who is going to pick up the tab this time?

Societal Collapse

What does 'societal collapse' mean? Essentially, it refers to the disintegration of complex human societies. Public services disappear and as food and water become scarce, the government loses control. There are many causes of collapse, including climate change, environmental degradation, inequality and oligarchy, complexity and external shocks like war, natural disasters, famine, and plagues (283). Our current crises include many of these causations. Our climate is rapidly warming, our ecosystems are being polluted and destroyed, income inequality is at its worst level since the 1920s, our economic system is only understood by a small group of people, and is largely in the hands of artificial intelligence, and large groups of humans have insufficient food due to droughts and flooding. There have been many known collapses in human history. Some have occurred over a number of years, such as the Romans, while others have happened abruptly, like the Mayans. History is full of such examples and it's worth looking at them if we are to avoid making similar mistakes.

One of the most famous examples of societal collapse is the Easter Islanders in the Pacific Ocean. Easter Island holds the crown for being the most isolated habitable piece of land on Earth. It covers an area of just 165 km². The closest habitable island is 2,000 km away, and the closest continent is South America which is more than 3,200 km away. The islands are so named as they were "discovered" by a Dutch explorer on Easter Day in 1722. Jacob Roggeveen encountered a baron wasteland with no trees or shrubs more than three meters high. There were no native animals bigger than insects. The people living there numbered around 2,000 and had no firewood to keep themselves warm on the windy winter nights. They only had several boats on the island and each boat could hold only two people each. They were flimsy craft that took on water easily. The people were not aware of any life away from the island, although they told stories of visiting a reef that was more than 418 km away. This piqued Roggeveen's interest. How was it possible that they could have travelled those distances in small leaky boats? And how could they have got to Easter Island in the first place? Investigators have been fascinated with the mystery of Easter Island ever since (284).

The other mystery was the huge stone statues erected along the coast. There were more than 200 of them standing on huge platforms 152 m long. The stones, called Moai, weighed as much as 82 tonnes. It appears that all the Moai were constructed in a single quarry and transported around the island. At the quarry, many Moai were found abandoned mid construction and these were even bigger and heavier than the Moai erected on the coast. Some were almost twenty meters high and weighed in at eighty-two tonnes. Many Moai were found mid transportation on roads that connected the quarry to the coast (284). How did the Easter Islanders move these huge statues without draft animals, thick timber, thick ropes, wheels, or any other power source?

The answer seemed to be that original Easter Islanders were part of a much more complex society than exists there today. So, what happened to them, and why did they pull down the Moai sometime between 1770 and 1864? Many people have tried to guess the answer, from the Moai being erected by advanced American Indians to alien life forms. Even though Easter Islanders were known to speak Polynesian, people could not accept that they had come from Asia. But, from Asia they had come, and Polynesian they were. DNA extracted from skeletons proves this. The islanders grew bananas, taro, sweet potato, and sugarcane. They also had domestic chickens which were from Asia (284). So, what went wrong?

From pollen analysis and radiocarbon testing, the history of Easter Island before the appearance of humans is that of a beautifully rich and vibrantly complex ecosystem. For at least 30,000 years before humans arrived, Easter Island was a subtropical forest of trees, bushes, shrubs, herbs, ferns, and grasses. They had Hauhau trees which produced rope, and its most common tree was the Easter Palm which grew to twenty-five meters in height and almost two meters in diameter. The Palm also produced nuts and its sap is used by Chileans to make sugar, syrup, and wine. The island's trees were pollinated by various species of native birds and the island was likely the richest breeding ground for twenty-five species of sea birds as there were no natural predators. It appears Easter Island was paradise until the first human settlers set foot ashore (284).

It is likely these Polynesians arrived by boat from Asia sometime between A.D. 400 to 700. Large scale deforestation was underway within a few hundred years and Moai construction peaked between 1200 and 1500. Few statues were constructed after this. The population was thought to number around 7,000, with some estimates much higher. Their diet differed to other Polynesians in that they ate much less fish. They did not have abundant corals surrounding the island and few places to fish. Instead, they chopped down trees to build ocean going canoes and they harpooned porpoises. From bones found in their sites, 1/3 of their diet was made up of porpoise. They also ate nesting birds which used Easter as a breeding ground, as well as domesticated chickens in small amounts (284).

As the population rose, competing tribes began making ever larger and more complex Moai to keep up with their neighbours. This is why the largest Moai found were not yet completed. They used ever more land for their own gardens and by 1400, the last Easter Palm had been cut down. The trees were under attack on a number of fronts. Human narcissism on one side. On the other, humans were wiping the bird population out so pollinators were diminishing, and finally the rats humans had brought inadvertently to Easter, were chewing on their nuts which could not then germinate (284).

After the last tree was cut down, canoes could not be made so porpoises could not be caught. The lack of trees meant that rich soil was eroded by rain and wind and crop yields decreased. Moai construction was abandoned, and people resorted to eating the domestic chickens, who until now, were eaten sporadically. The rats were also eaten for a while, and once the chickens were all but extinct, finally Easter Islanders started eating each other. There were no trees left for cooking fuel, so the islanders used sugarcane scraps and grass to fuel their fires (284).

As food became scarce, the chiefs, bureaucrats and priests who had kept control were unable to do so any longer. They were replaced by warriors and by 1700, the population had dwindled to as little as one tenth of its former size. To remain safe from each other, people began living in caves. Around 1770, rival warring clans started to topple each other's Moai until in 1864, the very last Moai was pulled down and desecrated. Within 300 years of cutting down their last tree, the islanders' demise was complete. A once flourishing society had collapsed into famine, war, and cannibalism (284).

The example of Easter Island is a microcosm of what we are doing now. The large-scale deforestation in order to graze cattle, industrialized fishing that depletes our oceans of trillions of fish each year and destroys ecosystems and deprives sharks, whales, dolphins, turtles, and penguins of their only food source is replicating the mistakes made by Easter Islanders on a global scale.

It's not too late for us, but with each year that passes, solving our problem becomes harder and harder. The scale of the damage we are inflicting on our ecosystem is astonishing, and the only difference between the Easter islanders and us, is that they had no history books from which to learn about previous mistakes. We, on the other hand, have no such excuse. We know what we are doing, and we continue anyway.

If we carry on business as usual, we are being warned time and time again that we risk societal collapse on a planetary scale. All our economies are tied to each other. The supply chains are complex, as are our food systems. If collapse happens in one country, that will have a cascading effect and no nation will be unaffected. As with the Easter Islanders, the main problems we face are human narcissism, and food and water shortages. We are building ever bigger houses and buildings, driving ever bigger cars, shopping with ever bigger trollies and creating ever bigger populations. All these things add up to a planetary Easter Island situation, and like the Easter Islanders, we also have no escape route.

As our population rises, our water runs dry, we run out of forests to chop down, fish to kill, and our crops fail, our leaders too will be unable to control society. Public spending will cease, and people will be left to fend for themselves. The reports coming out are in agreement about this, the only way they differ is the time frame. The Pentagon puts the date when the U.S. military can no longer operate at 2040 (285). Anglia Ruskin University's Global Sustainability Institute used a scientific model that agreed with the Pentagon's 2040 timeframe (286). A report from Breakthrough – National Centre for Climate Restoration is slightly more optimistic. They believe we have an extra ten years until our societies collapse. The two main authors are a retired Australian Admiral and the former chairman of the Australian Coal Association (287).

A 2014 research paper from Melbourne Sustainable Society Institute analysed historical data to update The Limits to Growth (LTG) report of 1972. The original report predicted that sometime mid-century, we would witness both economic and environmental collapse should we continue business as usual (BAU). After using data from the past forty years, the updated report found that, *a relatively rapid fall in economic conditions and the population could be imminent (288)*.

In 2020, Future Earth published their Risks Perceptions Report. The report surveyed 222 prominent scientists from fifty-two countries. They looked at five risks, climate change, extreme weather, biodiversity loss, food crises and water crises. Almost three-quarters (72%) of these scientists identified climate change to be a key driver of global systemic crisis. More than a third of these scientists believe that taken together, these five risks would worsen each other and lead to *global systemic collapse (289) (290)."*

Another study in 2020, published in Scientific Reports, stated that based on current resource use and population growth, there was a

90% likelihood that we will face complete collapse of our civilization between 2040-2060.

In May 2022, a U.N. Office for Disaster Risk Reduction report by Thomas Cernev stated that under business as usual policies that push us past the Earth's nine boundaries, there is a real risk that *total societal collapse is a possibility.*" According to Will Stefen of the Stockholm Resilience Center, in an interview with Jamie Morgan for the journal Globalizations, at our current rate, we will have passed through eight of the nine planetary boundaries by 2027, with the only remaining boundary not breached being the ozone layer. Business as usual therefore is creating the conditions necessary for the possibility of total societal collapse as early as the end of this decade.

Some reports go a step further in stating that collapse is already, not only underway, but also too late to prevent (291) (292). It's easy to look at these and sink into a deep despair. This will only ensure the worst arrives though. Now is not the time for people to give up. We owe it to our children, and their children, and also to ourselves. Who wants to be retiring after working for forty-five years, just as society starts to breakdown?

It's not all doom and gloom. We can take a glimmer of hope from the example set by Japan in the Tokugawa era, lasting from 1603 to 1867. From 1467 onwards, Japan was gripped by civil wars. These lasted until 1615 when Tokugawa Ieyasu stormed Osaka castle and his remaining enemies committed ritual suicide. This brought about the end of the war and long-lasting peace, population growth and prosperity (284).

The unintended consequence of this was that Japan embarked on a mammoth construction project. At first, it was the leading Daimyo (autonomous warrior barons) that made huge castles and temples to compete with each other, much like on Easter Island. Just the three largest castles built by Ieyasu required clear cutting of 26 km² of forest. Then, shortly after, around 200 castle towns appeared, and urban construction became the dominant cause of deforestation. Thatch was used for the roofs, and during the long cold winters, when wood was also burned for fuel, many towns were burned down. These towns then needed to be rebuilt and the deforestation continued at pace. After the most infamous Meireki fire in 1657, when more than 100,000 perished in the flames, half of Edo, present day Tokyo, needed to be rebuilt. The timber was also transported by boats which were also made from wood (284).

This construction boom peaked around 1570–1650 as timber became scarce. Unfortunately, by 1710, most accessible forest had been cut on the three main islands of Honshu, Shikoku, and Kyushu, as well as southern Hokkaido which was not yet part of Japan. This timber was necessary to supply housing for Edo, which by 1720 had become the most populous city on Earth. This mass tree cutting resulted in flash flooding and soil erosion (284).

At this stage, Japan was on the verge of going in the same direction as the Easter Islanders, but the Meireki fire and the resulting lack of wood served as an alarm call for the Daimyo. Nine years after the fire, they called for civilians to plant seedlings, and they instigated top-down edicts discouraging consumption and started to build up reserves (284).

Population also levelled out between 1721 and 1828 with a rise of only just over a million from 26,100,000 to 27,200,000. Population control, necessary because of scant resources, coupled with the top-down controls led to the creation of woodland management by 1700. Today, 67% of Japan is covered by trees (284).

The reason only a glimmer of hope is to be found here is that if you delve a little deeper into Japan's situation, you will find they just exported their deforestation to poorer countries. Starting in the 1950s, to fuel their post WWII recovery, they began importing timber from the Philippines. Due to forest depletion, imports peaked in the late 1960's and were banned in 1986. The Philippines ended up a net importer of timber as it lost most of its tree cover. Due to the resource depletion in the Philippines, Japan moved south to Indonesia. In the 1970s, Indonesia became the number one provider of timer for Japan until it too banned the export of logs in 1985. Due to these bans, Japan then focussed on Indonesia's neighbour, Malaysia. In the late 1980's,

90% of Japan's timber imports came from forests in Sabah and Sarawak. Pressure was mounting on Japan as global attention was concentrated on rainforest destruction, and in 1993 Sabah banned exports. Although exports from Sarawak declined during the 1990s, it continued to be the number one source of tropical logs for Japan until 1999. In that year, Japan was the number one importer of tropical timber, accounting for around 25% of global trade in round-wood equivalents that include plywood (293). Japan continues to rely on foreign supply to meet its domestic demand, even though much of its land is covered in trees (294). Japan released guidelines for timber purchasing in 2006, but the guidelines only apply to government bought timber which accounts for just 5% of imports. In 2018 it was the fourth largest importer of wood products, after China, the U.S. and the E.U. Much of these imports are alleged to be logged illegally in Russia, Sarawak, and Indonesia, or finished in China before being imported to Japan (295). While Japan was able to save itself at the last minute, it did so simply by outsourcing the problem. Unfortunately, we have no planet to outsource to.

A modern-day equivalent of Easter Island can be found in the Pacific Island chain of Kiribati. Here, beginning in 1900, an Australian mining prospector, Albert Ellis, arrived on the island of Banaba, and began removing phosphates for use in man-made fertilizers. By 1980, around 90% of the island's surface had been stripped bare by the British Phosphate Commission (BPC). The company, which was jointly owned by Australia, New Zealand, and the U.K., removed around twenty-two million tonnes of land, and the caves which were a source of clean water for centuries were destroyed or contaminated. Around 6,000 people have had to leave Banaba, mostly for the Fijian island of Rabi, with only about 300 remaining. The island can no longer sustain even this small number with no rains for more than a year, the population has been reliant on a desalination plant for drinking water, growing crops, and washing. Unfortunately, the desalination plant broke down in November 2020 and people were forced to drink contaminated water and sea water to survive. This resulted in widespread disease and sickness. All the crops died, and fish were the only source of sustenance until a ship arrived from the Kiribati capital of Tarawa in March 2021. The ship brought bottled water and equipment to fix the desalination plant but local elder, Roubena Ritata says *"Desalination plants are not a solution, how long until this one breaks and we're back in the same situation? What we need is the rehabilitation of our island."* The islanders are seeking support from the Australian and New Zealand governments to help them return to a method of capturing and collecting water from the underground network of caves. Decades ago, the British government awarded the island A$10 million on the condition it took no further action. The elders are acutely aware that relying on support from faraway places is unsustainable as the climate crisis gets worse. A local scholar, and associate professor at the Australian National University, Katerina Teaiwa is clear in her assessment of what happened on Banaba:

"They came in, had a big party, made lots of money and left (296)."

Ms. Teaiwa's assessment of what happened on Banaba could easily be used to describe what humanity has done to our planet. Clearly, lessons have not been learned. Rich countries hoovering up the resources of developing countries is exacerbating the destruction of the natural world and the ecosystems that we rely on for life. If we are to continue to live on our planet, we will need to drastically reduce consumption or face the real possibility of complete social collapse.

Following images:

Algeria slashes food prices amid riots: Magharebia is licensed under CC BY 2.0.

Easter Island Ahu Tongariki: Ndecam is licensed under CC BY 2.0.

Burj Khalifa aka Burj Dubai: Joi is licensed under CC BY 2.0.

Nuclear Power Stations

As with most technology, nuclear power stations are a prime example of a double-edged sword. Since the first power plant was switched on in Russia in 1954, their number has risen to 440 in thirty-two countries around the world. They produce around 2553 Terawatt hours (TWh) of power which is roughly equivalent to the entire power generation of India, Japan, and Mexico combined (297). In total, around 10% of the world's energy needs were met by nuclear in 2020, and with fifty-five plants being built, this will rise to around 15% (298).

Regarded as a miracle producer of clean energy, it wasn't long until the first accident occurred in Kyshtym, Russia in September 1957. There was another accident at the U.K.'s Sellafield plant in 1957, but it wasn't until 1961 in Idaho that nuclear power claimed a life. Three people died in the accident and in a total of ninety-nine accidents, forty-two people have lost their lives since (299).

Whilst nuclear power plants produce clean energy (minus their waste), in a warming world they could become an existential threat to human existence. There are two things to consider as the world warms. Firstly, sea level rise, and more alarmingly, societal collapse, as was discussed previously.

Why is societal collapse more worrying than sea level rise? Sea level rise is happening slowly, and given time, power plants can be moved or fortified against the rising tides. Societal collapse on the other hand can and usually does, happen quickly.

Let's look at sea level rise first as this is more certain than societal collapse. Nuclear plants need vast amounts of water to prevent the core and spent nuclear fuel from overheating. That's why 40% of the world's nuclear plants are located along the coastline, and that number rises to 66% if you include plants under construction. This places them in danger's way as sea levels rise (300).

Research shows that at least 100 U.S., European and Asian nuclear power stations are built just a few meters above sea level (301). Scientists predict that we will definitely see sea level rise of at least a meter by 2100, and possibly two meters. These projections do not include the impact of any tipping points.

Flooding is an existential threat to these power plants as it can knock out the electrical systems and send the plant into meltdown. This happened on March 11th, 2011, in Fukushima, Japan, following the Tohoko earthquake and subsequent tsunami. The tsunami breached the inadequate sea walls and inundated the site resulting in the power generation being cut. The diesel backup generators were also flooded, and the uranium core melted through the container into the ground, and expelled plumes of radioactive materials into the atmosphere. Around 165,000 people were evacuated, and many have still not returned, ten years later. The cost to taxpayers was over $200 billion and the clean-up is expected to last until the middle of the century, at least, and could cost half a trillion dollars (300). Fukushima was a man-made disaster due to the closeness of Tokyo Electric Power Company (TEPCO) and the government agency responsible for regulating it. The disaster could have been avoided, but scientific advice regarding the height of the sea wall and the backup power generation were not followed (302).

The worrying concern for U.S. citizens is that following the meltdown in Fukushima, the U.S. Nuclear Regulatory Commission (NRC) carried out an evaluation of U.S. nuclear facilities and found that fifty-five of the sixty-one facilities were at risk of flooding hazards they were not designed for (300). They also stated that the Fukushima disaster did not violate any U.S. safety standards.

One of those affected is Turkey Point, just a few miles south of Miami, home to millions of people. Turkey Point became operational in the early 1970s, long before people began to factor in sea level rise to risk assessments. Being in Florida, it was, however, designed to cope with hurricanes. Reactor vessels are elevated twenty feet above sea level. During Hurricane Andrew in 1992, storm surges were only ninety-one cm, but the highest storm surges were sixteen km north of the plant and reached five meters. Stronger storms such as Katrina saw storm surges of 8.5m. It gets more worrisome as other vital equipment is even closer to sea level. The back-up diesel generators, which were the cause of the Fukushima meltdown, are only 4.5m about sea level and have open louvers that could allow water to enter. Unfortunately for the people of Miami and the surrounding areas, this isn't the worst news. Turkey Point's cooling ponds are just a few feet about sea level. And at just one foot of sea level rise, the ponds begin to flood. At two feet of sea level rise, they are inundated, and at three feet, the whole plant is cut off from the mainland and can only be reached by boat or plane. According to the director of the Nuclear Safety Project for the Union of Concerned Scientists, Dave Lochbaum, the Nuclear Regulatory Commission, which oversees the safety of nuclear power in the U.S., when carrying out risk assessments, only past natural hazards are taken into account, and potential future hazards are not considered. If this all sounds quite irresponsible, the best is saved for last. Florida Power & Light are planning on spending $20 billion on two further reactors at Turkey Point. The plan for these reactors was passed in 2016 after a seven-year environmental review, and the report did not mention sea-level rise (303).

It's not just Americans that need to be worried. In the U.K., the planned $25 billion Hinkley Point C nuclear station is to begin generating electricity in 2028. Hinkley Point C is being built in the Bristol Channel which has one of the highest tidal ranges in the world. A previous plant on the same location was shut down for a week in 1981 due to serious flooding from a spring tide and storm surge, so the designers are building a 900 meter long 12.1 meter high seawall to ensure this will not happen in the future. Unfortunately, the plans were drawn up in 2012 before scientists became aware of the extent of Greenland flooding or the awareness that Antarctica was also beginning to melt. Projections of sea level rise have gone from 30 cm in the next fifty years to more than a meter. This is well within the lifespan of Hinckley C which is not projected to be decommissioned until well into the twenty-second century and spent fuel will be stored there for even longer (301).

Asia in particular will be heavily affected by sea level rise and many of the planned nuclear power plants sit merrily along the coasts predicted to be underwater sometime this century. People all over the world, shown in the map below, where the one-hundred operational or projected nuclear power plants are built just above sea level should be concerned as scientists forecast that climate change will affect coastal nuclear plants faster and harder than governments or nuclear regulators have foreseen (301).

Now, while we can be assured that the seas are going to rise, and that nuclear power plants will be in harm's way, the seas may rise slowly enough for nuclear power plants to be safely decommissioned. This is not to say that there is no risk from storm surges and extreme weather events. The more pressing concern, however, is the societal collapse discussed earlier. Let's take a look at the impact such a collapse could have on nuclear power stations around the world.

It's worth going back to 2011 in Fukushima in the weeks and months that followed the meltdown. The magnitude 9.0 earthquake off the north-east coast caused cars to bounce up and down in Fukushima Daiichi's car park. More than 6,000 workers were present, many of whom, had families living nearby. They headed for an earthquake proof evacuation building that had no windows. Less than an hour later, the tsunami breached the insufficiently high sea wall and swamped the plant, ripping through the reactor buildings. The men inside could not see the damage the tsunami was causing, but they did notice the plant had been plunged into darkness after the power was cut to four of the six reactors, meaning the nuclear fuel rods could not be cooled. To make matters worse, the back-up generators had been flooded. The prime minister Naoto Kan was privately making preparations to evacuate the greater Tokyo area of thirty-five million people, believing that the situation was out of control.

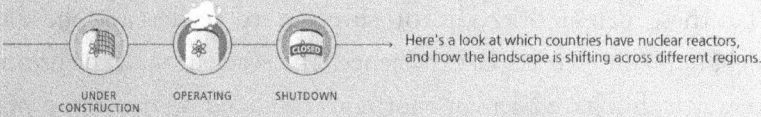

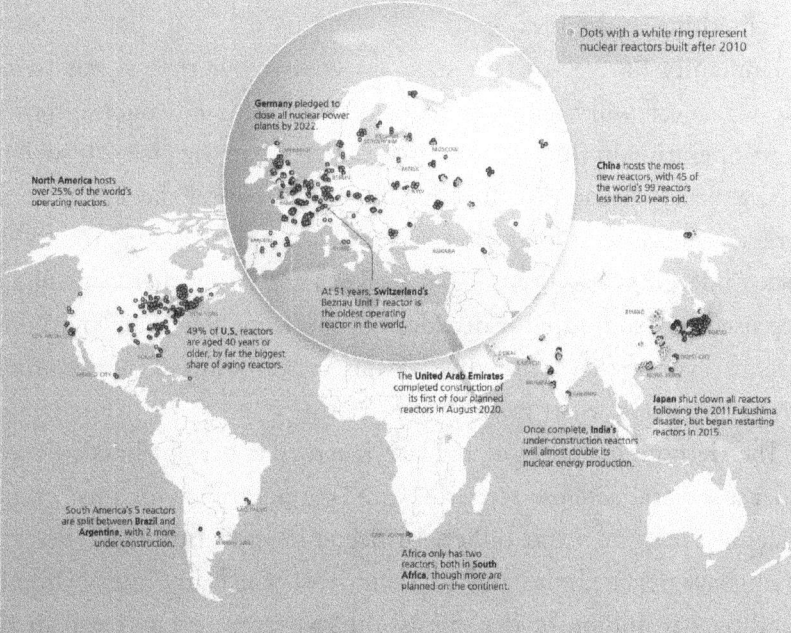

Mapped: The World's Nuclear Reactor Landscape
https://www.visualcapitalist.com/

News of this never reached the workers at Fukushima Daiichi, and in the following days, the workers did everything in their power to get a grip on the catastrophe. They were up against the odds, as aftershocks continued and hydrogen explosions hit three of the reactors, injuring several workers. TEPCO continued to spray coolant water on the overheating fuel rods night and day.

Workers laboured around the clock with minimal sleep, and a lack of food and water. This went on until December 2011 when the government declared the reactors had reached a stable state known as a cold shutdown. By this time, the world's media, who were covering the story night and day, had nicknamed the group of heroes at Fukushima Daiichi, the Fukushima Fifty (304).

What these men sacrificed in 2011 surely averted a disaster, the scale of which, humanity has yet to experience. The evacuation of the greater Tokyo area would have been an enormous task, and no one could plan for something so complex.

Humanity has yet to face such a scenario, but that is not to say that we never will. Take into consideration the 450 nuclear power plants operating in thirty-one countries than span the globe. If societal collapse was to occur, probably due to food and water shortages, would workers choose to stay and control fuel rods and labour around the clock to decommission the site in a safe manner, and then continue to keep used fuel rods safe for decades not knowing how their loved ones were, or even if they were alive. If the answer was yes, where would they get food and water from?

The decommissioning of nuclear reactors is extremely costly, and under a societal collapse scenario, who is going to pay for this? France is facing a cost of €200 billion to decommission fifty-eight reactors which have reached the end of their lives. Germany, meanwhile, is spending €38 billion to decommission seventeen reactors and in the U.K., decommissioning estimates vary from €109–€250 billion (305).

As with many large industries where profits are accrued by private individuals, the costs are almost exclusively borne out by society. Nuclear energy in its current form, is clearly not the answer to our long-term energy needs, and it might just be what makes our planet inhabitable.

Dear Indy,

The latest accident happened three days before you were due to be born on 11th March 2011. A magnitude nine earthquake hit off the coast of Tohoku in northern Japan. The earthquake triggered a huge tsunami which killed 15,894 people with thousands more missing. The tsunami hit a nuclear power station in Fukushima and knocked out the power system resulting in a meltdown of the uranium core. It was the biggest nuclear disaster since the infamous Chernobyl accident in 1986, coincidently the year your mother was born.

We were living around fifty km south of Tokyo, and we think due to the stress, you didn't fancy making an appearance quite yet. The weeks that followed the meltdown at Fukushima were the most stressful of our lives. Your mother had decided which midwife she wanted to deliver you, so we had to stay in Kanagawa. She was watching Japanese TV for news of the crisis, and they said we didn't need to worry, that everything was fine. I was watching the BBC, CNN, and Al Jazeera, and they were telling people to leave the area. The German and New Zealand governments requested their citizens leave. Many foreigners left the country. They were labelled flyjin by locals in reference to the name of gaijin, which means foreigner. I thought that was quite witty.

We were living really close to the U.S. base, Camp Zama, and each night helicopters were flying overhead. I was terrified. It took us three days to get petrol because the queues were just too long. Even if we wanted to leave, we couldn't as we had Takai and Tao, who being a dog and cat, wouldn't be allowed on trains or busses. The shops were empty of many foods, and bottled water ran out. Your mother was given some by the hospital as she was pregnant, but that was only enough for her. Each day, I had to check the government website to make sure the tap water was safe for drinking. Finally, at 8am on March 29th, 2011, you made an appearance and finally our stress levels started to come down.

It wasn't the ideal entrance to life on Earth, and I hope it wasn't a harbinger for things to come. You weren't quite born in a crossfire hurricane, but you came close!

TWO

Ecological Collapse

We have looked at how our changing climate is impacting our world, and how it will affect us all going forward, but the climate isn't the only crisis we are facing. On the other side of the same coin, is the destruction of our ecosystems. We are in the middle of an unprecedented sixth extinction. You may be asking yourself, if there have been five mass extinctions already then how can this be unprecedented? The answer is that this isn't so much an extinction event as an extermination event. The five previous extinction events, including that of the dinosaurs, were not caused willingly by a single species, they were the result of natural causes such as an asteroid hitting Earth. Unfortunately, we cannot excuse ourselves from blame. Through our short-sightedness, we are exterminating the natural world that we all rely on for life. In this chapter we will look at different parts of the ecosystem that are being destroyed and discuss how we are to blame.

Insect Apocolypse

The loss of large, majestic animals like tigers, and polar bears is disturbing to watch. We have grown up learning about them and they are part of our world, but it is not the loss of these well-loved species that will impact humans the most. It is the loss of the smallest creatures on Earth that will create the biggest hurdle to human survival. Approximately one-third of all plants worldwide are pollinated by insects (306). Insects may be small, but their accumulated mass outweighs humanity by seventeen times. There are a baffling 12,000 variations of ants, 20,000 types of bees, and an eye-popping 400,000 species of beetle. If you take a 30cm^2, 5cm deep section of healthy soil, you will find around 200 unique mite species, and they each have a slightly different job to fulfil in the ecosystem. Unfortunately, in what has been termed the insect apocalypse, half of all insects have been lost since the 1970s. A total of 40% of one million known species are facing extinction, and they are going extinct eight times faster than mammals, birds, and reptiles. And these losses are speeding up.

The U.K. has seen the biggest decline in insect numbers, but it is worth noting that this may be due to the U.K. being the most researched country. In the last century alone, twenty-three bee and wasp species have vanished from the U.K., 46% of U.K. butterflies have died out, and on U.K. farmland, in just nine years from 2000-2009, an astonishing 58% of butterflies disappeared. Fifty percent of all known species in the U.K. could have vanished since the 1970s, and the number might actually be much higher. The loss of insects has knock on effects for the entire ecosystem as they are essential nutrients for other creatures, they act as pollinators for crops, and they are excellent nutrient recyclers. One such impact of the insect apocalypse is a 93% decline in spotted flycatcher numbers since 1967 (307).

The U.K. isn't alone in the mass die off of insects. Puerto Rico has witnessed a staggering 98% loss in ground insects in the past thirty-five years. On the U.S. mainland, in Oklahoma, a half of bumblebee species that were present in 1949, were not present in 2013. The number of honeybee colonies dropped from six million in 1947 to just two and a half million in 2019. The staggeringly beautiful monarch butterfly population has declined by 90% in the past twenty years. That amounts to the loss of 900 million individuals. The rusty-patched bumblebee has dropped by 87% over the same period (308). The chance of seeing a bumblebee in North America has dropped by half since 1950 (309). Professor Paul Ehrlich witnessed first-hand the complete collapse of checkerspot butterflies on Stanford University's Jasper Ridge reserve between 1960 and 2000 (310).

Across the Atlantic, continental Europe has seen similar losses. The German Entomological Society discovered that when they measured simply by weight, the overall abundance of flying insects had dropped by 75-82% in the past two years. And the frightening part of this research is that it was carried out in nature reserves. Insect numbers outside these reserves could be much less. France is seeing the knock-on effects of the insect demise. Eight in ten partridge species have disappeared from French farmland. Nightingales have declined by 50%, and turtledoves by 80%. In total, half of all farmland birds have vanished in the past thirty years (310).

Disturbingly, over the past thirty years, 25% of insect species have disappeared and the decline is accelerating (311). We are now losing approximately 2.5% of insect mass each year. If these declines continue at this pace, we will lose a quarter of all species in the next ten years, and then half of all species in fifty years, and within a hundred years, we will have lost all insect species on the planet (308). The wondrous firefly is also on the brink, with more than 2,000 species at risk of extinction (312).

The problem for our insect friends is multifaceted. Firstly, the rapid growth of the human population has played a significant part in their demise. In 1950, just 30% of the world's population lived in urban areas. This had risen to 55% by 2018 (313). By 2050, this number will rise to 68%, and by the end of this century, around 85% of humans will live in cities. From a population of just over 2.5 billion in 1950 to over 7.96 billion in 2022, and a projection of around 10 billion by mid-century, our increasing urbanisation is having a clear impact on the survival of insects.

It is not just urbanisation that is pushing insects to the brink, our use of artificial light is also having a massive impact on our little helpers. For years, farmers have been using lights to deter insects, but due to our sudden expansion in numbers and territory coupled with the fall in energy prices, light pollution now affects a quarter of the planet's land surface (314).

We have all experienced watching moths come together in unison around a light bulb, although, these occurrences are not as frequent as they were. Moths are not like humans warming themselves up by sitting around a fire in winter. They use the moon as a navigational tool. As the moon is so far away, they use it to align themselves in relation to the moon's position. Moths mistake the bulb for the moon and their navigation is affected as the new light sources are much closer than the moon. As moths fly past the human light source, they realign their position accordingly. This results in a somewhat comical image for humans of moths repeatedly winging their way into what seems to us to be a dangerous object. We are correct in this thinking as one third of insects trapped in the orbit around these lights either dies of exhaustion or being eaten. As if the situation wasn't confusing enough, cars have exacerbated the problem by making these lights mobile. It's estimated that vehicle lights are responsible for more than 100 billion insect deaths per summer in only one country: Germany (314).

Artificial lighting has also been found to disturb insect breeding. Some insects rely on the polarisation of light to find a water source to breed, but artificial lights confuse some insect species into laying their eggs on asphalt where they face certain death. When you consider that some species only live for a day, if species lay their eggs in a poor location, the species is endangered in just 24 hours. Conversely, some insects are repelled by light. They face the opposite problem in that in areas of heavy light pollution, these species stop looking for food (314).

The second cause of the insect apocalypse is related to the climate crisis. Increased temperatures and heatwaves, which are only going to increase in severity, duration and frequency, impact insects greatly. In Puerto Rico, insect numbers have dropped precipitously to between 1/8 and 1/16 of their numbers in the 1970s. Researchers here analysed the collapse in numbers and their findings seem to point to the rising temperatures as the main cause of this massive drop. The island has seen temperatures rise by 2°C on average in the past several decades. Scientists are not so concerned by the long-term warming as they are by the temporary heatwaves that accompany it. All living beings have a limit to temperature tolerance, and it seems many species are already beginning to reach theirs (315). This, once again, is creating problems for other species reliant on the insects for their food. Puerto Rico has recorded sharp synchronous declines in lizard, frog, and bird populations since the 1970s (315). It may seem naïve to link the declines in insect numbers and animals who prey on them, but research from Puerto Rico shows that birds who eat fruits and seeds have seen no such decline in their numbers, whereas birds reliant on bugs for their dinner have (316).

Research from the University of Stellenbosch in South Africa concluded that we can expect to see a further 6% decline in insect populations if we reach 1.5°C of warming with losses of 18% forecast for 2°C, and a staggering 49% at 3.2°C. These forecasts are for our warming atmosphere alone. They do not take into account the effect of light pollution, or farming practices which will be discussed next.

The third problem for our arthropod chums is industrial agriculture. This can be broken down into two parts. Firstly, with our burgeoning population comes the need for more food. This leads to the felling of forests, and the creation of large swathes of land that host endless

mono crops like soy and corn. In areas with long histories of agriculture, to enhance profit margins, old hedge rows have been dug up and fences erected in their place. The vast majority of new farmland from 1980-2000 was at the expense of forests. During this time 55% of new agricultural land came from intact forests while a further 28% came from disturbed forests (317). This leaves insects with few places to live.

To put things into perspective, at the beginning of the agricultural revolution 10,000 years ago, just 4% of the Earth's habitable land was used for farming. This doesn't include glaciers, deserts, beaches, or sand dunes. Fast forward to 2022, and over half of the world's ice-free habitable land is now used for agriculture. 37% of our habitable land consists of forests, 11% shrubs and grasslands, only 1% fresh water and 1% for urbanisation and other human infrastructure. Today, of the 28,000 species categorised as threatened on The International Union for Conservation of Nature (IUCN) Red List, 24,000 of them are threatened mainly by agriculture (318).

The second way in which industrial agriculture is helping to exterminate insects is the over-use of synthetic pesticides. In a perverse happenstance, the prediction of the good Reverend Thomas Robert Malthus in 1798, that unregulated population growth will always outstrip food supply, has yet to come true. It's perverse because the way in which humans have overcome this greatest of challenges has led to arguably an even greater problem.

Beginning in 1939, DDT (dichloro-diphenyl-trichloroethane) began to be used by the U.S. military to combat malaria, typhus, and other insect-borne diseases. Troops were doused in it, and soon the deadly scourge of typhus was no longer a threat for U.S. servicemen. In Naples, Italy, the U.S. carried out tests on more than a million Italians with body lice that spread the disease. The DDT killed the lice and saved the city from a dangerous epidemic (319).

After the war finished, its producers quickly found a peacetime use for the chemical. It was discovered to be useful at killing insects in crop and livestock production such as the Colorado potato beetle which was wrecking crops in the U.S. and Europe. It was also used to kill mosquitos that spread malaria, preserve Arizona vineyards, West Virginia orchards, Oregon potato fields, Illinois cornfields and Iowa dairies (319). Additionally, DDT was sprayed on crops, around businesses, homes, and gardens to kill insects like mosquitos and other crop destroying insects. It was extremely successful and soon spread to other countries [143].

In pictures from the 1940s and 1950s, American housewives can be seen spraying DDT all around their homes, children are seen playing in the fog of spray from municipal spraying trucks, and the media were calling it "magic" and a "miracle". It was so popular that in 1948, its creator Paul Hermann Müller, was awarded the Nobel Prize in Physiology or Medicine for saving hundreds of thousands of lives (319). We will see in the next chapter how receiving a Nobel Prize isn't always a sign that something is a good idea in the long-term.

Not all U.S. citizens though, were as enamoured by this fast-working killer, and many citizens were extremely worried by the effects. Soon the unfortunate reality of what had been done, was made apparent. After the publication of Rachel Carson's seminal book, Silent Sprint in 1962, people became aware that DDT wasn't just killing the bugs intended but was also killing everything higher up the food chain and extinguishing entire ecosystems. It was also affecting human health. Due to the commotion her publication caused; the Environmental Protection Agency (EPA) was launched in 1970. As to be expected, the industry went into overdrive to discredit her findings, but President Kennedy vindicated her findings after a committee review (320).

The U.S. Department of Agriculture, whose remit it was to regulate pesticides before the EPA came into existence, started looking closely at DDT in the late 1950s and 1960s. The EPA didn't take much time once created to take action on DDTs, and within two years of its creation, large scale use of DDT was banned (321). Today, the only allowable use of DDT is to kill mosquitos that carry malaria. The disease killed 627,000 people in 2020, with the majority of deaths occurring in Sub-Saharan Africa (320).

As early as 1945, National Geographic warned that DDT was killing many beneficial and harmless insects, and that flowers and trees dependent on pollinators could also be killed. The same article forecast the demise of birds and fish as well. This will be discussed in a later section. Unfortunately, DDT was still seen as a miracle chemical in that it killed an enormous range of insects without affecting human health. That DDT was benign for humans was not strictly correct and since DDT was banned, research has continued. Some of the effects on humans include breast and other cancers, male infertility, miscarriages and low birth weight, developmental delay and nervous system and liver damage. It turns out that it wasn't only insects that were affected (319).

DDT was the early precursor for synthetic pesticides, and unfortunately, their use has continued. In research published in August 2019, it was found that neonicotinoid pesticides had increased the toxicity of agriculture in the U.S., and this was a driving force for the insect apocalypse we are witnessing. The study assessed data from 1992 until 2014 and was published in the journal PLoS One. The findings suggest that the use of pesticides has moved from predominantly organophosphorus and N-methyl carbamate pesticides to a mix dominated by neonicotinoids and pyrethroids. Neonicotinoids especially, although used in lower doses per acre, have been found to be forty-eight times more toxic to insects and to remain longer in the environment. This had already led to the European Union (E.U.) banning the substance in 2018. These pesticides are particularly harmful to bees who are essential for biodiversity, and the food that sustains us. Although banned in the E.U., Neonicotinoids are currently used in 120 countries around the world on over 140 different crops. They attack the insects' central nervous system resulting in paralysis and eventual death (323).

The loss of such key insects is having massive knock-on effects, and these will be discussed later. It is clear that the use of synthetic pesticides is resulting in the widespread loss of insects, and this carries a great risk to our ecosystems at large and fundamentally the survival of our species. What is abundantly clear is that to save our insect friends, intensive agriculture needs to be massively transformed, and pesticide use has to be drastically reduced in favour of more environmentally friendly farming practices. Additionally, large areas would benefit from complete rewilding which would allow insects to flourish, and with them the entire ecosystem. The scale of this problem is gigantic, and to offer a brief glimpse of the enourmity, Dino Martins, an entomologist at Kenya's Mpala Research Centre and explorer for National Geographic, has this to say:

"No insects equals no food, [which] equals no people. (322)"

The time for radical change is evidently upon us, and unfortunately the use of chemicals in our food systems do not end with the destruction of insects. They go much deeper than that, right under our feet.

demeureduchaos.org

Abeille Monsanto
Abode of Chaos is licensed under CC BY 2.0.

Monsanto
AraXilos is licensed under CC BY-ND 2.0.

Dear Indy,

The closest most urban city dwellers ever get to pesticides and herbicides is when they douse their garden once or twice a year. That was always the case for us, although we've rarely had a garden in order to douse.

That all changed when we moved to Osakikamijima in 2020. We are now face to face with the very real threat these chemicals pose to our existence. After moving into our dilapidated former mandarin farm, we are surrounded by active commercial farms, and they love to spray deadly chemicals throughout the year.

When we first saw them spraying, without informing us, we asked them politely to tell us before they spray so we can keep the cats and dogs inside and close our doors and windows. This doesn't stop us seeing the results of this excessive spraying, but it does give our furry friends some basic protection. The birds and bees will not be so lucky.

The wildflowers we planted along the border between our garden and the mandarin farms have all died and the edible weeds that line the mountain road have all turned from luscious green to lifeless brown. This is why we have to stop the dogs feasting on them when we go for walks. This was a favourite snack for them but eating them each day could cause health problems for them if the chemicals build up in their tissue. As the toxins are passed through the food chain, this is known as bioaccumulation.

It's not the farmers' fault. They all know the problems, but Japan Agriculture (JA) insists on the chemicals being sprayed at certain times if they are to buy the produce. In fact, the farmers don't use chemicals on the food they grow for personal use. They know the risks. These chemicals are not only wiping out wildlife, but they pose a serious risk to human health too. That's why we aren't using any chemicals on the food we grow. Let's hope the world moves away from this disastrous food system soon. Otherwise, we will find feeding ourselves becomes a very difficult thing to do. No pollinators equals starvation.

On a positive note, apparently I became the first person in the village to see a firefly in the past two years. The game isn't over yet!

Soil

It isn't only the insects above land that are being affected by industrial agriculture. The problem is much worse, and the insect apocalypse has an equally pernicious partner in crime, soil degradation. Soil degradation is the term used for soil which has seen its ability to grow crops diminished. Soil isn't something that we think about regularly. It is largely reserved for children and green fingered gardeners. Soil is so important for our lives, however, that maybe it deserves a place in discussions at the dinner table.

Amazingly, the brown stuff contains at least a quarter of all global biodiversity and is essential for providing clean water. It also helps to prevent flooding and drought. Significantly, it plays a vital role in ecosystem nutrient recycling as organisms decompose and provide sustenance for other wildlife (324). Every single handful of healthy soil contains billions of microscopic organisms. If we were to take just a gram of soil, it would host up to a billion bacteria cells. Overall, around eleven million species of organism live in the soil and less than 2% of these have been named and classified. It is also a depository of many possible life saving medicines. Vitaly for humans, healthy soil provides healthy nutrients to plants, and these in turn provide healthy nutrients to human beings (325). In total, 95% of global food supplies are reliant on soil (326).

In the fight against runaway climate change, soil management will play a huge part in the success or failure of human beings to keep temperature rises within manageable levels as three times more carbon is stored in the soil under our feet than is in the atmosphere today.

Research from 2017, found that better soil management has the potential to store more than 1.85 gigatonnes of carbon annually. This is more than all the carbon emitted by every plane, train, car, ship, motorcycle, helicopter and truck each year (327). Additional 2020 research has shown that better soil management and land use could allow the soils to sequester 23.8GtCO2e per year and research published in 2021 increased this amount to an astonishing 34.5GtCO2 annually (328) (329).

Since the agricultural revolution, 133bn tonnes of carbon has been emitted from the soil, and this number has increased rapidly since the beginning of the industrial revolution. Around a quarter of all manmade carbon emissions to date have come from the soil under our feet, and interestingly, around a quarter of our carbon emissions are absorbed by the soil annually (327). If the rest of this carbon is released, we can declare the fight against climate change to be over.

If we want to see the impact of losing topsoil, we can look at history. According to Roman tax records, Syria and Libya used to grow significant amounts of wheat, but as farmers degraded the soil over time, today these areas have barely any soil to grow crops. The effect of topsoil destruction is passed on to future generations as soil forms extremely slowly at the rate of one ccentimeter every 300 years, so once it's gone, it does not reappear in the lifetime of a generation, or even three (330). To be brief, according to soil expert, Prof Jane Rickson, from Cranfield University, U.K.:

"The thin layer of soil covering the Earth's surface represents the difference between survival and extinction for most terrestrial life. (331)"

We have seen how important soil is for our existence, but why should we be concerned about the earth in our gardens? At present, according to Volkert Engelsman from the International Federation of Organic Agriculture Movements:

"We are losing 30 soccer fields of soil every minute, mostly due to intensive farming (1)."

According to the United Nation's Food and Agriculture Organisation (FAO), soil degradation is a bigger obstacle to human survival than global warming, species loss or any other environmental crisis. They warn us that twenty-four billion tonnes of fertile topsoil are lost each year, and that 25% of the Earth's surface is degraded already. This is enough land to feed 1.5 billion people (332). The World Wildlife Fund (WWF) states that half the world's topsoil has disappeared in the past 150 years (333).

Mozambique, as we have seen, was hit by intense storms and droughts in 2019. As if this wasn't enough, they also had to deal with the loss of 40,000 hectares of cultivated land due to the use of pesticides (334). Further north, soil degradation in Ethiopia is believed to cost the country US$4.3 billion annually.

The problem of soil degradation isn't shared only in developing countries. It is estimated to cost Italy around €900 million a year (335). In Iowa, U.S.A., the depth of topsoil has diminished rapidly since 1900 when the depth was 35-45cm. By 2000, this depth had declined to 15-20cm. More worryingly, is that in the still-fertile fields of Iowa, more than fourteen million tonnes of soil were lost in 2014 due to a series of storms. The state university analysed eighty-two sites in twenty-one counties and discovered that from 1959-2009, soil structure and levels of organic matter had degraded, and acidity had increased (336).

In the U.K., 2.2 million tonnes of topsoil are eroded every year and 17% of land has already been eroded, with 44% of arable land at risk. About 18% of organic matter in the arable topsoil was also lost between 1980 and 1995. Experts warn that organic matter levels are so low that in the long term, food production might not be sustainable (337). Organic matter is essential for healthy soil. It provides nutrients and habitat to organisms and helps soil hold water. It would be possible to spend many more pages dedicated to soil degradation, and it is clearly an existential threat going forward. However, the FAO summed up the problem facing humanity due to soil degradation in a much more succinct way. They released an astonishing statement at a forum marking

World Soil Day in 2014, where Maria-Helena Semedo stated that if soil degradation levels continue apace, all of the world's topsoil may be gone by 2074 (1). The problem is global in scale, and unless it is tackled head on, it could combine with other factors to make life on planet Earth very inhospitable.

Erosion of soil is also exacerbated by the climate crisis. Areas that have traditionally experienced erosion will likely see much more in the future. Rainfall distribution and intensity are affected and with greater rainfall in shorter periods, erosion is likely to increase. Sub-Saharan Africa is already suffering from the erosion of its soils, with erosion accounting for 77% of overall degradation; 22% of arable land is at risk. Almost 70% of Uganda's land has been degraded by erosion, areas of grazing land have been devastated in Tanzania, and Malawi loses twenty-nine tonnes per hectare of land each year to erosion. The problem is not contained to Africa: 7.2% of Europe's total crop land is impacted at a cost of €1.25 billion ($1.4bn) annually (338).

So, how did we get into this mess? As was the case with the spread of DDT, the initial cause of this problem was extremely well intentioned. It can be traced back to March 25, 1914, in Saude, Iowa. On this day, a boy was born who would go on to save the lives of hundreds of millions of people, and in the process win the Nobel Peace Prize. Norman Borlaug was born into a hard-working family of Norwegian ancestry. He grew up poor and helped his father farm potatoes and raise cattle. Every autumn, his family handpicked half a million ears of maize on their family's forty-acre plot. In 1932, Borlaug decided to attend the University of Minnesota where he was offered room and board in return for working as a waiter in a diner. He duly became shocked by the effects of the Great Depression. Buildings were abandoned and streets were filled with the homeless victims of the dustbowl. Many people wrapped themselves in newspapers in the absence of blankets. A lot of these homeless were former dairy workers: kicked off their land after struggling to pay off debts. During World War I, the government had encouraged dairy farmers to increase yields to help feed the soldiers. High prices were paid for this milk, and in turn farmers bought more cattle and increased production. They also invested in tractors and pasteurization systems which placed them in debt. Unsurprisingly, after the war, demand plummeted, and farmers were left with high levels of debt, and no way to pay it off. They ended up selling milk at a loss, and eventually the banks foreclosed on them, and many ended up in Minneapolis. Fast forward to 1933, and Wisconsin was hit by massive discontent as people struggled to survive. Riots broke out and the young Borlaug witnessed this first-hand. There and then he decided that he wanted to make his life's work about ending hunger forever (192).

Borlaug studied forestry in Minnesota, but he ran out of funds in 1937 and began working as a forest fire fighter to improve his finances before taking a job with the U.S. Forest Service after his graduation. It was in university, that Borlaug would be introduced to the problem of black stem-rust fungus that was decimating wheat crops. This had been a problem for millennia, and Borlaug's family had actually experienced it in 1878 when they were driven out of business. Later in 1904 and 1916 it had led to widespread misery in the U.S. and northern Europe. Almost a third of the wheat harvest was lost in 1904 which forced the bread prices up. The spores from the rust were carried high into the air by the slightest of winds and spread for miles. On just a single acre of wheat with moderate stem rust, more than fifty-trillion spores were evident. The rust was such a problem historically that Romans were in the practice of sacrificing rust-coloured dogs to the Gods to appease them. In this rust, the young, and driven Borlaug, had found his cause (192).

During cold winters in the U.S. and Europe, the fungus cannot survive the cold, but in the warmer climates of Africa and Mexico, the spores are carried north by constantly blowing winds. It is in these winds that spores are brought north to meet the warming spring weather and settle on wheat crops. In 1914, the Rockefeller Foundation had succeeded in preventing the spread of cotton pests such as boll weevils. Decades later, the U.S. government was keen to assist Mexico in raising the wheat crop, but due to massive negative feelings towards the U.S., the government turned to the Rockefeller Foundation to help work under the radar. The foundation appointed three scientists to work on the problem, one of whom was Borlaug's professor from Minnesota, Elvin C. Stakman. Due to Mexico being the origin of stem rust in the U.S., Stakman was extremely interested in trying to eliminate it south of the border. Starting in 1931, work began in earnest. Soon, the scientists saw with their own eyes the desperate situation of the Mexican people who had little food, few clothes and had poor housing. At that time, the Mexican harvest had dropped by a third since the 1920s, even though they had planted almost a million more acres of wheat. To make matters worse, the population had risen by five million. The scientists worked hard to report back to the Rockefeller Foundation and a few months after they finally sent off their report, the Japanese bombed Pearl Harbour. In times of war, it was prudent for the U.S. to have a peaceful situation close to home. This led to the Rockefeller Foundation stepping up their efforts. Top of the list was improving maize production, but they also knew that if they could breed stem-rust resistant varieties of wheat in Mexico, that could cut off the rusts' path to wheat north of the border, thereby increasing yields, reducing hardship and improving food supplies for the troops (192).

Back in the states, Borlaug was obtaining his master's degree in 1941, after studying fungal disease for his dissertation. Sadly, he took no ecology, soil science or hydrology classes. If he had, the world may very well be a different place today. In October of that year, Stakman called Borlaug to his office to invite him to accept a job at Du Pont chemical company in Delaware. His wife made sure he accepted the job as no other offers were on the table. In December they drove to Delaware and Borlaug began working on a variety of jobs for the war effort. He couldn't have been too enthused by his new line of work because when Stakman offered him the chance to head up the stem-rust team in Mexico in 1943, he readily accepted. In September of the following year, he left for Mexico. He travelled the country looking at different varieties of wheat and farming techniques that seemed to prevent stem-rust. Scientists back in the U.S. sent many varieties for him to test in the Mexican soils. During this time, Borlaug was further concerned to see the poverty that existed all around him and the lack of healthy soil in which to grow wheat (192).

He continued his work at breakneck speed, cross breeding thousands of foreign varieties in the hope that one would be resistant to stem-rust. A full decade after arriving in Mexico, Borlaug finally had some success in producing a short variety that not only fended off stem-rust, but also produced more grain. The only downside was poor taste and low protein content. It took another five years of work to improve the taste and amount of protein, but by 1962, they had succeeded in producing a prolific short variety of wheat that was not only disease-proof but could also grow in soils that were rich or poor anywhere in the world. In this line of work, Borlaug and his team had found the Holy Grail. There were only two things necessary for the wheat to thrive and the harvest to be large: water and fertilizer. He believed that pouring on water and fertilizer would be the answer to worldwide hunger in growing larger than ever quantities of wheat (192).

Up until this point, the work of Borlaug is hard to criticize. It was from this moment forward that the problems we experience today, can be traced back to his influence. Borlaug offered a package that could be used in any country on any continent, seeds, fertilizer, and water. Each would need to be adjusted slightly to suit each environment, but by and large the package remained the same, seed, fertilizer, and water. In 1968, the term 'Green Revolution' was used to describe Borlaug's package as it had led to crops in Mexico growing from 345kg per acre to almost 1,143kg in just twenty years. Harvest had tripled on Borlaug's watch. Similar results were coming in from India and Pakistan, and his discovery certainly averted famines and saved many lives. People around the world rightfully applauded his incredible dedication and effort. In 1970, in gratitude for his many years of work, he was awarded the Nobel Peace Prize (192).

Sadly, Borlaug's package had two problematic features: water and fertilizer. Water was increasingly in short supply and these new crop varieties required more water than usual. Additionally, the fertilizer favoured by U.S. corporations was synthetic and this led to long term declines in soil fertility, reduced genetic diversity and soil erosion. To buy this package, subsistence farmers got into debt, and many were displaced in favour of larger operations. The discovery of Borlaug also set in motion the wheels of the larger genetically modified seed companies to plant larger than ever monocrops that required fertilizer by the truck load. He also threw his full support behind the large-scale use of pesticides, even going as far to endorse the use of DDT in 1971, and in doing so denigrate the Silent Spring author, Rachel Carson. In response to concerns from environmentalists about the use of chemicals in our food system, Borlaug had this to say:

"Because of unwise legislation that is now being promoted by a powerful group, of hysterical lobbyists who are provoking fear by predicting doom for the world through chemical poisoning, then the world will be doomed not by chemical poisoning but from starvation (339)."

Unfortunately for Borlaug, his legacy may not be the one he worked long and painstakingly for. His miracle discovery, while providing incontrovertible short-term benefits, has led to long term degradation of our soils which ironically may be unfit to grow crops on the one hundredth anniversary of his Nobel Peace Prize.

As they say, hindsight is a beautiful thing, and in hindsight, chemical companies were not ideal bedfellows for a man with such lofty ideals. As Borlaug struggled under intensely difficult conditions to provide adequate food for the poor, his partners were locked in a war of manufactured doubt. As concerns arose about the safety of their products, their spin went into hyper drive. The same PR companies employed by the tobacco industry, and more recently, the fossil fuel industry were purchased to provide "evidence" that spraying chemicals on the food that we eat was safe. Multinational behemoths like DowDuPoint, Bayer, Monsanto and Syngenta insisted that their products were not only safe, but necessary to feed our growing population. In the fifty years following the end of WWII, insecticide use increased ten-fold in the U.S. In the same period, crop losses almost doubled (340). This hardly sounds like a necessary product.

The end of the war also coincidently saw a new market opening up for producers of nitrate. At the war's end, the U.S. had ten large-scale nitrate factories making bombs for the war effort. With the advent of peace, these companies looked elsewhere for a market for their goods. They settled on agriculture, and especially, the new hybrid strains that had been developed. By the mid-2010s the U.S. was responsible for 12% of the planet's nitrogen-fertilizer use, much of which was designated for corn which covers 30% of U.S. farmland. Whilst nitrogen is essential for good soils, the excess nitrogen that comes from the synthetic kind seeps into streams, rivers and eventually oceans, in turn causing ocean dead zones. It also emits nitrous oxide, which is 300 times more potent than carbon dioxide, and more importantly for this section, destroys organic matter in soil (341). Research is suggesting that a simple shift to more diverse crop rotations would eliminate the need for artificial fertilizers while not affecting food production. Unfortunately, this is not happening. Fertilizer use is growing rapidly. Around 191 million tonnes were used in 2019, up from 46.3 million tonnes in 1965 (342).

The industrialized use of pesticides on our food is affecting the earthworms in the soil too. Earthworms in the soil under crops sprayed with pesticides only grow to half their usual weight and fail to reproduce as efficiently as earthworms where no pesticides are used (343). In the U.K., two out of five fields have either too few or no worms at all, according to Rothamsted Research. In turn this has led to a drop in song thrush populations (344).

Agrichemical companies have long led us to believe that their products are vital to feed the burgeoning world population. However, it is becoming apparent that organic farmland helps to reduce soil degradation, conserve soil moisture and living organisms, and just as importantly, actually produces equal or increased crop yields (345).

A 2021 study published in Frontiers in Environmental Science highlighted the problem pesticide use has had on our soils. One of the co-authors, Nathan Donley, from the Centre for Biological Diversity stated:

"Study after study indicates the unchecked use of pesticides across hundreds of millions of acres each year is poisoning the organisms critical to maintaining healthy soils" ... *"Yet our regulators have been ignoring the harm to these important ecosystems for decades (346)."*

It seems that just as with DDT, short-term profits have been prioritized over long-term health, and the clock is now ticking down to 2074. Even if we stop degrading our soils, they may still come back to haunt us as the atmosphere warms. It has been suggested that if we hit 2°C of warming, then the soils will begin to spill some 230GtC into the air, and potentially tip us into abrupt climate change (347).

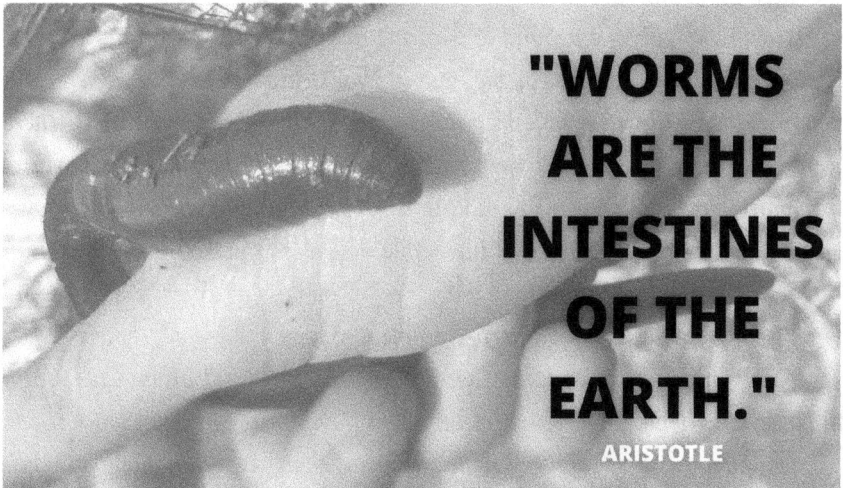

"WORMS ARE THE INTESTINES OF THE EARTH."
ARISTOTLE

Dear Indy,

Let's talk soil. That lovely brown stuff in the garden that we love so much. When we moved into our new farm, our garden was covered in black plastic sheets. They had been laid down to stop excessive weed growth as the farm was empty for over twenty years. A new garden had grown on top of the sheets, and this shows that nature will always find a way. It's not really the planet we are trying to save. It's us! The planet will rebound eventually, and life will go on without us. In fact, the planet might be quite happy to see the back of us, well those of us participating in the global economy anyway.

After pulling the black sheets back and in the process, doing my back in, we found that while there was life above the plastic, there wasn't much life underneath. In fact, the soil seemed completely void of life. Where there should have been earthworms a plenty, we found none. This was the same everywhere the sheets were placed.

When we finally got around to planting the Habaneros, Carolina Reapers, and Trinidadian Scorpions, in the land at the back where no plastic was laid down, as I was tilling the soil, I was confronted with an existential crisis. There were beautiful earthworms every which way I worked. Fortunately, I wasn't using any machinery, so the casualties were minimal, but the sheer numbers of big fat juicy earthworms was incredible. It was so amazing to see them all there but at the same time, it made my job of missing them as I used my pitchfork much harder. It was a nice problem to have.

Unfortunately, this isn't the case everywhere. Fields all around the world are increasingly not home to worms. It seems the culprit here is not black sheets to prohibit weeds, but the very same chemicals that we put on our food. They seep into the ground and our little worm friends don't much like them either. People, including you, might think worms are gross but they are fantastic little friends to us and without them, we will find growing food becomes a lot harder, if not impossible. In addition to Save the Whale and Save the Tiger campaigns, maybe we should have Save the Worm campaigns too. If they can thrive, we might be able to as well.

Plants

Much is made of the disappearance of wild fauna, and rightly so. Losing animal species is heart wrenching for many humans to witness. By focussing intently on the demise of animal species though, are we ignoring the impact climate change is having on the life-giving flora that provide not only our nutrition and oxygen, but also many of our medicines, clothing and building materials?

All the calories we consume today come either directly from plants or indirectly through consuming the flesh and secretions of animals who got their nutrition from plants. Additionally, plants on land and in the oceans produce the oxygen we rely on for life. Plants have also long been used to cure human ailments, with evidence from clay tablets dating back more than 5,000 years that Sumerians were using hundreds of different plants, including opium. The ancient Egyptians were using more than 850 plant medicines, and the Romans more than 1,000 (348).

One of the most commonly used drugs today is Aspirin. It was first synthesised in 1899, but as far back as the ancient Egyptians, it was made from its natural ingredient: willow tree bark. Several anti-cancer drugs are also derived from plants (349). The widespread use of Quinine in anti-malarial drugs owes its existence to the Cinchona calisaya tree in South America, and morphine likewise is derived from the object of European peace: the poppy (350). In total, a quarter of modern pharmaceuticals, including two thirds of all cancer fighting drugs, are derived from tropical forest plants (351).

Our clothes are also dependent on plant life. Cotton and hemp are derived from plants, and even synthetic fibres such as rayon are manufactured from cellulose, mainly found in plants. The wool that we take from sheep is also reliant on the grass that sheep consume for energy. Many of our traditional buildings also use plants, from bamboo to adobe and thatch and hemp to baled straw. The plants we tend to take for granted have served humans, and our evolutionary descendants extremely well for millions of years.

Since the first land plants appeared on Earth around 470 million years ago, they have been instrumental in helping to drive Earth's climate. The importance of plants in our atmosphere can be summed up with a story about the humble moss that dates back to the origin of land plants in the Ordovician period. It seems that the emergence of shallow rooted plants like mosses and liverworts in this period led to a mini-ice age and mass extinction. The mosses were able to extract nutrients from the bedrock they attached to which allowed the rocks to suck CO_2 out of the atmosphere. CO_2 levels, incredibly, fell by around sixteen times their previous levels and the world rapidly cooled before entering an ice age as a result. Unfortunately, for our water loving ancestors, one of the chemicals leeched from the rocks was phosphorous and this entered the oceans in turn triggering a massive algal bloom. Creatures fed on this bloom and used up the oxygen which suffocated oxygen breathing animals and led to a mass extinction of marine life (352).

Plant life has flourished on Earth in the years following the Ordovician, and they continue to play a vital role in regulating our ecosystems and sustaining life on Earth for all creatures great and small. Today, we are fortunate that around 435,000 species of plant, give or take a few, call Earth home (353). While fulfilling essential roles in our ecosystems, they also provide humans with beauty, grace, and heavenly scents.

We are still discovering new flora at the rate of 2,000 species a year. Sadly, many of the species being discovered are already threatened with extinction. These plants are known as the "living dead" in that they number so few that they are either unable to reproduce or the large animals needed to disperse their seeds are extinct. A total of 571 species are known to have become extinct in the time since 1750, and extinction rates are thought to be 500 times higher today than prior to the industrial revolution. Due to our limited knowledge, the true number is probably much higher. More plants have disappeared than birds, mammals and amphibians combined (354). Around 37% of the world's plant species are very rare; they've only been seen five times or less, and these species are extremely vulnerable to climate change (355).

A report that emerged in late September 2019, laid bare for all to see the situation we have created for ourselves. The International Union for Conservation of Nature (IUCN) report estimates that 58% of trees that are endemic to Europe are at risk of extinction, with 15% considered critically endangered (356). One of those at risk is the horse chestnut, a tree more relevant than any other for children growing up in the U.K. The tree produced every single conker that was ever used in the playground game of conkers. The 'conker' tree has provided seeds of joy to British children since the 1850s, and its demise will be much lamented (357).

Another species of tree that is heading for extinction is the African baobab. There are nine species of baobab tree and they can grow to around thirty metres in height as well as circumference and can live for thousands of years. The trees have been used by humans as a meeting place to discuss problems and find solutions for millennia. In addition, their leaves are used in traditional medicine, the fruit is high in nutrients and the seed's oil is used by the cosmetic industry. It is unlikely the species will make it to 2200 (358).

Another famous species of tree that is at risk is the Joshua, popularised by U2's 1987 album The Joshua Tree. The tree has survived in the Mojave Desert for around 2.5 million years, but due to drying conditions, drought and wildfire, the species is threatened with extinction. Joshua trees can live to 300, but their young and seedlings cannot store enough water to survive droughts like the 376 weeklong drought that lasted from 2011 until 2019. The best-case scenario for the Joshua tree is that one of every five will survive the next fifty years, but failure to stabilize temperatures within this time frame will likely push the tree over the edge (359).

Even more famous that the Joshua is the Giant sequoia. This is the largest single-stem tree on our planet and can live to be more than three thousand years old. They have very strict climate requirements and only exist in a small area in the Sierra Nevada Mountains. They live at between 1520 and 2130 meters above sea level. Their bark can grow to become almost a meter thick, and their branches can measure two and a half meters in diameter. The trees are renowned for their girth, but they do grow over ninety meters. Giant sequoia are well regarded for their hardiness, but they have come under attack recently from bark beetles linked to the climate emergency. The trees are fabled for their indestructible nature, but the beetles, coupled with the prolonged drought conditions and subsequent wildfires mean the trees face an uncertain future. Contingency plans made for the 2050s are being utilized now, such as planting seeds at higher elevations and thinning the forest floor to avoid wildfires (360). The giant sequoia is not alone in this challenge though, with around 40% of all forests in the U.S. at risk of attack by pests. Around six billion trees are standing dead in the U.S. west. During droughts, trees rely on hydrated soils deep underground, but as these deep soils eventually become dehydrated, the trees become stressed and frail. Due to ill health the trees fail to photosynthesize and when the bark beetles appear, the trees are unable to cope (361). The situation is turning forests from carbon sinks to carbon emitters and as rainfall and wildfire patterns become more unpredictable, more and more trees will likely be affected (362).

While humans can switch on air conditioners when it gets too hot, animals are faced with having to physically move in order to adapt. Plants, have neither option available to them. In order to deal with rising temperatures, plants grow taller to help them cool themselves off. Their stalks become taller and leaves narrower. This helps to make the plant more unstable and the proportion of substances like proteins are reduced (363).

There is one other option, it seems, open to plants. Miles Silman, a forest ecologist from Wake Forest University has been carrying out research into future projections for forest evolution. His research led him to Manu, Peru. Beginning in 2003, Silman laid out seventeen plots each measuring 10,000 sq. meters (2 1/2 acres). Each plot was at a different elevation that varied in temperature. The elevation at the first plot near the top of the ridge was 3,450 meters, whereas the lowest plot was near the Amazon basin at sea level. The temperature alternated by several degrees. Silman and his students measured every single tree over ten centimetres in diameter. The trees were tagged, identified, and given a number. The results were extremely interesting while also concerning for many of the species being recorded. Ninety percent of the species found in Plot 1 were not found in Plot 4 which is 762 meters (2,500 feet) lower down the ridge. This is not unusual as trees in the tropics usually have a narrow altitudinal range, whereby they live best at one elevation. The first re-census was carried out just four years later in 2007. The results showed that on average, trees were moving up the ridge at a pace of 2.4 meters per year. This average masked huge differences in the ability of species to flourish at different elevations, and temperature. The Schefflera arboricola from the ginseng family was chugging up the mountain at 100 feet (30 meters) a year, while at the other end of the spectrum, Ilex trees hadn't moved at all (364). The study was replicated by nature in Japan. Here, the tree line on the iconic Mount Fuji has been steadily climbing the past forty years and now sits some thirty meters further up slope than it did in 1978. The scientists from Niigata University's Sado Island Centre for Ecological Sustainability believe this is due to the 2°C rise in temperature at the mountain's summit over the past fifty years. This lengthened the growing season (365).

For the trees unable to move their range, the future looks extremely bleak. As temperatures rise, they will be unable to move, and they will likely find conditions incompatible with life. Others more able to prosper at different temperatures, may flourish. However, the overall outlook is not good. Research from the University of York discovered that with minimum warming, around 9-13% of all species would become extinct by 2050. Maximum warming would lead to between 21-32% of species dying off. Taking the median number would result in around a quarter of all species becoming extinct by the middle of the century (364).

In the Earth's most naturally abundant areas, like the Galapagos and Amazon, up to 80% of all plant species could be lost by 2100 unless we take urgent climate action. Even if we meet the Paris Climate Agreement 2°C target, then 45% of the Amazon's plant life could be lost, 45% of southern Africa's Miombo woodlands are at risk, and south-west Australian plant species could drop 40%. If temperatures were left to rise to 4.5°C, then areas like south-west Australia could see losses of almost 75%, southern Africa's Miombo woodlands could experience more than 80% declines and the Amazon almost 70% (366).

How can this be? Many climate change deniers enjoy making the claim that CO_2 is good for plant growth. According to them, we should be increasing CO_2 in the atmosphere because this increases plant growth and plant growth is good. Let's assume that CO_2 wasn't a pollutant responsible for tens of thousands of deaths every year, shouldn't increased CO_2 levels actually increase plant growth? It seems that like with many things, more is better, until it isn't. For example, the more money people earn correlates positively with happiness, right up until you earn $75,000. This makes sense as a lack of money can increase stress levels as people struggle to pay bills. The more money you earn over and above $75,000 does not equate to any increase in happiness, however, it actually leads to a decrease. Once a stable lifestyle is experienced, more money doesn't equal more happiness (367).

The same effect is evident with plants and CO_2 it appears. Whilst it is true that increased temperatures do lead to increased CO_2 levels, and increased CO_2 levels increase the amount that plants are able to carry out photosynthesis, scientists are unsure what will happen if plants grow more quickly. Around 30% of our greenhouse gas emissions are currently absorbed by foliage and trees on land. Tropical and boreal forests are the biggest carbon sinks as they account for more than half (53%) of this amount. Due to deforestation in tropical forests, the amount of carbon absorbed there has actually begun to fall. It is presently rising in the northern boreal forests in Canada and Russia which has helped to increase the overall sink by one billion tonnes of carbon from 1992-2015. Due to what is termed the 'CO_2 fertilisation effect', the more CO_2 emitted by humans, the more is available to aid photosynthesis. This seems to be helping plants to grow faster and store more carbon. What scientists are unsure of is that if plants are growing faster, will this lead to them dying younger. Professor Anja Rammig, a researcher of land-surface interactions at the Technical University of Munich is concerned that the increased carbon sink the boreal forests have become may reverse itself in the coming decades (368).

In further studies conducted in California, a report found that higher temperatures were harmful to plant growth. The only variable that resulted in increased productivity was the addition of nitrogen. When increasing temperature or CO_2 level, yields actually declined. A further French study found that corn yields which had been rising steadily for decades were now beginning to slow, and yields were hit hard during heat waves. Maize yields are likewise found to decrease for every day above 32°C. Wheat is also affected with a 1°C rise in temperature leading to a 5% decline in harvest (369).

Further testing on CO_2 has found that certain plants could become more poisonous at increased levels. Cassava and sorghum, which hundreds of millions of people are dependent on for sustenance, both produce cyanogenic glycosides which release cyanide gas if their leaves are crushed or chewed. About 10% of all plants and 60% of crop species produce cyanogenic glycosides. Researchers grew these plants at various CO_2 levels of 360 parts per million (ppm), 550ppm and 710ppm. With each increase in CO_2, the amount of cyanide relative to protein rises. The study concluded that anyone reliant on these crops for survival would be at risk of cyanide poisoning, especially during droughts. Not only were the leaves becoming dangerous, but the number of tubers also dropped by 50% with increased CO_2 (370).

As our ecosystems continue to unravel around us, some may be inclined to reach for a bottle of something inviting to reduce anxiety levels. That something nice may not be wine in our warming world. If global average temperatures reach the dreaded 2°C (3.6°F), then areas suitable for growing grapes could be reduced by as much as 56%. With current targets, it's more likely we will be heading to 4°C (7.2°F) than

2°C, and 4°C would shrink suitable areas by 85%. The good news for wine enthusiasts or eco anxiety sufferers, depending on your reason for reaching for the bottle, is there are around 1,100 varieties of grape planted today. This gives us hope as it will be possible to grow different varieties in different areas. The bad news is that even after utilizing various grapes in different soils and climates, wine grapes are very sensitive, and losses are still considerable. At 2°C, even after altering grape type, losses would still equal 24%. In warmer conditions, losses would be considerably higher. By swapping grape variety, losses at 4°C would drop from 85% to 58%. Whichever way we look at it, wine consumption is going to have to drop, either by a quarter or maybe by half. As the crisis unfolds, we may have to change tipple if we wish to take our minds off of things at the end of a long day (371).

The worrying trend to follow is the role of plants in what are known as 'co-extinctions', when one organism becomes extinct because it depended on another unfortunate species that has disappeared. Researchers studying this impact simulated 2,000 'virtual Earths' to understand the probability of life coming to an end on our planet. They included scenarios such as multiple atomic bombs, large asteroid impact, and runaway climate change. What they found was that virtually all species, regardless of how tolerant to change they were perceived to be, were affected, as they were also reliant on other species that were less tolerant and had perished. The co-author of the study, Professor Bradshaw, stated that failure to include 'co-extinctions' in future modelling results in underestimating species loss by as much as ten times. As if this wasn't worrying enough, the situation can be summed up neatly according to Dr Giovanni Strona of the European Commission's Joint Research Centre:

"Another really important discovery was that in the case of global warming in particular, the combination of intolerance to heat combined with co-extinctions mean that 5-6 degrees (9-10.8°F) of average warming globally is enough to wipe out most life on the planet."

It is clear that rising temperatures will not increase all plant growth past a certain level, the plants themselves may become less nutritious, or even more poisonous, and with everything on the planet connected in a web of life, plant extinctions may drive or be driven by the demise of other organisms leading to what is being called the 'extinction domino effect' that could lead to the end of all life on Earth (372).

Baobab by Noel Feans is licensed under CC BY 2.0.

Deforestation

"When humanity wins its battle against the forests, this victory will be humanity's greatest defeat!"

This quote from the Turkish poet Mehmet Murat Ildan perfectly illustrates our current situation. We are locked in an ongoing battle against our unbelievably patient life source, Earth. Should we win this battle for dominance, will not our celebration be muted as we purvey the destruction left behind? In place of unparalleled beauty, what will we build instead?

There are more than 60,000 species of tree and they are under unprecedented threat from climate change and deforestation. Trees are the longest living organism on Earth. There are trees alive today that were inhaling carbon dioxide and exhaling oxygen when humans first learned to write 4,600 years ago. The same trees will have stood around, and taken in, everything our historians have noted down in all of written history. The Great Basin Bristlecone with the ID WPN-114 Prometheus in Nevada, USA has been going about her business slowly but surely for a staggering 4,900 years (373). Trees, like humans, never die from old age. As we get older, we become more perceptible to sickness and death. Unlike humans though, trees never stop growing. There is a physical limit to their height as trees cannot send water from their roots to the top layer of leaves, making it impossible to carry out photosynthesis. Trees get around this by growing wider, and they actually grow faster the older they become (374).

Typical hardwoods consume around 379 litres of water a day and absorb around 150 kg of carbon dioxide per year. They also lower the air temperature by evaporating water in their leaves, help to improve water quality through filtration and also prevent against storm flooding (375).

These trillions of trees cover about a third of the planet's land area. There are thought to be around three trillion trees, with 1.4 trillion in the tropics, 700 billion in boreal areas like Canada and Russia, and 600 billion in temperate regions. They are home to more than half the world's land-based plant and animal species. More than one billion rural people rely on forests, with many people completely dependent. Forests have long acted as carbon sinks. It is estimated that forests around the world are responsible for removing 2.1 billion tonnes of carbon each year. Forests also act as natural aqueducts by moving up to 95% of their absorbed water to places where it is needed. Individual trees act like fountains as they suck water out of the ground and release it through their leaves into the atmosphere as water vapour. The trillions of trees carrying out their task create rivers in the sky, and the rivers take that vapour hundreds or thousands of miles away from where it falls, and the cycle continues. They hold water in the soil, prevent erosion, and help to cool the planet by producing a cooling effect. Currently, around 13% of the world's forests or about 5.24 million sq. km are managed to conserve their abundant biodiversity (376).

Unfortunately, forests are under relentless attack from human beings. Since humans first started cutting down trees, it's estimated that 46% have been felled. Around 17% of the Amazon has been destroyed in just the past fifty years, and things are speeding up. Since the mid-90s, a forest the size of South Africa has been destroyed. A staggering football field of forest is felled every single second. That's sixty football fields of forest lost every single minute, 3,600 each hour or for people who think on a daily basis, 86,400 football fields a day (377). In the time it has taken you to read this far in the paragraph, eighteen football fields of forest have been slashed and burned. Between 2000 and 2010, thirteen million hectares of forest were lost each and every year. To put this into context, this area is similar in size to the nation of Greece. The situation for tropical forests is even worse as half have been lost in the last century alone. In a single year, 2017, an area the size of Bangladesh (158,000 square km) was destroyed (378). By 2030, unless the situation is reversed, we are likely to see another 1,700,000 sq. km of forest lost. There are eleven deforestation hot spots around the world that account for 80% of the planet's tree clearing, with ten of them being tropical (379). We will take a closer look at these a little later.

When we think of deforestation, we tend to think of the tropical forests like the Amazon or those in Southeast Asia. This is not surprising as more than half of the trees left on planet Earth are in these tropical zones. Ten of the eleven deforestation hotspots are tropical forests and 80% of the land projected to be lost by 2030 is in the tropics (379). They are also home to much loved species like orang-utans and leopards whose demise humanity follows closely. We tend to ignore the state of forest loss in our own countries as this occurred before our time. It is both out of sight and out of mind. Unfortunately, not understanding the historic state of deforestation around the world blinds us to the reality of our current plight. It is too easy to look at the ongoing attack of Brazil's forests by their climate change denying President Jair Bolsonaro and conclude that deforestation is someone else's problem. Though, a quick review of the historical literature provides a more detailed view of the problem of deforestation.

Let's take a look at Europe where the industrial revolution took hold. When the ice from the last ice age in Europe finally retreated around 12,000 years ago, forests reappeared and spread quickly in areas that they had first inhabited some 400 million years previously. The forests were seldom damaged by humans until Europeans switched from hunter gathering to agriculture around 9,000 years ago. At this time, Europeans began to convert forests into crop and grazing land as well as use the wood for fuel. As the population rose, more and more forest land was cut down.

With the advent of the iron age, large quantities of fuel were needed to supply furnaces to produce pig iron from iron ore. The Romans' source of fuel was charcoal, and this came from the lush forests that once lined the Mediterranean. They went about clear cutting an astonishing 250,000 sq. km of forest to fulfil their needs. This is approximately twice the size of present-day Greece (380).

Socially, Europeans viewed an abundance of forests as underdeveloped. Mr. Bolsonaro is still making the same argument some 9,000 years later. Clearly, we are slow to learn. As our ID WPN-114 Prometheus was beginning to flourish in Nevada, 80-90% of Europe was still covered by forests, but in the last 5,000 years, population growth led to widespread deforestation of the continent. By the 18th century, just as the industrial revolution was getting underway, forests covered only around 10% of land. It was at this point that the concept of conservation began to take hold. Unfortunately, new forests that were planted were usually monocultures that did little to address biodiversity. Soon, forest managers recognized that regeneration of forests was not a quick fix but could take 100 years or more. The forests began to reappear until the First World War increased demand for wood. Hot on the heels of the First World War came the second, and again, forests took a direct hit from both military assault, and military needs. After the war, much effort was spent on reforestation and within a couple of decades, forests sheltered around 42% of European land. Essentially, in less than the life of two generations of tree, Europeans had cut down around 38% of "theirs" (381). Today, much of the deforested land is used to graze cows and sheep. 71% of the EU's farmland is used to feed livestock and 63% of all arable land is given over to feeding animals. This amounts to 1,250,000 sq. km of European land and not only land is provided for the industry. The EU also gives animal agriculture between €28-€32 billion a year in subsidies. This amounts to 18-20% of the total EU budget (382).

In the United States, a similar pattern of deforestation was witnessed. In 1630, forested land was around 1,023 million acres (46%). Around a quarter of those forests were converted to agriculture. Nearly two thirds of this conversion happened in the second half of the 19th century, during which around thirty-four square kilometres of forest were felled every single day for fifty years. By 1910, total forested land accounted for 754 million acres, or 34%, and by 2012 this had risen to

766 million acres. Today, around 10% of forests in the United States are preserved (383). In the USA grazing takes up roughly a third of all land (2,646,644 sq. km) and cropland 1,582,320 sq. km. Around 32.5% of cropland is devoted to growing animal feed so animal agriculture in the contiguous USA accounts for 41% of total land use (384).

As Europeans began their outward quest for land, Britain took the proverbial bull by the horns and went land stealing crazy. After arriving in Australia in 1606, the Brits took a bit of time to warm up, but once their axes were sharpened, they didn't disappoint. Although 75% of Australia is inhospitable, the coasts were abundant in almost untouched flora and fauna. After Captain Cook's arrival in 1788, deforestation in South-eastern Australia occurred immediately. Any areas with fertile soil were felled first, mostly for wheat production or grazing sheep. Deforestation in Western Australia rapidly increased between 1920 and the 1980s, and most recently, Queensland has become the number one deforestation hotspot. Between 1991-1995, of the 1.2 million hectares cleared, 80% was in Queensland. By the 1980s, 38% of Australia's forested areas had been lost (385). This was all achieved by a population numbering just twenty-five million. Today, Australia has the unenviable title of being the only developed nation to be considered a deforestation hot spot.

Deforestation has many impacts, some of which are less obvious than others. We are all aware of the impact on wild animals and forest dwelling humans and the role of forests in drawing down carbon, but the geographically distant impacts tend to be less understood. The World Resources Institute published an article in 2018 that argued that, through disruption of the water cycle, any large-scale deforestation of the three major tropical forest zones of Southeast Asia, Africa's Congo basin or the Amazon could pose a risk to important breadbaskets like the U.S., India, or China. It has long been ignored that the forests themselves are making their own rainfall. While much of the rain in coastal regions is evaporated from the oceans, areas inland rely on trees in the forest itself to produce around 1/3 of rainfall. If the forests disappeared, these areas would be like deserts. Air that moves over areas with lots of vegetation produces twice as much rain as air than moves over little vegetation. Forest loss in the Amazon is set to reduce dry-season rainfall by 21% by 2050 (386).

When Jair Bolsonaro won the Brazilian presidential election in October 2018, he did so by promising rural citizens that his government would open up the Amazon to business interests like cattle ranching, mining and logging. He did not disappoint. In July 2019, around 1,345 sq. km of forest were cleared. This is equal to the size of greater London and was a third more than the previous July record. Scientists are concerned that the Amazon is in danger of reaching a tipping point, after which, the forest will start to become a savannah. What is happening in Brazil today is a far cry from just a decade ago. After introducing regulations to protect the forest, deforestation declined by 80% between 2006 and 2012. However, since Bolsonaro encouraged ranchers and miners to commercialize the Amazon, the amount of forest cleared rose by 13%, which was the highest increase in a decade (387). The situation is so severe that one of Brazil's leading climate scientists is forecasting that we have only fifteen years left to save the forest. At present rates of clearing, around half of the Amazon may turn into savannah by 2035. At the moment, around 17% of the forest has been felled, and if this exceeds 25% then the tipping point may have been crossed. This would convert the Amazon from being a carbon sink to becoming a carbon source. We will likely see emissions surge as forests disappear and their carbon is released. This will in turn raise temperatures further, and the cycle continues without human interference. This is one of the tipping points humanity has to do all it can to prevent. The best people to protect the forests are those who have flourished as part of the ecosystem: indigenous tribes. Bolsanaro understands this and is trying to undermine their legitimate land rights. The roughly 300 tribes who inhabit the Amazon occupy around 13% of the land in Brazil. Bolsanaro believes that they are standing in the way of progress, and he is sticking to his election promise of allowing mining on their lands (388). Other countries like Columbia have decided to act now to protect their share of the forest. They are are on track to reduce deforestation by 30% by

August 2022, and for a country that is 35% covered by forests, this is an important target. Without action from Brazil though, the forest's future looks grim (389). And the repercussions will be felt by humans in Brazilian cities, neighbouring countries and as far away as the United States where around 50% of rainfall in the Midwest evaporates from the land, including in the Amazon (386) (390).

Across the Atlantic Ocean, around 15% of the planet's forests are found in Africa. Many people rely on the forests for sustenance and our closest relatives, the bonobo and chimpanzee call these forests home. They are not alone as around 1,000 bird species, and 400 mammals including elephant, gorilla and forest buffalo also reside in the relative safety of the canopies. African forests also store around 171 gigatonnes of carbon and support an estimated 100 million people. Unfortunately, the fast-rising human population in Africa is leading to the forests becoming extremely fragmented. In the past 100 years alone, around 90% of the forest coverage in West Africa has been wiped out. In the Congo Rainforest, the planet's second largest, between 2002-2019, 5.6% was destroyed, according to Global Forest Watch. At current rates, by 2030, around 30% of the Congo Rainforest, will be gone, and by 2100, there may be little more than a few chunks of forest left here and there (391).

Nearly 8% of all life is found on the Australian continent. That's around 130,000 species of flora and fauna. Most of this life is found in the forests of the east coast. Close to 70% of the forests in Queensland and New South Wales are considered cleared or disturbed. The iconic koala, among many others, has become a vulnerable species due to the loss of forest cover. The runoff of soil and rain increases between 40-100% due to deforestation and this sediment is polluting the Great Barrier Reef. The unprecedented wildfires experienced in summer 2019/2020 are also influenced by deforestation. In areas of tree cover, temperatures are generally cooler, but due to tree cutting, temperatures rise, and this helps to dry out the forests and make them more combustible. Trees also help to produce water through transpiring. Just one tree can transpire hundreds of litres of water in a single day. One hundred litres has a cooling effect equal to two air conditioners (386). Multiply that by millions of trees and you understand the importance of protecting trees. Former Australian Prime Minister Scott Morrison liked to claim that Australian carbon emissions (1.3% of global total) can't possibly have exacerbated the fires that have engulfed the nation. They are way too insignificant to have played a part according to him. Mr. Morrison is either not very intelligent or a rather disingenuous human being, or possibly both. Australia just so happens to be the biggest exporter of coal on the planet and the third biggest exporter of fossil fuels, after Russia and Saudi Arabia in that order. Australia does not include its exports of fossil fuel in its own emissions. These are added on to the nation that actually burns the fuel. If these exports are included, then Australia's emissions increase to 5% (392). This figure is forecast to increase to 13% by 2030 as Australia continues to export its excess of black gold. This would see Australia become the largest exporter of both coal and natural gas. As an emitter, Australia is tied in fifth place with Russia when exports are included. Australians are already the biggest emitters on a per capita basis without including exports, but when you include exports, Australians stand out like a sore thumb (393). Regardless of Australia's contribution to emissions, their own tree cutting is leading to increased fire risk and these fires have already doubled Australia's annual emissions (394).

Just to the north of Australia, the great forests of Southeast Asia stretch from Papua New Guinea through to Indonesia, Malaysia, Thailand, Vietnam, Cambodia, Laos, and Myanmar. Just as in Europe, people here rely on the forests for food and wood. Population levels were so low that this use was sustainable. From the eighth century onwards, as the population shifted towards permanent agriculture, it was possible to feed more and more people and the population soared. This continued until the fifteenth century when large areas of Sumatra, Borneo, Vietnam, and Malaysia were deforested for commercial agriculture including pepper, sugar, cloves, and coffee. Trees were also felled to fuel the export logging trade (395). Around 50% of tree cover in Southeast Asia has already been lost. Vietnam and Thailand lost around 43% between the end of the U.S. war in 1973 and 2009. The U.S. military destroyed around 19,942 sq. km, or six percent of Vietnam's total land area through the long-term use of defoliants like Agent Orange. More than seventy-two million litres of herbicides were dropped and around thirteen million tonnes of bombs (396). Between 2005 and 2015, around eighty million hectares (800,000 sq. km) of forest were lost, or around eight million hectares (80,000 sq. km) a year. 62% of this loss was inflicted on Indonesia, Malaysia 16.6%, Myanmar 5.3%, and Cambodia 5%. Deforestation in the region is the world's fastest (1.2% annually), and most biodiversity loss occurs here (397). It is forecast that the area will lose around 42% of its biodiversity and 3/4 of its original forests by 2100 (398). Animals at risk are some of the most beloved on Earth, including Orang-utans, Asian elephants, and the Indochinese tiger.

Deforestation, like the climate crisis, is an existential threat to human survival. We are reliant on the forests for the very air we breathe, water we drink, the medicines we take, many of the clothes we wear and, in many cases, the food we eat. Is it possible to reverse this current situation and protect the world's forests before a point of no return? The first thing we need to do is explore why these forests are being felled, and then look to see how we can first slow down, and then, if possible, reverse the damage our species has inflicted over the past 10,000 years.

There are numerous causes of deforestation. These are known as deforestation 'drivers'. Each forest has a set of specific drivers, and no two forests are the same. The specific areas at risk are the Amazon, the Atlantic Forest and Gran Chaco, Borneo, the Cerrado, Choco-Darien, the Congo Basin, East Africa, Eastern Australia, Greater Mekong, New Guinea, and Sumatra. As much as 420 million acres (1.7 million sq. km) of these forests could disappear by 2030.

Let's take a look at the largest drivers of deforestation around the world. While there are a number of drivers, there are just four commodities that stand out. These are beef, soy, palm oil, and wood products. Of these, grazing cows is by far and away the number one driver across all tropical forests. The production of beef generates double the deforestation of soy, palm oil and wood products combined. These are the overall second, third and fourth largest drivers.

In South America alone, around 2.71 million hectares, equivalent to the state of Massachusetts, is cleared for grazing every single year (399). This is five times more than any other driver in the area. Additionally, as animals have said goodbye to green and sunny fields and instead moved into dark, cramped, and dirty factories, the food they eat has also changed. The old justification for eating animals: whereby they are converting grass, which humans can't digest, into protein that we can, has largely fallen by the wayside as animal feeds have become more and more exotic over the years. Today, grass is not on the menu, soybeans are. Soybean production is the second biggest driver and it has grown fifteen-fold since factory farming began in the 1950s (400).

In the world's largest rainforest, the Amazon, the main driver is also animal agriculture. Here, according to the United Nation's Food & Agriculture 2006 report 'Livestock's Long Shadow', 70% of deforested land is being used to graze cattle. This is an unbelievable statistic, and is truly shocking to many people who have been, for years, misinformed of the true cause of Amazon destruction. To make things even worse, the same report found that much of the remaining deforested land is being used to grow soybeans for worldwide use in factory farms (401). Other drivers such as mining, logging, and mega infrastructure projects pale in comparison when placed alongside the true cause of rainforest destruction in South America: animal agriculture. It is clear that if these majestic forests are to be saved, and with them, ourselves, then we need to change our diets.

To the east of the Amazon, along the coast lies the aptly named Atlantic Forest. This is considered the most at threat of all Brazil's eco-systems. It once stretched along the coast of Brazil, Paraguay, Uruguay, and Argentina. Today, it has been broken down into patches and pro-tected areas (402). Only 15% of the original 1.1 million square kilometre forest remains after centuries of colonialization and development. This is not surprising when you consider 75% of Brazil's human population call the region home (403). Whilst the usual culprits like animal agriculture, mining and logging are pervasive in the Atlantic Forest, the main driver here is eucalyptus plantations. Once native to Australia, the eucalyptus tree was seen as a fast growing and low-cost wood that could be made into a variety of paper products. Today, eucalyptus plantations surround vulnerable patches of Atlantic Forest. This, together with hunting, has led to the demise of 70% of the forest's biodiversity. The trade-off has been financial with Brazil's Suzano, the world's largest eucalyptus-derived pulp producer, making record profits of $2.1 billion in the first quarter of 2022 (403). Whilst in the Amazon, the pressure is on to slow down and stop deforestation, the situation in the Atlantic Forest is very different as there isn't much forest left to protect. Here, massive restoration efforts are needed if the forests are to make a comeback, and along with them, all the species that call the Atlantic Forest home.

Another of South America's tropical forests is the Cerrado. Today, it is considered more of a savannah than a forest. The area was first settled in the early 20th century by ranchers who grazed cattle at very low densities in order to copy the grazing habits of native animals that had been plundered by hunters. Brasilia, the capital of Brazil, was built in Cerrado in the 1950s. The forests were on call to provide timber and fuel for this development, but within a decade soy and cotton were being grown in the acidic soils. By the 2000s, forested land had been converted into monocrops of soy with between 50 and 70% of the Cerrado disappearing (404). The Cerrado was first in line for the chop, as the Amazon destruction didn't get underway until the 1980s. In an attempt to protect the Cerrado, the World Wildlife Fund (WWF) created the Cerrado Manifesto in 2017. This attempted to get multinational corporations to voluntarily agree to eliminate the conversion of native vegetation for agriculture and livestock. The project was successful in getting McDonalds, Tesco, and Walmart to sign up, but had less success with the world's biggest commodity companies like Cargill, Bunge and Amaggi. The reason could be that these three companies are already signed up to the 2006 Amazon Soy Moratorium. It seems they were fine to leave the Amazon alone, as they knew they could exploit its lesser known relative, the Cerrado. And they aren't quite finished with her yet (405).

In the northwest of the continent lies the Chocó-Darien which runs along the Pacific coast from Panama in the north through Columbia and into Northern Ecuador. The forest houses some of the highest number of native plants in all South America. Jaguars, ocelots, giant anteaters, tapirs, and tamarins call the jungle home. The region covers around 187,500 sq. km and it lays claim to be the wettest place on Earth with some areas getting up to sixteen meters of rain a year. The region has lost up to 1/3 of its forest cover with a variety of drivers apparent in the region. Infrastructure programs such as the Inter-American High-way are impacting the forest which still has around 66% preserved in a natural state. Another major driver is firewood with more than 50% of wood for fuel coming from the forest. Although the Chocó-Darien is not in as perilous a situation as the other South American forests, as the population continues to increase, the forest will come under more and more pressure as the century unfolds (406) (407).

The last forest in South America that we will focus on is the Gran Chaco. Straddling Bolivia, Paraguay, Argentina and Brazil, Gran Chaco is the second largest forest in South America. The area is inhabited by 3,400 species of plants, 500 species of bird, 150 mammals and 220 species of reptiles and amphibians. Like elsewhere, deforestation has been playing out in Gran Chaco over decades, but it has speeded up recently. Between the years 2000 - 2012, more than an acre of forest was cut every single minute (408). Around 85% of the forest has been lost and it is forecast that by 2030, millions of additional acres will disappear.

In Argentina, the Gran Chaco has been decimated by demand for cheap animal products around the world. It is considered one of the most deforested areas on the planet. Since genetically modified soya beans were authorised in 1996, close to a quarter of native forests have been felled. 80% of Chaco deforestation occurred in Argentina. About 7.2 million acres of forest were cut between 2010 and 2018 with an area the size of the capital, Buenos Aires, wiped out in June 2018 (409).

Most of the soya grown on deforested lands is exported to key markets including Russia, the Middle East, Australia, Asia, and Europe. With Soya products making up 31% of all exports, it is difficult to see the government changing direction. Especially, with the peso dropping by 50% as they borrowed $57 billion from the International Monetary Fund in 2018 (410).

From looking at the forest losses in South America, it appears the number one driver in the region is animal agriculture, with grazing and animal feed being the main causes. Paper production in the Atlantic Forest is a major concern, and the Chocó-Darien is afflicted by human encroachment and fuel needs. In the following pages we will look at the situation on other continents.

Let's take a look at Africa, our ancestral home. A quarter of the continent is covered in forests. Unfortunately, around the same figure (27%) of the total energy needs in Africa are supplied by these forests. Between 70-90% of Africa's population relies on fuelwood for cooking and surprisingly, around 45% of the world's renewable energy supply is sourced from wood. In sub-Saharan countries, wood is being consumed 200% more than the annual growth in trees. Deforestation in Africa is happening at twice the global rate. More than 40,000 sq. km are felled every year (411).

In West Africa, the second largest forest on Earth stretches majestically up from the Congo River which also happens to be the world's second largest basin. The Congo Basin, as the forest is known, covers two million sq. km of Cameroon, the Central African Republic, the Democratic Republic of Congo, Gabon, Equatorial Guinea, and the Republic of Congo. The Republic of Congo is one of six countries that is home to 60% of the planet's remaining forests. Twenty percent of the world's tropical forests are found here. As a biodiversity hotspot, it claims the continent's prize with 11,000 plants (1,400 of which are potential fights against cancer), 600 tree species and 10,000 animals including 400 species of mammals: such as leopards, hippos and lions, primates comprising our closest relatives, the chimpanzees and bonobos, as well as gorillas and forest elephants, amphibians, fish and more than 1,000 species of bird (412). This amazing array of life has attracted humans to live in the region for over 50,000 years and the forest supplies food, fresh water, and shelter to around seventy-five million people. More than 150 different ethnic groups live here, and the population is increasing by 1.7 million annually. This population increase is putting strain on the forest's ability to thrive, but demand from overseas is also threatening the future of the forest. More than 15,000 sq. km of forest are being lost each year. That is forest roughly half the size of Belgium lost each and every year (413). Much of this destruction is to extract resources such as timber, rubber, palm oil, diamonds, and oil. These extraction projects require roads and other infrastructure which in turn attracts more people who need more land for food.

The incredible species that call the region home have also helped to adapt the forest to be a climate champion. Whereas the Amazon is home to few large mammals, the Congo Basin is home to some of the biggest animals on our planet. The presence of these magnificent species means the density of small trees has remained low compared to a high density of taller trees that evade the reach of hungry animals. This has led to old growth trees in the Congo Basin storing around thirty-nine billion tonnes of carbon in their thick trunks and vegetation (414). Unfortunately, these very trees, like teak, are highly prized for their timber and at the current pace of deforestation, this wonderful forest and all its inhabitants, immortalized by Livingstone and Joseph Conrad's Heart of Darkness, face an unsure future.

On the east coast of Africa, stretching all the way from Mozambique in the south to Somalia in the north lies the East African forest. The forest totals around 1,340,000 sq. km and is home to an amazing cornucopia of wildlife including African elephant, black rhino, white rhino, lions, and wild dogs. Around 21% of Africa's and 4% of the global forest cover is found here. Sudan has 46% of the forest's tree cover, followed by Tanzania (29%) and Kenya (13%) (415). One of the major problems in the area is grinding poverty. People here are among the poorest on the planet and rely on the forests for fuel and food. As the population increases, demand on the forest also goes up. In addition to the local needs, international demand for cheap wood from Europe and Asia, especially China, is wholly unsustainable and often illegal. More than 90% of Tanzania and Kenya's forests have been cut down, but the worst deforestation has been seen in Mozambique, Tanzania, and Zambia. Information on deforestation here is hard to calculate as governments do not collect accurate data. From the data that is available, East Africa lost around 55,000 sq. km of forest between 2000 and 2012 and is expected to lose a further 120,000 sq. km by 2030 unless action to protect the forests is taken as a matter of urgency (416). If we are to ensure the survival of Africa's most enigmatic flora and fauna, and in turn provide sustenance for the fastest growing human population on Earth, then rapid action is needed to reverse deforestation.

Asia, home to 60% of the human population, has been developing at breakneck speed for the past several decades. Fuelled initially by Japan's post war reconstruction, the mantle has since passed to South Korea and Taiwan, and more recently to the largest country in the region: China. The financial hubs of Hong Kong and Singapore are increasingly investing in Southeast Asian economies such as Vietnam, Thailand, and Indonesia. This growth has been a blessing for many Asians who have known nothing but abject poverty. The growth, unfortunately, has not been as positive for the area's rich biodiversity.

On Borneo, the planet's third largest island, human population growth, rising GDP and pressure from multinational corporations has led to around one third of the rainforest that straddles the Indonesian/ Malaysian border being burned to the ground between 1973 and 2012 (417). During the 1980s and 1990s, much of the rainforests were shipped to Japan, the United States and Europe in the form of garden furniture, paper pulp and chopsticks. Today, the forests are unrecognisable from the impassable and virtually unexplored jungle that was ruled by human head-hunters until the early 20ᵗʰ century. The "protected" lowland forests have experienced a staggering 56% decline since the mid-1980s. Today, the island's biggest threat is the ubiquitous Elaeis guineensis, better known as palm oil trees. More than six million hectares of forest, or 14%, were felled between 2000 and 2017. About half of this land was converted to palm oil plantations, with 92% converted within a year of the trees hitting the ground. Plantations increased in number by a staggering 170% or 62,000 sq. km in the same period and 88% were harvesting palm oil while 12% was for pulpwood. More than 2/3 of these plantations were on the Indonesian side of the border (418). The good news for these forests is that the expansion of plantations into old-growth forest is slowing down due to moratoriums on new plantations in primary forest. As the international community scrutinizes the deforestation on Borneo, people are beginning to avoid the use of palm oil products. This is easier said than done as palm oil is in everything: from lipstick to pizza dough and shampoo to cookies. It is even finding its way into factory farmed animal diets as well as cat, dog and even goldfish food. It is the most common vegetable oil on the planet and for good reason. Oil palms produce around 6-10 times more oil than temperate oilseed crops like rapeseed, soybean, olive, or sunflower. This efficiency is what has been driving the destruction of Borneo's forests, but what is likely is that if we were to rely on less efficient oils then deforestation elsewhere would probably be far higher. In fact, if oilseed crops replaced palm oil, then an extra 500,000 sq. km of farmland would be needed to produce the same amount of oil (419). The choice for humanity is stark. Do we continue to increase palm oil plantations at the expense of the ten primate species, 222 species of mammal, 350 species of bird, 150 reptile and amphibian species and 15,000 species of plants (420), or do we either source our oil from less efficient sources that result in more need for farmland, or perhaps do the unthinkable, and change the way we consume? Recent statistics show the world is taking one of the latter two paths as palm oil expansion has slowed since 2012 with much of the new plantations being planted on land formerly cut down for an alternative crop (421).

Unfortunately, it isn't only palm oil that is leading to the grand old forest's destruction. The old foe of logging still persists. Demand from South Korea, Taiwan, Japan, India, and the Middle East ensures that forests are felled to make plywood and home interiors. Japanese logging companies, which were discussed in an earlier section, have been a long-time customer of the forests, but fortunately its custom is dwindling by the year. The Japanese market for logs has dropped by 90% since 2007 when Japan imported 576,576 m3. In 2017, Japanese imports had fallen to 40,689 m3 (422). There was still time for the Olympics to get in on the destruction of Borneo's forests though. More than 134,000 large sheets of tropical plywood from the island were imported to Japan to help build concrete moulds for stadium construction for the 2020 Tokyo showpiece. In many cases this plywood was sourced from orangutan habitat, and 80% of the giant apes are now living precariously in unprotected areas (423). The future of this wonderfully diverse hotspot is at stake, and many challenges lie ahead, but the fact primary forest clearance is slowing down is positive. Unfortunately, as is usually the case, deforestation is being exported elsewhere. In this case, to neighbouring Papua New Guinea.

Like elsewhere, the island of Papua, east of Indonesia, had been covered by forest since the end of the last ice age. Humans arrived as far back as 60,000 years ago from Southeast Asia. They shared the forests with 744 types of bird, 395 amphibious species, 289 species of reptile, 2858 varieties of fish, 269 types of mammal and 11,544 varieties of vascular plant (424). This is an amazingly diverse biosphere that provides its inhabitants with food, water, protein, building materials, medicines, and warmth. Around half of the forest's species are yet to be classified and a third of all species found here are unique to Papua. The island has a long history of annexation, colonialization, and independence. In 1975, Papua New Guinea (PNG) declared independence and joined the United Nations later the same year. From its independence until 2002, the forest lost around 15% of its tree cover (425). Since 2002, deforestation of the island's canopy has exacerbated, with an average of 81,265 hectares of forest being cut down every year until 2018. Of this, an average of 39,665 hectares were primary forest (424). By 2014 1/3 of the commercially viable forests had been logged (426). Unfortunately, as regulations reduced supply in Borneo, PNG picked up the slack. From 2014-2018, a total area of 700,311 hectares was destroyed, 357,311 hectares being prime forest. The average jumped from 81,265 hectares a year to 140,062 hectares. In total, a staggering 48.8% of the forest has disappeared this century. The annual loss of forest cover amounts to 2.1% (424). The largest driver in these forests is logging. Trees are sent to China where they are turned into furniture and flooring before being sold around the world. The journey from soil to shopping cart is more than 14,000 km. According to the NGO Global Witness, the PNG government is currently abusing a land-leasing scheme to remove people from their land. In all, they state that 12% of the country has been handed over to foreign-backed companies for logging or cropland (426). If deforestation continues at current pace, then by 2030 there may not be any forest left in PNG.

Further east in the Indonesian archipelago, stands the only place on our wonderful planet that you can find elephants, rhinos, tigers, and orangutans living together. It is also home to 15,000 known plants, 201 mammal species and 580 types of bird (427). It is the sixth largest island and because of it, Indonesia earned the nickname "Emerald of the Equator." Welcome to Sumatra. More than half of the sixth largest island in the world was covered by forty-four million hectares as recently as 1985. Through a lack of governance and land grabbing, deforestation has gone unchecked at a rate of 1.8% a year. By 2016, only around 25% of this forest survived and only 8% of Sumatra is virgin forest. The key drivers have changed as the decades have progressed. The 50s and 60s saw rice and smallholder rubber and coffee plantations emerge on deforested land, and this gave way to logging in the 70s and 80s before palm oil and pulp production took over a decade later. In recent years, the level of forest loss has declined, partly due to police enforcement of forest laws and partly because there are few primary forests left to be logged. Sadly, huge pressure is on the island to allow the removal of more and more trees to make way for palm oil, acacia, and mining operations. With demand for vegetable oils set to increase by 46% by 2050, it is hard to see much future for these majestic forests, and their inhabitants (428). Yuyun Indradi, political forest campaigner with Greenpeace Southeast Asia in Jakarta surmises:

"This is the fastest, most comprehensive transformation of an entire land-scape that has ever taken place anywhere in the world including the Amazon. If it continues at this rate all that will be left in 20 years is a few fragmented areas of natural forest surrounded by huge manmade plantations. There will be increased floods, fires and droughts but no animals (429)."

The emergence of Asia as an economic powerhouse has had no bigger impact than on the Greater Mekong, home to seventy million. Stretching all the way from the South China Sea in Vietnam, the forest incorporates most of Vietnam, Cambodia, and Laos as well as the east of Thailand, part of Myanmar and also Yunnan and Guangxi in China. The Greater Mekong is around 200 million acres which is similar in size to Texas and Arkansas combined. Unlike these two states though, the Greater Mekong is one of the most biologically diverse areas on Earth. There are around 20,000 species of plants, 1,200 species of bird, 800 reptiles and amphibians and 430 mammals, including the Asian tiger and Asian elephant (430). Since 1997, over 1,300 new species have been discovered. Unfortunately, the forests are being lost faster than species discovery. Between 1973 and 2009, almost a third of the forest was lost. Cambodia lost 22%, Laos and Burma 24% and Thailand and Vietnam 43% (431). The Mekong is known as the 'rice bowl' of Asia and as pop-ulation increases, more and more land is required to grow the region's staple crop. This is the largest cause of Greater Mekong deforestation. The Mekong River, which at 4,828 meters, is one of the world's longest, also provides around 25% of the world's freshwater fish catch. In total, more than 1,100 species of fish live in the Mekong, and this makes it the second most biodiverse river on Earth. With rising sea levels, the Mekong Delta will be inundated by saltwater and with a rising popu-lation, the future of this important area is in jeopardy. Within the next couple of decades another third of its forest cover will be lost if trends continue (430). The Greater Mekong Forest may be as little as 10% of its original size by 2030.

On the world's largest island, as was mentioned, deforestation has blighted human history. The first human settlers here slashed and burned forests and caused the extinction of many of the giant fauna that had evolved only here. Tree cutting has continued since Europeans settled the continent and around 5,000 sq. km of virgin bush are destroyed each year (432). Australia is home to an astonishing 1,250 plant species and 390 animal species that are threatened. 77% of all these plants (964) and 73% (286) of animal species are endangered as a result of deforestation and forest fragmentation. The WWF predicts that by 2030 around 60,000 sq. km of land will be cleared unless land clearing laws are changed (433).

In the state of Queensland where deforestation is most rampant, the number one driver is livestock grazing which accounted for a whopping 91% of all clearing in 2014-2015. Smaller drivers include forestry (5.4%), cropping (1.7%), mining (1.1%), infrastructure (0.3%) and settlement (0.7%). Together, these drivers were responsible for the loss of 2,960 sq. km of forest in that single year (434). A year later, 3,950 sq. km of forest disappeared. To put the scale of the damage into context, this is roughly the size of Sydney or 1,500 football fields per day. For those who visualise on a per tree basis, around 400 million trees were cut down in a single year. 91% of this land clearance was for beef burgers, steaks, and milk. The clearing, as mentioned earlier, is having considerable consequences. For example, in 1990, the year after the climate crisis became public, emissions from deforestation accounted for a quarter of all Australia's emissions. By 2012, this had dropped to 6% of the total although the emissions hadn't dropped (433). Another of the consequences that is coming home to roost now is an increase in temperature and a decline in precipitation. The clearing of deep-rooted native vegetation leads to a drop in cloud cover and with it, rainfall. This has led to the warmer and drier climate Australians are having to face up to today as record temperatures have led to record droughts which have led to record wildfires (434).

As we can see from the various forests discussed, the drivers vary from forest to forest. In Africa, population increase coupled with demand from international commodity markets and logging interests is attracting more and more people into the forest and as a result more and more land is cleared. The situation in South America differs by forest, but the main drivers here are animal agriculture with cattle grazing and animal feed production the largest cause. The great forests of Southeast Asia also have a variety of drivers apparent which include large scale palm oil plantations, subsistence farming for the growing population and international demand for timber. And finally, the east coast of Australia can also blame the burgeoning demand for burgers, beef steak and dairy products for the loss of their incredible biodiversity, and newfound droughts and wildfires.

As the great forests of Southeast Asia, Australia, Africa, and South America are felled at record speeds, it is difficult to watch. Knowing, not just what these forests mean for everyone who lives in them, but also for those of us who do not, is like watching our life support system being smashed to bits with baseball bats, and there being nothing we can do to stop it. When President Macron of France criticized President Bolsonaro of Brazil for overseeing a huge uptick in tree cutting, Bolsonaro was quick to taunt back that no one complained to Europeans when they were felling their great forests in order to increase their GDP. Just under 60% of Brazil's land area is covered in forests. Can Brazilians (60%) take British (13%), Americans (34%), Australians (17%), Chinese (23%) and French (31%) seriously when they criticize them for deforestation? (435) It's rather a case of the pot calling the kettle black.

It is clear from looking at global land use that there is one industry that stands out from the crowd. 10% of the planet is covered by glaciers, 19% is barren land such as deserts, beaches, sand dunes and rocks, 37% is forested, 11% shrubs and grasslands, urbanisation equals just 1% and freshwater is 1%. Then we look at agriculture which uses half of all Earth's land. With a population of 7.96 billion and growing, it is difficult for land not to be used to grow crops for humans, but the amount of land actually used to grow food for humans is much smaller. Seventy-seven percent of this cropland is used to grow food, not for humans, but for animals that humans then eat. While more than ¾ of all our farmland (27% of total land available) is used by animal agriculture, only 18% of our calories come from eating animal products and only 37% of our total protein (318).

The corona virus that turned our world on its head in 2020-2021 is also the result of human activities. As we clear cut forests to satiate our growing demand for meat, wild animals come into closer and closer contact with humans. It is from these animals that three quarters of all newly emerging diseases originate. As deforestation increases in the following years, we will likely see more and more diseases like COVID-19 make the jump between species. Factory farms were introduced to reduce land use, but the feed still requires forests to make way for the mono crops we feed them. These filthy enclosures are also petri dishes for diseases which often spread quickly between animals who are stressed and sickly. Around a quarter of infectious diseases in humans arise from the way we treat animals in factory farms and the trend is towards larger factories that house higher populations of animals (436). China is now housing pigs in thirteen story buildings with one thousand pigs on each floor (437). As the world gets back to "normal", most of our focus is on preparing for the next pandemic, but very little attention is paid to actually preventing the diseases from occurring.

This all seems like a hugely detrimental trade off and if we are to avert the sixth extinction it is clear we will need to alter our diets if we are to avoid losing the world's last biodiversity hotspots.

Amazon burns vs cows grazing
Animalrebellion

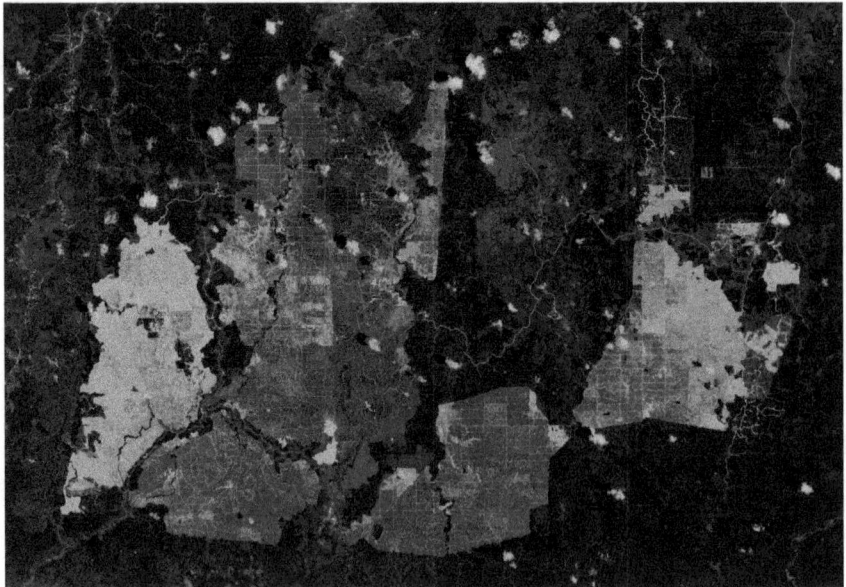

Palm oil plantations
europeanspaceagency is licensed under CC BY-SA 2.0.

Dear Indy,

When you read this, I wonder what will be left of the world's forests. Did we finally get our act together and protect the forests as if our lives depended on it, or did we carry on as usual?

If we carried on as usual then there will be little forest cover left by now. If we got our act together then maybe our planet is on a new path.

I remember first becoming aware of deforestation and its implications when you were a little boy. I guess I'd always known about the destruction of the tropical forests, as growing up in the 90s, it was discussed in school and on kid's TV shows. We were all led to believe that the major culprit was the logging industry and urban sprawl. It wasn't until I watched the documentary 'Cowspiracy' in 2015 that I learned the truth. I had been a follower of Greenpeace and World Wildlife Fund for years and I trusted them to tell us the truth. Unfortunately, commercial interests were evidently more important to these organisations than helping to preserve the forests they were tasked with protecting.

What I learned from watching this documentary and following the links to the facts on their webpage was that the main cause of Amazon destruction was not logging or mining, but cows. An astonishing 70% of Amazon destruction was caused not by cutting down timber for furniture or houses or digging out rare minerals for use in technology. It wasn't even for massive dams. It was for burgers and steaks. 70% of Amazon deforestation to date was to graze cows who would then be exported around the world as an answer to the world's desire for cheap beef. Most of the rest of the forest loss was to grow huge tracts of soy. This soy wasn't even grown for human consumption. Guess what it was used for? Yep, you got it! Instead of the rich biodiversity that was the Amazon rain forest, it is being used to grow feed for factory farmed animals. Since the 1950s, the animal agriculture industry has been moving animals inside in order to cut costs and increase profits for investors. These animals no longer touch the grass outside. They no longer feel the sun on their backs, and they no longer smell the Earth. Today, they just 'are' in a dark, damp, cramped existence. And what do they eat? They eat soy, grain, and increasingly palm-oil kernels. All their food is grown elsewhere and imported. The food replaces some of the most biodiverse land on our planet. The system is hugely unsustainable. The next day we became vegan, and a month later we visited your uncle Chops in Australia. What we saw there ensured that we would never return to eating animals. On a ten-hour drive north from Sydney to Brisbane, we saw nothing but fields of cows over and over and over. Four hours inland, and back down to Sydney. We expected koalas and kangaroos everywhere. All we got were cows.

Wild Animals

"Love the animals. God has given them the rudiments of thought and joy untroubled".

This simple message from Fyoder Doestoyevsky at first glance seems like common sense, after all, we named ourselves Homo sapiens. The sapiens part means 'wise' and 'intelligent'. Although, anyone using such lofty adjectives to describe themselves might certainly be lacking in hubris. Just ask my compatriots from 'Great' Britain.

If we look at the way in which humans treat non-human animals today, and in fact, the way we have treated non-human animals since we first stood on two legs, it is easy to see why Doestoyevsky felt compelled to speak out.

In this chapter, when the term animal is used, it is used to denote non-human animals who human animals, for millennia, have tried at all costs to distance themselves from. In previous chapters we have looked at the impact our species is having on our climate, and also on the natural world. The way humans are being impacted by the climate crisis and ecological breakdown is dwarfed by the impact our actions are having on our fellow Earthlings. Not only are the climate crisis and ecological emergency severely affecting many species ability to survive, human indifference to animal suffering further compounds the awful reality of life on Earth for our evolutionary brothers and sisters.

Planet Earth in 2022 is a grim place for animals to live. We have already filled the seas with plastic, hunted animals to extinction, destroyed their homes for our own pleasure, and driven them to the very brink. In simple terms, since the 1970s, humans have wiped out 50% of all animal populations (438). This is just the tip of an extremely large iceberg as a further 27% are at risk of extinction (439). The background extinction rate is between 1,000-10,000 times the norm and only five times in history have so many species been lost so quickly. At the high end, as many as 274 species become extinct every day (440). Humans can hold their heads up high as they have become the first animal to cause a mass extinction on Earth. Quite the achievement for a hardy band of warriors who couldn't even use stone tools until 2.5 million years ago. Unfortunately for humans, these extinctions are part of a wider problem that if left unaddressed will likely lead to the eventual demise of the wise and intelligent one itself.

In the following pages, we will look at the myriad ways in which humans have driven, and are driving animals to the brink of extinction. Rather than looking at each of the causes in order of their magnitude, let's take a journey through human history to see how we got into this mess.

Our species' journey began around six million years ago when a female ape had two daughters. One of these daughters became the grandmother of chimpanzees, and the other is our great great granny. Millions of years later, in East Africa, a number of species related to us roamed the forests. These groups all belonged to the group *Homo* which is the term used for humans. About two million years ago, some of these bands of early humans walked out of East Africa and headed off in search of excitement. They headed in different directions, some headed to North Africa, some went to Europe and some even walked all the way to Asia. All these early humans had to learn to survive in very different climates from jungles and deserts to snowy peaks. They all evolved different characteristics to help them. The humans that hung around in East Africa were known as Homo rudolfensis while the hardened trekkers who made it out to East Asia were known as Homo erectus. Those that preferred the snowy winters of Europe and Western Asia were called Homo neanderthalensis. These early human groups settled and began populating their territory. They adapted to their terrain and evolved to be part of nature, much as animals today have. On the island of Flores in Indonesia, a group of humans got cut off by rising waters and found themselves stuck on an island with very little food. The humans with the bigger frames found it difficult to find enough food to survive, whereas the smaller ones coped better. Over time, this resulted in the emergence of a dwarf human called Homo floresiensis. Wherever humans established themselves, they adapted brilliantly to their surroundings. Some hunted animals, while others gathered plants (245).

Our early ancestors survived mainly on a diet of nuts, seeds, and fruits. They supplemented this, when possible, by chasing after small animals and devouring insects which were high in protein and low in danger factor. Our omnivorous behaviour came around somewhat opportunistically. As is clearly evident, we don't have the teeth necessary (carnassials) to rip flesh from bone or the claws needed to take down an animal in flight. So, we had to rely on larger carnivores, like lions, to do the hard work. Then, we had to wait until the scavengers showed up to fill their boots before we could emerge from the shadows cautiously to crack open the bones with stone tools to extract the nutritious marrow and finish off whatever flesh had been left behind by the wasteful hyenas and jackals (441). This may sound surprising to many who have never needed to skulk wearily through bushes in order to avoid being eaten by the planet's dedicated carnivores. However, this was reality for our early descendants. They were not the top of the tree species we are today.

In fact, they were just a bunch of middle-of-the-road survivalists who until this point were fulfilling their own niche much like all the other species that surrounded us. Things for the animals took a turn for the worse when some intellectual Einstein of the day chanced upon a magic trick of huge importance to the future of life on Earth: fire. Although this enabled us to keep warm, cook food and frighten away predators at night, it also offered us an offensive weapon. We didn't waste time in putting this new tool to devastating effect in burning down thick vegetation to create grasslands where hunting could be achieved more easily. This was in effect the start of our attack on planet Earth. The war was underway.

Fast forward to 50,000 years ago and you will find our fore-fathers, Homo sapiens, who had recently left Africa themselves for the adventure of a lifetime, mixing it up with other species of Homo, in some cases it seems, mixing it up real good. Research into the human genome in 2010 had some far reaching and beautiful results. Modern day Europeans and Middle Easterners have between 1-4% Neanderthal DNA, while Australian aborigines have up to 6% Denisovan DNA. It seems from our DNA that sapiens occasionally bred with other forms of Homo, but not very frequently. More commonly, early sapiens fought with and defeated our early cousins. Today, as we know, there is only one type of human that still walks the globe. And we don't walk any-more, we stride across the globe without a care in the world (245).

After becoming the champions of the Homo wars, we spread quickly around the planet, bow and arrow in hand. We crossed the straight between Indonesia and Australia about 45,000 years ago with seafaring technology in the bag along with fire on demand and our bows at the ready. This was the first time that any mammal had ever crossed from the Afro-Asian landmass to the long-isolated continent of Australia. What we found there were giant koalas, two-meter-tall kangaroos, marsupial lions, five-meter-long snakes, and huge flightless birds. These animals had evolved without humans for millions of years and were unprepared for this new hostile tool laden visitor. This allowed humans to instantly jump from mid-table to top-of-the-league protagonists, a jump never before or since witnessed. What ensued was the absolute slaughter of all the giant fauna and in many cases the burning of di-verse flora. Of the twenty-four fifty kilogram plus animal species living on the Australian continent, twenty-three were driven to extinction. In total 90% of Australia's megafauna disappeared. One fact of relevance to Australians today is that until humans arrived in Australia, eucalyptus trees were scarce, but due to their resistance to fire, they emerged as the most common plant on the continent. This enabled koalas to flourish for another 44,000 years, until modern day Europeans arrived to begin the slaughter once again. This was the beginning, if you will, of our ruthless attack on all things non-human (245).

Around 16,000 years ago, when sea levels were very low, humans walked over the land bridge from Siberia to modern day Alaska. The journey must have been incredibly challenging with temperatures of minus 50°C and no sunlight at all in winter months. After arriving in Alaska these early humans had to wait for 2,000 years in the frozen north until the end of the last ice age when the glaciers blocking their way finally melted. Being cooped up in the ice-covered lands of Alaska must have been so challenging that to celebrate, they went on an ecologically disastrous rampage across the North American continent. The wildlife they encountered here was similar to the giants secluded on Australia. There were mammoths, mastodons, bear sized rodents, massive lions, horses and camels, sabre-tooth cats and eight-ton sloths. Suddenly, these humans had gone in every which way they could, decimating animal populations along the way. Again, they had lucked out because these animals had not evolved with humans and had no idea these seemingly innocuous little fellows were lethal. Within 2,000 years, humans had reached the southern tip of South America, and on their passage, they had wiped out thirty-four of forty-seven large North American mammals and fifty out of sixty in South America. Just in case some of you are thinking the arrival of humans and the disappearance of these giant fauna is a coincidence, the giant sloths, whose final remains only date to 12,000 years on the mainland continued for a further 5,000 years on the islands of Cuba and modern-day Dominican Republic. They only succumbed 7,000 ago, 'coincidently' when humans first made their way across the Caribbean. No, this was no coincidence. This was the continuation of what we had started in Australia. This was the march of man for complete dominion on the only known habitable planet, the third rock from the sun (245).

As the agricultural revolution got underway 10,000 years ago, there was an estimated population of between one and ten million (442). As the advent of agriculture freed up humans to create pottery and other artefacts, it is possible to say that our love of all things shiny began here. This also freed up people to have larger families as they didn't have to continuously move in search of food. For millennia our descendants continued to evolve in their respective corners of the planet. Each area afforded its human population varied pros and cons from availability of food to regional temperatures and the fluctuating ability to trade over distances. From the 1400s, as an arms race gripped European elites, our species continued its outward expansion across the seas. The British led the way, but were joined by the French, Dutch, Belgians, Germans, Italians, Portuguese, and Spanish. Europeans were merciless in their drive for economic and military power.

The Spanish and British were locked in a battle for maritime supremacy and in 1492, as we will see in part three, the Spanish crown hired the Italian explorer Christopher Columbus to sail off across the Atlantic in search of other people's wealth. After arriving in the Caribbean, Columbus and his men went on a rampage, imprisoning many, and slaughtering many more. From a population of around eight million in 1493, the population had fallen to around three million in 1496. These humans were likely related to the seafarers responsible for the extinction of the giant sloth thousands of years earlier. By the time Columbus returned home, there were only 100,000 native people left (443). It wasn't just the gilded armour and human indifference that enabled Columbus to wipe out the Indians so effectively. They also had the assistance of bacteria. For the proceeding millennia, Europeans had been living with domestic animals such as pigs, chickens, goats, cows, and sheep. In this time, humans had gotten used to the diseases spread by animals. These diseases had spread across the entire Eurasian landmass. Humans on the American landmass, however, had had no such contact with these animals and with them the diseases they carried. When Columbus arrived in Hispaniola with seventeen ships, 1,500 soldiers and hundreds of pigs and other domesticated animals, the animals were taken ashore. Within a day, the Europeans had been taken ill, Columbus included. It appears the pigs, who had been kept isolated at the bottom of the boat, were carrying swine flu. For the Europeans, this was an inconvenience but for the local population it was a disaster. Within four months, a third of the Indians were dead. As more and more boats, humans and animals arrived on the formally isolated islands, more and more Indians were decimated by European diseases like smallpox and flu. By 1516, of the original one million inhabitants of the now Dominican Republic and Haiti, only around 8,000 or 9,000 were left alive (442).

The Spanish spread throughout Central and South America and everywhere they went, Native Americans disappeared. Out of an estimated population of between eleven and twenty-five million, only two million Aztecs survived. The diseases accompanied the Spanish at every turn and between 70% and 90% of people died after sustained European contact (442).

At this point, you may be wondering what this has to do with wildlife, and that is a valid point, so let's refocus. This part of human history is euphemistically referred to as the innocuous 'Columbian Exchange'. It led to massive changes for humanity, not least dietary. Until this point in history, no Europeans had ever laid eyes on a potato which has since become the staple crop of many European countries. Likewise, banana trees had never been grown in soil in the Americas and India had no chili peppers. Imagine Indian food with no chilies. Africa had no peanuts. If you visit the Congo rainforest today, you will probably taste the staple crop, Cassava. This plant is originally from South America. If you were then to visit the Amazon rainforest to visit the Yanomami tribe, you might enjoy a meal of plantains which were domesticated in Africa. In short, the Columbian Exchange allowed humans around the world to import a plethora of crops best suited to their local environment. Great news you might think, and for crop yields, it certainly was. Animals didn't get such a good deal, however. This increase in crop yields encouraged people to have more and more children and soon deforestation expanded. The incredible array of biodiversity on Earth was now in jeopardy as humans undid the work that hundreds of millions of years of isolation had created. Our network of ships began to reconnect the seven continents that had once been the supercontinent of Pangaea. As the biological diversity of Earth was standardised, the Anthropocene was beginning to emerge.

While native people often saw themselves as custodians of the land as opposed to owners and made attempts to conserve rare animals and use the land sustainably, the Europeans saw everything as a renewable resource. When they first explored the Americas, Africa, and Asia, they didn't see animals, they saw fur, feathers, and skin. When they looked at forests, they saw furniture. When they saw elephants, who were protected by the Ikoma tribe in Tanzania, they saw ivory. It is unsurprising then, that the powers of Europe wasted little time in carving up these lands for their own enrichment. By 1800, Europe and her colonies covered just over half the globe (444). Britain, at one point had appropriated 25% of the globe's land mass for a nation of just 242,495 sq. km.

The nations of Europe begged, stole, and borrowed, but mostly they stole. To develop their own economies still further, they enslaved native people, stole their resources, and killed wildlife with impunity as if there was an endless supply. While it is possible to argue that humans had been hunting animals for many millennia, this was not hunting for necessity, but hunting for sport. These animals were not eaten by the shooters themselves, but instead their skin was removed and stuffed by taxidermists and sent home where it was mounted on the wall. This crazed behaviour was in large part down to Christianity with the long-held belief that God had made man in his own creation and given us 'Dominion' over all the other animals. It led unsurprisingly to the killing of animals for sport in the 1880s. The whole sorry state of affairs can be traced back to a single man named Roualeyn Gordon-Cumming in 1848. After a five-year sojourn to Southern Africa, he returned home to the U.K. with thirty tonnes of acquired baggage, including animals as trophies. Within two years, he had opened his collection to the public and soon thereafter began the white man, and woman's love affair with killing animals for fun (445). As an example, between 1875 and 1925, the British hunted 80,000 tigers in India. One man, George Yule, ended the precious earthly experience of over 400 tigers. On a royal visit in 1911, King George V and his group of 'sportsmen' rode out on more than 600 elephants. By the end of the day, thirty-nine tigers, eighteen rhinoceroses, four bears and a leopard had been murdered for enjoyment. This was their way of celebrating the birth of Jesus, their saviour. It was Christmas day. The current tiger population is less than 4,000. The British didn't only hunt tigers. They also aimed their high velocity rifles in the direction of cheetahs, leopards, bears and elephants. Cheetahs are now extinct in India and tigers, rhinos and elephants are endangered. In North America, American bison were driven to the very brink of extinction by European settlers who reduced their population from an estimated thirty million animals to less than 100. There are stories of settlers firing their rifles out of moving trains as they travelled west. They did this in less than the lifetime of some humans alive today (446). In total, some 1.2 million animals have been killed in the name of 'sport' (447).

Today, Europeans are aghast when they see native people killing wild animals in foreign countries, yet it was their descendants who pushed many of these beautiful creatures to the edge of extinction. An estimated 500 men were shot dead for suspected poaching in Mozambique in 2017. If they had been born a hundred years earlier with white skin, they would have received a hero's welcome on their return home with the animal well and truly stuffed (446). If they had been born white today, as long as they had enough money, they would have been welcomed in Africa as "conservationists". It is truly Orwellian double speak that fun loving killers are considered conservationists and not the psychopaths they clearly are. One of the chief absurdities is that the twenty-sixth president of the United States, Theodore Roosevelt, is considered: *"one of the most powerful voices in the history of American conservation."* Apparently, he, *"cherished and promoted our nation's landscapes and wildlife (448)".* He sounds like a fantastic fellow. The following might come as a shock to anyone unfamiliar with the animal loving conservationist champion. On one of his many hunting safaris in Africa, he produced quite a few orphans. One such unlucky rhino calf refused to leave her mother's side after Teddy and his son Kermit had shot her mother. Now, while many would take pity on a child who had just witnessed the murder of their mother, Teddy and Kermit decided to pelt the poor animal with sticks and clods of earth. The freshly produced orphan, knowing no better, refused to leave her dead mother and good old Teddy shot her in the buttocks with his rifle. This time, she left her mother so Teddy and his son could tear of her skin and remove her horn. After killing another rhino mother, he describes what happened next, *"we shot the calf, which when dying uttered a screaming whistle, almost like that of a steam engine."* Other courageous stories of conservation involve a pair of elephants shot, with one being downed and the other escaping. Teddy talks of wishing to pursue the escapee, only to give up as he not only wished to remove the tusks of the dead elephant, but also remove the skin, and time was of the essence (449). Again, those rushing to defend Roosevelt will contend that he was a product of his time. Sounds like a reasonable claim. His epic safari which resulted in the deaths of 11,400 animals took place in 1909 (450). While he was orphaning children in Africa, Thomas Edison, Nikolas Tesla, and Albert Einstein were abstaining from eating animals for ethical reasons. Clearly, not so much a product of his time as people would have us believe. In fact, people have been abstaining from meat for ethical reasons as far back as Pythagoras in 495BC.

If the plan, as hunters argue, is to hunt these animals in order to protect them, then it isn't working. In the past one hundred years, the number of lions in Africa has dropped from 200,000 to just 20,000. There were an estimated twenty-six million African elephants in 1800. By 1923, the population had dropped to ten million with elephant after elephant being downed in order to turn their tusks into billiard balls for western elites (451). In 1933, with clear evidence emerging of the almost complete collapse of the Bison, and the obvious decimation of wild animal populations in Africa, the first Conference for the Protection of the Fauna and Flora of Africa was held. Soon after, nation states began to set aside land for wildlife reserves. Sadly, this didn't stop the unconscionable slaughter of these majestic animals for sport. It still continues to this day and will be discussed later.

While, up until now, Europeans had been working hard to eradicate wild "resources", it wasn't until the industrial revolution in the 18th century that humans really stamped their authority on the planet. Until 1700, the human population rose at an average of 0.004% meaning that in 1700, there were 600 million of our descendants tilling the soil and working as artisans of various kinds. Over the next century, this number began to rise more steeply and by 1803, Earth had welcomed the one billionth human baby (452). The population at the outbreak of World War I was roughly 1.5 billion, and at the war's end, the population had been depleted of around sixteen million. It wasn't just humans who suffered though. A total of sixteen million animals also lost their life in the world's first Great War. This number includes 484,143 horses, mules, camels, and bullocks killed in service of the United Kingdom. The horrors of war are unimaginable to those of us fortunate to have never experienced them, but try to imagine the stress these animals must have endured. While humans understood the absolute dreadfulness around them, the animals were completely in the dark. In Germany, 30,000 dogs also died, slugs were used to detect mustard gas, whales who were mistaken for submarines were bombed from the air and 175,000 whales were killed for rifle oil, fuel, and trench stoves (453). To the east, Russian soldiers were chowing down on European bison which was a keystone species in the boreal forests. At the outset of the war, Russian elites had already reduced the bison herd to only 400, but by 1918, the starving soldiers had finished the species off (454).

The impact on natural resources in wartime is significant with forests in Belgium, the U.K. and France decimated by First World War fighting with French forests taking the brunt. They underwent a thorough change from diverse ecosystems in 1914 to barren monocultures in 1918. The U.K., when faced with U-boats sinking their supply of logs from Scandinavia, Russia, and Canada, embarked on a native deforestation program that resulted in 1,800 sq. km of deforestation. The Ottomans obliterated the cedar forests of Lebanon, and the infrastructure for future extraction was put in place in Africa. In addition to the logging of forests to provide timber for trenches, new and more destructive weapons were in use, including chemical weapons. Many of the chemicals used in the war poisoned the soil and led to a huge area known as the 'Zone Rouge' in France being declared off limits (455). As well as men, animals were also killed by chlorine and mustard gas, and in response to the battleground destruction he witnessed, the French writer *Henri Barbusse* had this to say:

"Fields of sterility" where "frightful loads of dead and wounded men alter the shape of the plains" and "everything appears turned over...full of rotten-ness and smelling of disaster. Where there are no dead, the earth itself is corpselike."

Other soldiers told of the land losing its nature and becoming something artificial (454).

By 1928, the human population finally hit two billion. It had taken 125 years to double in size, and by the start of the next World War in 1939, another 300 million had been added. In the years between world wars, in most cases, technological advancements had made animals redundant, but the flip-side was that the new weaponry was capable of destruction on a scale never before imagined. The destruction in the First World War pales in comparison with what occurred a few decades later. Europe again tore itself apart and around 3% (70-85 million) of the world population lost their lives. Regardless of technological break-throughs, animals would not be spared. In a single week at the begin-ning of hostilities, around 750,000 British pets were sent to slaughter due to concerns over the availability of food. Pets would receive no rations therefore their owners took them to the vet to be put down. As the bombs rained down on the U.K. in 1940, more and more people ended their pets' experiences (456).

As we discussed earlier, WWII was responsible for giving the world the DDT that went on to cause so much environmental damage around the world. It was also a harbinger for our blossoming chemical romance. In fact, the war brought so much environmental devastation that many of the problems we face today originated in those horrific six years. Not only was the war fought with guns, tanks, and bombs, but a war was also fought with shovels, hoes, and bulldozers. This second, and lesser-known war, was against nature herself. To provide logistics for the war effort, an army of soldiers and prisoners of war hurriedly built roads, railways, ports, and airports. All this brought invasive species to new territories and formally untouched areas were polluted and destroyed. Food production also increased dramatically during the war. In some places, wheat acreages grew by 200%, 400% or even 1000%. This massive demand for food resulted in further losses to nature and set the world on the course of industrial farming that has caused, and is causing so many problems today (454).

The sheer amount of munitions and supplies necessary for such an enormous event as WWII are unbelievable. Around six million vehicles including tanks, eight million mortars, artillery, and anti-tank guns, 850,000 airplanes, 45,000 missiles and 55,500 ships were produced for the war effort and technology advanced at a fantastic speed. At the start of the war, most air forces were using propeller fighter planes, but by the war's end, jet planes had become the norm (457). In 1940, bombs dropped in the Battle of Britain were responsible for destroying half a dozen or so houses in one foul swoop. Just five years later, humans had the potential to take out entire cities with a single bomb.

When the war finally came to a close in 1945 with the atomic bombings in Hiroshima and Nagasaki, the world had changed forever. Two new superpowers emerged, and the cold war began between two war time allies. All the airplanes, ships, and tanks, not to mention bombs, needed to be disposed of. Kingman Air Force Base in Arizona was home to 5,500 rusting hulks, ships were sunk, bombs dropped into the depths of the ocean and jeeps left to be turned into public transport in the Philippines (457). At the end of fighting, economies needed to be rebuilt, cities needed to be reconstructed. In short, the war between people had ended, but the war on nature was about to speed up. It was in this period that the Japanese turned their attention to forests in the Philippines, as was discussed earlier. The impact the war had on animals is immense although records are hard to come by, as understandably, people's focus lay on the human cost. In the aftermath of the war however, the lives of both domesticated and wild animals were about to become unrecognisable.

Industrial farming was already in full swing in the U.S. due to farmers being supported by Great Depression era government subsidies to produce corn. It became so cheap that it was only considered worthy of feeding to animals. After the war, with the help of Marshall Plan money, industrial farming arrived in Europe and soon entire landscapes were replaced by monocrops as far as the eye could see. The price of raising animals for food dropped hugely and the effects can be seen all around us today (458).

In the aftermath of the war, many countries faced a shortage of manpower for the rebuilding effort. Britain turned to its colonies for help. Between 1948 and 1970, more than 500,000 people were brought to the U.K., and many of these people are now unceremoniously being returned to countries they left decades ago.

In the U.S., the focus was not on manpower, but on technology. The war effort had meant that many farmers had left their farms to work in factories with some states seeing 30% of farmers switch to factory work. With the end of hostilities, farms had a dearth of labour, and this initiated the transition to factory farming. Chicken production had already been shifted indoors in the 1920s after the discovery of vitamins meant that animal feed could be supplemented. Mrs. Wilmer Steele of Sussex County in Delaware lays claim to be the first human on Earth to raise broiler chickens indoors. In 1923 she had a flock of 500 chickens, but by 1926, the popularity of her chickens had increased so much that she was able to house 10,000 birds in a single building (459). The idea of factory farms for other animals in the 1940s also had support in and out of agriculture, but the war made people even more enthusiastic to use machinery and automation to do the job of humans (460).

Factory farms, or Concentrated Animal Feeding Operations (CAFOs), as they are formally known, didn't become the standard for other farmed animals until the 1960s. In the intervening years, the industry began trotting out the falsehood that factory farms were necessary to feed the world and provide food security. The reality, as is usually the case, was not about feeding the world, but about maximising profits. The dictionary definition of a factory farm is *"a large, industrialized farm, especially a farm on which large numbers of livestock are raised indoors in conditions intended to maximize production at minimal cost"* (461). This seems quite explicit and doesn't mention the necessity of feeding the world. The industry had simply realised that by shifting animals indoors, they could house 2,000 animals on a single acre instead of two. Rearing animals intensively this way would also allow the industry to control and intimidate farmers into purchasing expensive equipment that placed them in debt to the corporations who owned the animals. Farmers were simply paid to raise them to a certain weight by a certain date. Animal husbandry, which had existed since the agricultural revolution, was officially over, and billions of live sentient animals were pushed off the land and into dark cramped warehouses with no direct sunlight, no fresh air, no grass, and no room to play or socialise.

The horrors of factory farming are there for all to see, and the majority of people would prefer it if it didn't exist. Whether it is male piglets being castrated without any pain killers, or one day old male chicks being macerated, the industry is relentless in its drive for efficiency. Animals kept in cages barely large enough to allow any movement at all, geese with tubes slammed down their throats to force feed grain into their livers, baby male calves being shot in the head because they can't produce milk, mothers separated from their babies within forty-eight hours, skulls of piglets being slammed onto concrete floors because they are growing too slowly, live animals left to die in agony because treatment reduces profit, the list goes on and on. An entire book could be devoted to the sadistic nature of factory farming, and in fact many have been. Famed historian, Yuval Noah Harari, described it best in his book Sapiens: a brief history of humankind:

> *"If we accept a mere tenth of what animal-rights activists are claiming, then modern industrial agriculture might well be the greatest crime in history."* (245)

This is not a book about the utter cruelty of factory farming though. It is a book about the collapse of our ecosystems, so how is factory farming, and in fact, animal agriculture in general, related to this I hear some of you mutter. Well, the sixth extinction and animal agriculture go hand in hand. According to the World Wildlife Fund's 2017 report, Appetite for Destruction, animal agriculture is responsible for a staggering 60% of all species loss. This number is in addition to the roughly 3,000 farmed animals killed every single second of every single day. It isn't difficult to see how the breeding and slaughter of seventy billion farmed animals every year is impacting wild animals. For a start, 45% of the Earth's surface is used for livestock (462). This is in most part land that was formerly forested and home to flourishing ecosystems with a plethora of wild animals. Seventy percent of deforested land in the Amazon is now used to graze cattle and most of the remainder is used to grow soya beans. Many people seem to believe that these beans are intended for vegetarians and vegans, but the truth of the matter is that these beans don't make it into human mouths at all. They are exported around the world to feed the animals living in factory farms. In fact, human consumption of soya beans accounts for just 1% of the total with 98% of all soya beans being fed to farmed animals (463). This situation is mirrored in other South American rainforests. The deforestation of the East Coast Australian forests can also be blamed on our love for beef. 73% of land cleared in Queensland between 2013-2018 was to graze cows. In the Great Barrier Reef catchment area, that number rises to 94% (464). 964 plant species and 286 animal species are listed as threatened by the Australian government and 77% of plants and 73% of animal species on the list are threatened because of deforestation. To put things into context, humanity makes up only about 0.01% of the planet's total biomass. If you were to weigh all the mammals on the planet, humans would make up 36% of that weight, domesticated animals such as cows and pigs make up 60%, and wild mammals today account for just 4% (433).

Humans and livestock make up 96% of mammals on Earth

Humans
36%

Livestock
are 60%

Wild mammals
4%

Data from EcoWatch

Since the human aided extermination of wild animals began, wild mammalian populations have fallen by seven times. In the past century, humans have completely altered the biomass on Earth. It isn't just the use of almost half the land on Earth by animal agriculture that is leading to this 60% decline. Almost a third (27%) of fresh water supplies are diverted to farmed animals. To produce one kg of beef, 15,415 litres of water are used. This is in comparison to grain which uses just 1,664 litres of water (465). This has led to aquifers drying up and the knock-on effect to wild animals should leave no need for explanation, but its impact on humans is also worth noting. Around 1.1 billion humans worldwide don't have access to water while 2.7 billion suffer water scarcity for one month each year (466). This is all happening as seventy billion farmed animals are adequately fed and hydrated. How is this acceptable? The industry also produces vast amounts of raw waste, much of which sits outside, unlike the animals, in large slurry pits which produce ammonia, phosphorus and nitrogen. These pits routinely leak, and in times of large storms, they completely discharge their contents into river systems. Rivers and lakes in famously picturesque New Zealand are now so polluted by animal agriculture that 2/3 of its waterways are now unsafe to swim in. In 2016, 5,000 people became ill and four died as a result of sheep faeces contaminating the water supply. The impacts on New Zealand's biodiversity are alarming with three quarters of their native freshwater fish now at risk of extinction (467). There is a knock-on-effect as these rivers drain into the oceans resulting in what are known as dead zones. These are areas depleted of oxygen where nothing lives.

In Wales, a staggering 75% of all land is given over to sheep farming, but as George Monbiot points out, Wales actually imports seven times more meat than it produces (468). This doesn't seem to be a good trade-off. Looking out of the window in Japan, you can enjoy looking at thick jungle and listening to the birds all day, and at night, you can hear the wild boar foraging outside the fence. If you take the ferry to the mainland of Honshu and go into the mountains, you will be able to find monkeys swinging from trees or stealing farmers' food. Good luck finding somewhere like that in Wales.

If we were to take the U.K. as a whole, around 85% of the land is used for either grazing animals or growing their feed. With these numbers, you might think that eating animals would provide the bulk of our calories and protein. The reality is somewhat different with only 32% of calories and 48% of protein derived from animal products. Again, the trade-off here is extremely poor (469). You might think that using 85% of all farmland for animal agriculture would mean the U.K. was self-sufficient in animal products? Not the case. The U.K. provides only 61% of its own food with meat imports three times more than exports, dairy, and egg imports twice more and fish imports also double. This not only makes no sense, but it places the U.K. in a dangerous position when emergencies like the COVID-19 pandemic hit.

All around the world, in developed nations, self-sufficiency is extremely low, and while this is good for trade, it does little for our planet. Many might think this is inevitable, but according to research from Harvard University, the U.K. could convert its grazing land and land used to grow animal feed to forests, thus sucking CO_2 out of the atmosphere and providing habitat for wildlife, while still being able to grow enough crops to provide more than the recommended protein and calories for each person in the U.K. This carbon dioxide removal (CDR) would be equal to between nine to twelve years of U.K. carbon emissions (470).

The problem of animal agriculture, unlike coal power plants, is not going away any time soon. Meat production has quadrupled since factory farms took over the landscape while milk has followed a similar trajectory (471). The U.N. expects meat production to double by 2050 (472). The fifteen countries expected to see the greatest change from forest to grazing lands are expected to convert 3,000,000 sq. km of forest by mid-century. This is an increase in land use of between 30-50% (473). How will wild animals survive the coming onslaught if we are to double our meat production? The Amazon is already burning, and as was mentioned earlier, it may be close to a point of no return where it becomes a savannah. As the population increases, and incomes rise in less wealthy nations, demand for animal products rises. Without a concerted effort to change our dietary behaviour, how will it be possible to protect the very lungs of our planet, and with them, the wonderful cornucopia of wildlife that calls them home.

While the impact of animal agriculture is pernicious, the felling of great forests is underway elsewhere for cheap oil. In Indonesia and Malaysia, our great cousins, the orangutans, are losing their foothold in this world, here, not for meat, but for palm oil. The ubiquitous product finds its way into everything, from snacks we eat, to cosmetics we use, energy for our cars, medicines we ingest, and even, once again, animal feed for animals locked away in the dank, caged hell holes we call factory farms. In 2019 alone, 8,578 sq. km of jungle were burned to the ground. In just three months leading up to November 2019, not only did all wildlife either perish in the fires or the lucky ones became homeless, but 626 mega tonnes of CO_2 were emitted. For context, this is more than the entire annual emissions from Australia (474). These emissions continue the cycle of positive feedback loops as the planet gets warmer and drier, more fires occur and in turn raise the heat further.

Agriculture isn't alone in leading to the mass extermination of animals. Continued human conflict also puts added pressure on their ability to survive. Starting in 1977, the fifteen-year civil war in Mozambique ended with thousands of humans dead. Its impact on the non-human animal population was even worse with 90% of large herbivores living in Gorongosa National Park wiped out by poaching. The carnivores, including hyenas, leopards and wild dogs fared even worse as they disappeared completely. Once the war was over, the animal populations began to recover, but war is never far away in many African nations who are home to many of the last vestiges of large wild mammals on Earth.

Across Africa, during peacetime, animal populations are relatively stable, but once the gloves are off, populations suddenly decline. Impala populations in Uganda dropped by 4.5% per year during the country's numerous wars between 1982 and 1995. Hippo numbers decreased by 12.5%. Later, in Chad, the number of elephants dropped by a staggering 44% in just three years of civil war between 2006 and 2009 (475).

It is difficult to be optimistic when it comes to the future of our childhood story-book companions in Africa, India, and Asia. As water becomes even scarcer, and populations continue to rise, wars will inevitably increase and with them, the majestic animals we all love and cherish will likely disappear for good. As if everything we have done to wild animals over the years wasn't enough to make grown men weep, even as animals are added to the endangered list, human beings, the wise ones, consider the best way to preserve them is by killing them for sport. As we discussed earlier, hunting for enjoyment began during the British empire. Not to be outdone, the U.S. empire has continued the trend.

In nine short years starting in 2005, a staggering 1.26 million wild animals were imported to the US as "trophies" to be mounted on walls. More than 1,200 different species were shot for a giggle, including

5,600 African lions, 4,600 elephants, 4,500 leopards, 330 southern white rhinos and 17,200 buffalo (476). Many of these animals are listed as vulnerable, near threatened or endangered, but that doesn't stop the human desire to kill them.

Today, you don't even have to leave the comfort of the United States to get your killing kick. Wild African animals, and those from many other areas will literally be flown to a park near you so you can shoot them from the safety of a raised chair in what are called canned hunts. Animals, tired and hungry after a long journey from their homeland are released in specially arranged locations where a psychopath with a rifle will be waiting on a nice and safe elevated perch. As the animal searches for food that has been left out intentionally within sight of the killer, the demented 'sportsman' or woman will wait until the animal is in sight and then with a short movement of their middle finger, a life will be extinguished.

According to the Humane Society of America, there are now more than 1,000 canned hunt operations in over twenty-eight states (477). Forty-three percent of the murdered animals originate in Canada, with South Africa providing 32%. The others come from Namibia, Mexico, Zimbabwe, New Zealand, Tanzania, Argentina, Zambia, and Botswana, in that order (476). The owners of these kill zones use conservation as their excuse for the murder that takes place there, but if we are serious about wildlife conservation, surely murdering wildlife as if they are nothing more than a commodity is not the way. Zoos claim the same excuse for their profiteering from the incarceration of wild animals for amusement, yet, in an unlikely twist, they even provided animals to the canned hunting operations to be shot and stuffed (478).

We couldn't incriminate people in the United States for their blatant disregard for non-human animals without also highlighting the appalling ways in which animals are treated in the world's second largest economy: China. Traditional Chinese Medicine (TCM) has been practiced for thousands of years, and can include tai chi, aqua puncture, herbal remedies, massage, suction cupping and moxibustion where herbal leaves are burned near the body. These are all benign practices that many Chinese believe to help restore the balance of Qi which can slip out of kilter due to an excess of either yin or yang. Unfortunately, for animals, both wild and domesticated, TCM also includes the consumption of animal parts usually traded from Southeast Asia or Africa. To date, there is no scientific evidence to suggest that any of the following 'cures' are capable of anything more than simply providing the user with a placebo effect. Rhino horns are used to cure gout or rheumatism, and in some cases even cancer. Ground tiger bones are used to fight meningitis or malaria while tiger penis soup is believed to make men virile. Tigers can have sex up to twenty-five times a day. What isn't mentioned is that each encounter usually lasts between two and ten seconds. Men copying this practice may prove to be more irritable than pleasurable. Sun bears are kept in cages their entire life to produce bile from their liver which is considered the best medicine against a range of ailments from haemorrhoids to conjunctivitis and even hepatitis. These animals are forced to endure the ultimate form of torture. Sea turtle eggs are consumed in the false belief that their consumption will lead to a longer life (479). Not only are these 'cures' completely incapable of any real medicinal effect, but they can also lead to slow and agonizing deaths, like a snake bite victim being given powdered tiger horn that is made up of keratin. It's akin to a doctor in Sydney offering a snake bite victim some of his own powdered fingernail instead of anti-venom.

So, what of the animals used to supply this horrendous trade? The island of Sumatra was once home to 'swarms' of rhino and both Chinese and European accounts confirm the large numbers. Due largely to their hunting for their horns, which resemble the mythical unicorn in Chinese folklore, they have been almost wiped out. There may be as few as thirty left on the island. A similar story has played out on other Southeast Asian islands including Java and Borneo. The one horned Sumatran and Javan rhino was once found from India, through to Vietnam, Indonesia, and Malaysia, but the last rhino in Vietnam was hunted for their horn in 2011 and on Borneo, the final rhino is not in a healthy condition (480).

Tigers have faced a similar trajectory in Southeast Asia with reports in the 19th century of Sumatra and Java being 'infested' with man eating tigers. Even as recently as the 1980's, villagers in Cambodia feared going out at night because of the threat of tiger attack. At the turn of the 20th century, it was estimated there were 100,000 tigers in the wild, but this number has dwindled to just 3,200 with the Javan tiger being declared extinct in the 1980s and the last wild Chinese tiger being caught, killed, and eaten by a poacher in 2009 (480).

Asian elephants have fared little better. While their tusks are smaller than their African cousins, meaning they were slightly less of a target, their demise has been nonetheless partly driven by demand for their tusks. They used to live as far north as the outskirts of Beijing, but understandably, China is no place for marauding wild elephants. Today, there are fewer than sixty individuals in Vietnam and elephants in Cambodia have been found murdered with their tusks missing. In Myanmar, elephants are being killed for their skin which is used for clothes, jewellery and also medicine. The problem in China is so entrenched that in 2014, tiger skins, bear paws, otter skins and clouded leopard pelts were found in the Chinese Chamber of Commerce offices in the Cambodian capital of Phnom Penh (480).

As mentioned above, the Chinese wildlife trade was not reserved for mammals, nor purely for 'medicinal' purposes. Helmeted hornbills were also prized for their 'red ivory'. These amazing creatures are used to make jewellery boxes and other bric-a-brac. Pangolins, thought to be the origin of the COVID-19 virus, are the most traded wild animal, but macaque monkeys are also bought and sold before being cooked alive in Hong Kong and Guangdong for their brains. These are just some of the animals that are being driven to extinction by the wild animal trade, and as populations decline, the industry is turning to farming wild animals instead.

The collapse of wildlife around the world has not deterred Chinese wildlife vendors. They simply set up factory farms to house wildlife instead of pigs and chickens. China is home to an astonishing 20,000 wildlife farms, housing a vast array of fauna, including peacocks, civet cats, porcupines, ostriches, wild geese, boar, snakes, frogs, turtles, racoon dogs, geckos, Asian black bears, emus, musk deer, foxes, bamboo rats, mallards, ostriches, deer, white rhinos, crocodiles, macaques, black swans, antelope, and salamander. This isn't even a full list. The insanity of wiping out wildlife for momentary pleasure or imaginary benefits, and then continuing to torture and kill these animals who have lost their homes and freedom is a double smack in the face to non-human animals. If we can't amend our ways after witnessing the extermination of entire species, what hope is there for us to avoid the worst of the coming climate and ecological crises. Even a global pandemic that has taken the lives of millions of humans and brought our dear economy to its knees is not enough for the all-seeing, all-controlling Chinese dictatorship to close these hell holes down. The same can be said of corrupt governments around the world who continue to sanction the depravity of factory farms, even as outbreak after outbreak originate in them. These 'farms' are a ticking time bomb, and the feeding of huge amounts of antibiotics to healthy animals to promote growth and prevent disease will likely result in pathogens that are resistant to anti-biotics (481). This could send us back to the dark ages where we have no treatment for basic illnesses.

It will be incredibly difficult, going forward, for non-human animals to continue to live side-by-side with us unless we drastically change the way we perceive them. Whether it's zoos, circuses, factory farms or canned hunts, humans in the 21st century still regard non-human animals in much the same light as in the Victorian age. Their existence is given little consideration and, in many cases, like factory farms, we treat non-human animals worse today than at any period in the history of mankind. Never before has the treatment been as sadistic as it is today, with animals crammed into disgusting cages where they can experience nothing of the joy of being in order that a human animal can experience the joy of a momentary taste sensation. We are no longer in the business of killing animals. We have now diversified into torture. And the more sadistic our treatment of non-human animals becomes, the closer the sixth extinction will creep to human feet.

So far, we have looked at some of the ways in which humans are causing the sixth extinction, but we have left the elephant in the room to last. The climate crisis could be the straw that broke the camel's back, as rising temperatures, worsening wildfires, super-storms and diminishing supplies of food and water lead to a total collapse of animal populations. Coupled with deforestation, our warming climate is the largest threat to wildlife, and if ignored, will likely result in humans being left with only the animals we have domesticated as company. In total, 27% of all assessed species are at risk of extinction with a quarter of all mammals. Terrifyingly, 99.9% of critically endangered species and 67% of endangered species are expected to disappear from the wild within the next one hundred years (482). How will we explain this to our children and grandchildren?

"No comment" by AlphaTangoBravo / Adam Baker is licensed under CC BY 2.0.

https://mydreamforanimals.com/

Teddy Roosevelt on safari. https://commons.wikimedia.org/

LIST OF GAME SHOT WITH THE RIFLE DURING THE TRIP

	BY T. R.	BY K. R.
Lion	9	8
Leopard	—	3
Cheetah	—	7
Hyena	5	4
Elephant	8	3
Square-mouthed rhinoceros	5	4
Hook-lipped rhinoceros	8	3
Hippopotamus	7	1
Wart-hog	8	4
Common zebra	15	4
Big or Grevy's zebra	5	5
Giraffe	7	2
Buffalo	6	4
Giant eland	1	2
Common eland	5	2
Bongo	—	2
Koodoo	—	2
Situtunga	—	1
Bushbuck:		
East African	2	4
Uganda harnessed	1	2
Nile harnessed	3	3
Sable	—	3
Roan	4	5
Oryx	10	3
Wildebeest	5	2
Neumann's hartebeest	—	3
Coke's hartebeest	10	3
Big hartebeest:		
Jackson's	14	7
Uganda	1	3
Nilotic	8	4
Topi	12	4
Common waterbuck	5	3
Singsing waterbuck	6	6
Common kob	10	6
Vaughan's kob	1	2
White-eared kob	3	2
Saddle-backed lechwe (Mrs. Gray's)	3	1
Bohor reedbuck	10	4
Chanler's buck	3	4
Impalla	7	5
Big gazelle:		
Granti	5	3
Robertsi	4	6
Notata	8	1
Thomson's gazelle	11	9
Gerunuk	3	2
Klipspringer	1	3
Oribi	18	8
Duiker	3	2
Steinbuck	4	2
Dikdik	1	1
Baboon	—	3
Red ground monkey	1	—
Green monkey	—	1
Black and white monkey	5	4
Serval	—	1
Jackal	—	1
Aardwolf	—	1
Rattel	—	1
Porcupine	—	2
Ostrich	2	—
Great bustard	4³	3¹
Lesser bustard	1	1
Kavirondo crane	2²	—
Flamingo	—	4
Whale-headed stork	1	1²
Marabou	1	1
Saddle-billed stork	1²	—
Ibis stork	2¹	—
Pelican	1	—
Guinea-fowl	5	5
Francolin	1	2
Fish eagle	—	1
Vulture	—	2
Crocodile	1	3
Monitor	—	1
Python	3	1
	296	216
Grand total		512

"DATING BACK TO TEDDY ROOSEVELT, HUNTERS HAVE BEEN THE PILLAR OF CONSERVATION IN AMERICA, DOING MORE THAN ANYONE TO CONSERVE WILDLIFE AND ITS HABITAT."

Gale Norton

48th United States Secretary of the Interior from 2001 to 2006 under President George W. Bush.

List of Roosevelt's 1909 safari victims. Roosevelt via Archive.org

Oceans

The way humankind has impacted the land is there to see in all its glory. Every day, we can see the innumerable ways in which we have bent nature to our will, whether it's concrete cities, endless stretches of monocrops, the disappearance of insects, polar bears scrounging for food on rubbish tips or koalas and orangutans sitting on freshly cut down tree stumps. The state of our oceans, unfortunately, is harder for humans to see. The only hint of the damage we have caused can be found around our coastlines as wave after wave deposits more and more plastic on our beaches and shorelines. You have to rely on scientific studies to get an inkling of the perilous state our oceans are in. Covering over 70% of our blue planet, the oceans are home to 97% of Earth's water and 78% of the planet's animal biomass. With the mass melting of our ice caps, the amount of ocean is growing by the day. Unfortunately, the amount of biomass is going in the other direction. While large scale hunting of wild animals on land has largely ended and instead switched to models of confinement, the destruction of marine wildlife is speeding up. Though we can estimate fairly accurately the number of land animals killed for food each year, it is extremely difficult to put a figure on the number of fish and marine mammals hunted annually as these distant relatives of ours are not recorded by individual numbers, but by weight. By taking the average weight of fish caught scientists have estimated that the number of fish hauled from the seas to suffocate slowly on factory ships is between 0.97 - 2.74 trillion. That's 970 billion to 2,740 billion fishes ripped from their habitat each and every year. All to satiate just 7.9 billion humans. This mammoth figure does not even include fish caught illegally or bycatch (fish caught by accident) who are thrown back into the ocean to die. It is estimated that 30% of fish caught every year go unreported so the actual figure is likely much larger (483). With our population expected to hit almost ten billion by 2050 and rising income levels meaning fish consumption is expected to rise from 154 million tonnes in 2011 to 186 million tonnes in 2030, one could easily question whether this is at all sustainable (484). The answer, according to research published in the journal Science is definitely not. The report estimates that we will see fishless oceans by 2050 (485). Add to this mass annihilation of our very own oceans, the fact we have heated the seas to the equivalent of one atomic bomb per second for the past 150 years and you start to get a clearer picture of what we have done. The founder of Sea Shepard, Paul Watson, is fond of saying, that, *"if the oceans die, we die"*. Many will see this as hyperbole, but he is well founded in this prediction as more than half of our oxygen is produced by our once flourishing ocean eco-systems (486). In the following chapter, we will first look at a problem that certainly every human can get behind: plastic. Then, we will look at a few harder truths that people may not find complete agreement on, but nonetheless are the reality of our situation.

Plastic

T he word plastic originally meant something that was 'pliable and easily shaped'. Recently, it has come to refer to polymers which means something with many parts. Polymers are made up of long chains of molecules, one of which is our own strands of DNA. Polymers are found in nature with the most obvious example, cellulose, which is the material of the cell walls in plants. The American John Wesley Hyatt, motivated by $10,000 prize money, was working on creating a synthetic version of cellulose. His work was intended to replace the use of ivory and horn in billiard balls as we were becoming aware of the severe drop in elephant and rhino populations. Decades later, in 1907, Dr Leo Baekeland produced the world's first fully synthetic version of cellulose which he termed Bakelite. Bakelite was used as an insulator to help roll out electrification across the U.S. It was sold as *the material of a thousand uses* as it could be shaped and moulded to make almost anything. The success of Baekeland and Hyatt caught the attention of chemical companies, and while these two scientists were trying to solve an existential problem, these chemical companies were merely chasing a profit.

Fast forward to the 1940s, and the Second World War yet again, increased the demand for products. This time, for products that specifically avoided the use of natural resources that were scarce. Nylon, which had been invented in 1935 to replace silk was used to make parachutes, ropes, body armour and helmet liners. Plexiglas was used instead of glass for aircraft windows. During the war, U.S. plastic production rose 300%. After the war, plastics began to be used to replace steel in cars, paper and glass in packaging and wood in furniture (487). The future seemed amazingly rosy for humankind who could now rely on a copious material that was cheap, safe, and clean, but most of all versatile. It could be used in everything. And it was.

For the next twenty years, plastic could do no wrong. It wasn't until the 1960s that plastic trash was first noticed in the oceans. The 60s were a time of environmental reflection with the publication of Rachel Carson's Silent Spring and by this time, the amount of plastic waste began to worry people. The concerns continued to rise until the 1980s when the plastic producers themselves decided to introduce recycling as the solution to this product, made to throw away, but which never actually disappears.

Today, plastics are as ubiquitous as flowers and plants. They are an indispensable part of everyday life, from the water we drink, to the clock on your wall, the pen in your hand, collar on your pet, the garden hose, the light switch, PCs, speakers, rubbish bins used to hold the plastic packaging that fills them, vacuum cleaners, cars, watering cans, chairs, food containers, notebook covers, cell phones, pill bottles, and baseball caps. And, in a twist of imagination, they are even used to produce indoor plants that replace actual plants.

In 1950, the world produced fifty million metric tonnes of plastic. In twenty-five years, that number had doubled and in the next thirteen years it doubled again to 200 million metric tonnes. By 2020, the amount of plastic being produced had risen to 367 million metric tonnes (488). Now, this figure sounds a lot, but it's very abstract. To provide a clearer image, the total amount of plastic produced since the 1950s is 8.3 billion tonnes, 6.3 billion tonnes of which has become waste. This is still a little abstract, so let's compare it to the animal it was initially invented to preserve, the elephant. The total weight of all the plastic produced in the past seventy years is equal to one billion elephants (489). There are currently around 450,000 elephants roaming the Earth.

While the companies profiting from our throw away culture have led us to believe that merely recycling all our plastic is the answer to the problem, in 2017 it came to light that the vast majority of the plastic we place in separate recycling bins doesn't actually get recycled. In fact, only 9% of our total waste was recycled. A further 12% was burned and the rest (79%) was sent to landfill or discarded in nature, often accumulating in our oceans (489).

To understand just how gargantuan the problem of plastic is, let's take a look at the most salient item, the humble plastic bottle. Every single minute of every single day, one million plastic bottles are bought, that's 20,000 bottles every second. By 2030, another half a trillion will be sold. To put this in some sort of perspective, if we took all the plastic bottles sold in 2016 and placed them end to end, they would stretch halfway to the Sun. That's 480 billion bottles. The amount of plastic produced annually now amounts to roughly the weight of humanity (490).

While these Pet bottles can easily be recycled, less than half of the bottles in 2016 were even collected for recycling. The actual number of bottles that made it into another form was just 7%. The majority of these bottles either end up buried in the ground or clogging our oceans. The amount of plastic that makes it into our oceans is estimated at between five and thirteen million tonnes a year. Much of this plastic is ingested by fish, birds, and other marine wildlife. Albatross and Shearwaters have been found with an astonishing 3,000 pieces of plastic, weighing eight kilograms, in their stomachs. Plastic cannot be digested so it remains, and the birds eventually die from a lack of nutrition. (489) We have all seen the images of birds, and whales with their stomachs opened up to reveal huge amounts of plastic sitting inside. In further research by the Ellen MacArthur Foundation, there will likely be more plastic in the ocean by 2050 than there are fish. This isn't just a problem for marine wildlife as according to the WWF, around three billion humans rely on fish for their main source of sustenance. Various scientific studies have shown that this plastic is already making its way into humans. Research from Ghent University in Belgium calculated that people who eat seafood are ingesting around 11,000 pieces of tiny plastic each year. Plymouth University in England found that one third of all fish caught in the U.K. contained plastic. These findings, and many more, have led to the EU Food and Safety Authority calling for urgent research into the effects on human health (490).

It is easy to scapegoat plastic bottles and drinking straws because of their ubiquitous nature, but according to research published in Scientific Reports, 46% of plastic found in the Great Pacific Garbage patch was from the fishing industry. Abandoned ropes, oyster spacers, eel traps, crates and baskets joined discarded fishing nets in a plastic rubbish heap that is twice the size of Texas. These nets often trap dolphins, whales and turtles who end up starving to death as they are unable to free themselves to eat. Each year, it is estimated that more than 650,000 marine animals, including turtles, seals, whales, and dolphins are killed or injured in fishing nets. Some of these animals are caught in active fishing nets and then thrown back into the ocean to die a lingering death. Just as many, however, are caught in what are called 'ghost nets', abandoned nets floating silently through the ocean. These are just some of the 640,000 tonnes of fishing gear discarded either intentionally or by accident in the oceans every year (491) (492).

In recent years, it was discovered that much of the plastics being carefully placed in recycling bins in high-income countries were actually being exported to low-income countries where they do not have the facilities necessary to actually recycle them. This waste is then, either discarded, incinerated, or buried in return for a small profit for the importer and at huge cost to the entire planet. China and many other lower-income nations have now begun to not only shun the import of plastic waste but send it back to the producing nation to deal with itself, as should have been the case in the first place. Yet again, the industry has been misleading us to encourage continued use of their products (493).

Just twenty companies produce 55% of single-use plastic products, and many of them are the same fossil fuel companies pretending to be green. ExxonMobil is top of the list with Dow, Sinopec, Saudi Aramco. PetroChina, Lotte Chemical, INEOS, Total, Formosa Plastics

Corporation and China Energy Investment Group also part of what should rightly be called the plastic crime syndicate (494). The damage these companies have knowingly caused needs to be held to account.

Even if plastic does get recycled, it cannot be recycled in infinitum. As was mentioned, plastic is made from polymers and each time a plastic product is recycled, the chain of polymers gets shorter and the quality decreases. Plastics can at best be recycled between two to three times before the plastic has degraded to the point it can no longer be used. It must then be buried or discarded in nature where it will sit for centuries to come (495).

To emphasize just how gigantic the problem of plastic waste is, humans have yet to set foot on Mars, and yet the first plastic waste has preceeded our arrival. In June 2022, it was reported that a thermal blanket which accompanied NASA's Persevearnce Mars Rover was found discarded on the red planet. Let's hope this isn't a harbinger for things to come.

Albatross at Midway Atoll Refuge USFWS Headquarters is licensed under CC BY 2.0.

Marine Debris on Kure Atoll USFWS Pacific is licensed under CC BY 2.0.

Beach Trash

Justin Dolske from Cupertino, USA is licensed under CC BY-SA 2.0.

Fish

The practice of catching fish for food dates back as far as 40,000 years, as humans were first making their way from Europe to the eastern Mediterranean, or Levant. Fish bones have been found at shell middens, the waste heaps of humans, dating back to this time and cave paintings also depict fish being speared. Between 7,500 – 3,000 years ago, Native Americans on the Californian coast had developed the fishing hook, and we are still using varieties of it to this day. The human population 5,000 years ago was about five million, similar to the population of New Zealand today. As human technology has advanced and as the human population has increased, the number of fishes we take from the ocean has extrapolated beyond anything our ancestors could have imagined (496).

As the population was hitting 2.5 billion in 1950, humans were taking seventeen million tonnes of fish from the sea and another two million from fresh water. In just a decade this had risen to thirty-one million tonnes of wild oceanic fish and three million tonnes of freshwater fish. There were two new statistics to add with one million tonnes now coming from factory fish farms in the sea and a further one million tonnes coming from freshwater fish farms. The number of wild fish taken from the sea continued to rise unabated until it amounted to fifty-eight million tonnes from the sea in 1970 with a further five million tonnes now being removed from fresh water sources. The number of fish coming from fish farms, both fresh and oceanic remained unchanged. The human population now stood at 3.7 billion. By 1980, fish catches from the sea had begun to rise more slowly than previous decades but sixty-two million tonnes of our scaly distant cousins were removed from the oceans and a further five million tonnes from fresh water sources as neoliberalism took hold around the world. As wild fish catches in both oceans and fresh water slowed down, factory fish farms began to expand. In 1980, fish produced in factory farms had doubled in weight to two million tonnes from ocean and fresh water-based operations. The 1980s saw the human population continue to rise and by 1990 it was almost 4.5 billion. Oceanic fish populations were further reduced by seventy-eight million tonnes and freshwater catches depleted six million tonnes. Oceanic fish farms really took off in the 80s and by 1990, eight million tonnes of fish were now being sourced from fish farms in the open ocean. Fresh water fish from farms also rose to five million tonnes. While the 90s saw a slower increase for wild fish taken from the sea, tonnage still increased to eighty-five million tonnes with nine million tonnes of freshwater fish also depleted. The 90s saw a huge expansion of fish farming with freshwater farms taking over as the largest producer of farmed fish. Nineteen million tonnes of freshwater fish were produced in 2000 with a further fourteen million tonnes coming from sea-based farms. As the new millennia got underway, the human population had passed six billion for the first time, and as the human population exploded, the number of wild fish taken from the sea began to drop. By 2010, fish ripped from the oceans declined from eighty-five million tonnes to seventy-six million tonnes with eleven million tonnes now coming from wild freshwaters. The slack was taken up by fish farms both inland and oceanic with thirty-six million tonnes of fish now being produced in fresh water and a further twenty-two million tonnes in the ocean. The past decade has seen an uptick in wild oceanic fish with eighty-four million tonnes being taken by 2018 and twelve million tonnes coming from wild freshwater fish. Fish farms grew massively in the past decade and fish taken from fish farms inland and on the sea now account for eighty-two million tonnes, just two million tonnes shy of wild fish (497).

WORLD CAPTURE FISHERIES AND AQUACULTURE PRODUCTION 1950-2018

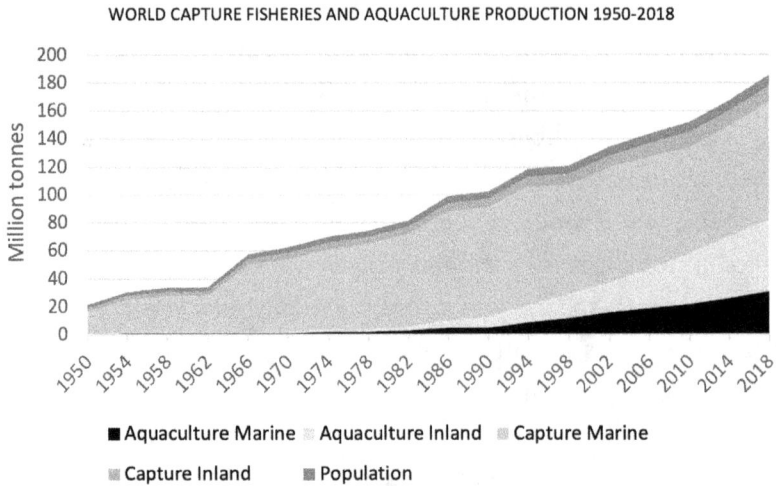

■ Aquaculture Marine ▨ Aquaculture Inland ▨ Capture Marine

▨ Capture Inland ■ Population

The State of World Fisheries and Aquaculture 2020: Sustainability in Action
Food & Agriculture Organisation

It is easy to assume that we switched over to farming fish because we wanted to conserve wildlife, but yet again, the reality is somewhat less warm and fuzzy. Beginning in the 1950s, and the onslaught of industrialized fishing, humanity has driven large carnivorous fish like cod, sharks, halibut, grouper, tuna, swordfish, and marlin to the edge of extinction. Less than a third of their population survive today (498). Most of these declines have occurred since the 1970s. The popularity of these fish encourages fisherman to go after these top-of-the-food-chain predators, rather than the smaller anchovies and sardines. The massive declines have led to these carnivorous hunters being farmed instead. Now, these fish who travel far and wide in the wild are crammed into circular pens just seventy meters across (499). Around the world, we have driven the largest wild sea life to dangerously low levels. You might think the 'wise one' would try to reverse the situation by introducing smaller quotas and preservation areas, but alas, what we are doing instead is going after the smaller pelagic fish in order to feed to the carnivorous fish we keep in these inhumane factory fish farms. Not content with the mass annihilation of the ocean's predators, we are now annihilating the other end of the ocean food chain. This is known in the industry as, *"fishing down the food web."* Of course, this has massive ramifications for the one third of carnivorous fish left and other wild-life who need the pelagic fish for their own survival. African penguins are now endangered because the small fish they need for survival are being dragged from the ocean by humans to be ground up into fish meal and fed to farmed salmon, pigs, and chickens. Around 20% of wild fish caught worldwide becomes fishmeal for animals reared in factory farms in China and the European Union. There are just 50,000 African penguins left in the wild and they could disappear forever by 2030 (500). They are not alone as whales, penguins and leopard seals are threatened by the overfishing of their food source: krill, in Antarctic waters. Krill populations have declined by 80% since the 1970s (501). As the sea ice in the Arctic melts, we can expect to see fleets of gigantic factory ships heading to the North pole and the fate of polar bears, narwhale and the rest of the Arctic ecosystem may be sealed.

Removing both top of the chain predators and bottom of the chain creatures like krill has huge debilitating effects on the entire oceanic ecosystem. Kelp forests, which are home to many unique species, are now threatened by an explosion in sea urchin numbers as their usual predator, sea otters, are removed from the ecosystem (498). As we remove the krill that whales, seals, and penguins rely on for food, we not only reduce the populations of all these animals, but we impact climate change too, as will be discussed in the section on whales.

In the Northeast Pacific, southern resident killer whales, made famous for many by the movie Free Willy are at risk of extinction from several threats: overfishing of their prey, the Chinook salmon, key among them. As the Chinook population decreases, so does the availability of food for the killer whales. The demise of the Chinook doesn't only affect killer whales because the salmon provide sustenance for bears who depend on their annual salmon run to lay their eggs for the next generation. As Chinook numbers drop, bears can be seen emaciated on riverbanks and as the adult fish die just days after laying their eggs, their decaying bodies contain the nutrients necessary to fuel the food chain that will sustain their offspring (502).

In Alaska, in 2020, the government took the unprecedented decision of closing the Gulf cod fishery. The number of Gulf cod has hit historic lows and Pacific fish populations in general are so low that the short-tailed shearwaters reliant on them for their own survival are now being found dead on beaches in Alaska, and as far away as Sydney, Australia. It is believed these fish are dying both due to overfishing and ocean warming. Low rainfall in Alaska has also led to the decimation of Alaskan pink salmon. Tens of thousands of salmon who were making their annual spawning trip were trapped in small tide pools where they eventually suffocated prior to completing their job of producing the next generation. (503).

One stated solution to the destruction of fish populations is to provide protected areas, similar to national parks, where populations can rebound. There are 178 of these Marine Protected Areas (MPAs) in British waters and they account for almost a quarter of U.K. territorial waters. In a 2019 study, it was found that of seventy-three MPAs analyzed, all but two of them were being dredged and bottom trawled. Boats from the U.K., the E.U., and Russia spent some 200,000 hours ripping up the seabed. To prevent this absolute devastation, Greenpeace took the unusual action of dropping massive boulders into the Dogger Bank MPA to prevent ships from trawling the area in search of cod, whiting and sand eels (504). In the same year, the U.K. government announced a global alliance of countries to protect 30% of the planet's marine environments by 2030 (505). If they take this project as seriously as they have the domestic MPAs, we can expect to wave goodbye to many species in the coming decade. Everything is ultimately linked, and humanity must be careful in the direction it takes if we are not to tip entire ecosystems into complete collapse.

The alternative to wild fishing is the fish farms that have proliferated but these further endanger wild populations by the spread of disease. The European Union buys more than 200,000 tonnes of frozen and untreated fish every year from the North Atlantic, West Africa, and South America. These fish are non-native and when dumped into tiny oceanic fish farms, they raise the risk of exotic viruses being spread to their geographic region, in this case the Mediterranean Sea (499). In Canada, large-scale fish farms have been found to spread the Piscine orthoreovirus (PRV) to wild fish in the area. The closer wild fish are found to fish-farms, the more likely they will be infected by PRV. In farmed Chinook salmon, the virus causes jaundice and anemia, but in its wild cousin, it causes kidney and liver failure. Successive Canadian governments have been made aware of this yet have failed to act; however there are plans to end open pen Chinook salmon farms by 2025 (506).

Fish farms are causing concern for other reasons too, including the introduction of invasive species, parasites, and chemical pollution (507). An example of the indifference we have to life under the waves can be found in Scotland, where in 2021, the government was pushing for a banned insecticide to be used in Scottish salmon farms. The insecticide was banned by the E.U. in 2018 because it was blamed for destroying bee populations when used on crops. It's manufacturer, Bayer, attempted to have the ban overturned but lost its case in 2021. Amazingly, the ban doesn't apply to use in rivers or the ocean where it is intended to be used to kill the lice that are prevalent on fish farms where salmon live on top of each other with little space to move freely (508).

According to the UN's Food and Agriculture Organisation (FAO), around 90% of global fish stocks are now fully depleted or overfished. The situation is expected to get worse as a further 17% rise is expected by 2025. Per capita consumption of fish is now estimated at twenty kg per year and since 2000, one fifth more fish stocks are at worrying levels (507).

Can anyone imagine this absolute destruction and disregard for oceanic life happening on land? These large carnivorous fish are the equivalent of the lions, tigers, cheetahs, and pumas that we all pertain to love. We are ripping them out of their habitat and then leaving them to suffocate in great pain on giant factory ships.

To justify this annihilation, humans have long told themselves the story that fish are unintelligent and do not feel pain. This ignorant belief has desensitized us to the images of fish flapping around in desperation and confusion as their oxygen supply is cut off. Even as mounting evidence suggests they not only feel pain, but are significantly more intelligent that we have believed; the head of the Scottish Fishermen's Federation, Bertie Armstrong suggests it is "cranky" to believe fish should be protected by any kind of welfare laws. He insists *"the balance of scientific evidence is that fish do not feel pain as we do."* Let's unpack this statement a little. Firstly, how on Earth can humans possibly know what other sentient beings can and can't feel? We have no idea what kind of pain dogs feel when stepped on or cats when sat on accidently. We can't know this. But no one would seriously contend that dogs being burnt to death in the Yulin dog "festival" do not feel pain. We can't know for certain they don't; but it is highly likely that they do. Any sentient being that has evolved for millions of years will have evolved to escape danger. The best way to guarantee that animals flee from situations that could cause harm is to develop the ability to feel pain. Otherwise, animals would continue to walk into burning forests, and pigs and chickens would happily walk to their death. They don't. Secondly, why does Bertie worry about the type of pain suffered? Why is pain only an important factor if it is the same as human pain? It is likely that pain in different species is experienced in different ways, but the type of pain experienced should not cloud our judgement as we cannot possibly know how that pain is experienced. Let's imagine alien lifeforms arrive on Earth and they are significantly more intelligent than humans. Maybe they would contend than we don't feel pain as they do, which would almost certainly be very true. Would that be a good justification for suffocating us on mass? These aliens may further contend from our willful destruction of the ecosystems that provide us with life, that we are a species without reason and our language may seem primitive to these beings who communicate telepathically. We would hope that they are a merciful species and as the English philosopher Jeremy Bentham said in his 1789 work An Introduction to the Principles of Morals and Legislation:

"When considering our ethical obligations to other animals, the most

important question is not, can they reason? Nor, can they talk? But, can they suffer? (509)"

Let's take a look at the evidence mounting up that suggests fish are extremely capable of suffering. Anatomically, fish have neurons called nociceptors. These are used to warn of potential danger such as high temperatures, high pressure, and even corrosive chemicals in their environment. In addition, the same opioids that mammals produce as natural pain killers are also produced by fish. The brain activity in fish is also similar to land vertebrates and in studies, when pins have been stuck into goldfish or rainbow trout, the nociceptors send a torrent of electrical activity to the brain region. These regions are not limited to the hindbrain and brainstem which are responsible for reflexes and impulses, but cerebellum, tectum and telencephalon, the very regions necessary for conscious sensory perceptions (510).

In further studies on rainbow trout, bright Lego blocks have been dropped into their tanks. Rainbow trout usually avoid new objects in case they are dangerous, but when injected with acetic acid, they are much less likely to display defensive mechanisms. Scientists believe this is due to them being distracted by the pain of the burning acid. The same fish when injected with both acid and morphine continue to avoid the potential danger. This suggests the morphine is dulling the pain, but of course the source of pain is still there, and this suggests the fish are experiencing pain and not merely responding reflexively (510).

The scientists weren't finished with what surely amount to tortuous experiments. Next, they injected acetic acid into the lips of rainbow trout. The fish then breathed more quickly, rocked back and forth at the bottom of the tank, and rubbed their lips on the gravel and the side of the tank. Other fish had salt water injected into their lips and showed no signs of any distress, but when fish had acetic acid and morphine injected into their lips, they showed some signs of distress but less than the fish who were given no morphine (510).

If you still aren't convinced that fish can experience pain, consider this experiment conducted by one of the world's foremost experts on fish pain, Lynne Sneddon from the University of Liverpool. In the experiment, fish were provided with two choices: a barren aquarium or an aquarium with gravel, a plant, and a view of other fish. The fish consistently preferred the decorated chamber. No real surprise there. But, when the fish were injected with acid and the undecorated tank was flooded with lidocaine to numb the pain, the fish abandoned the livelier tank in favour of the pain numbing water. To emphasise the result, the fish next had the pain killer injected directly into their bodies so they could take the pain killer with them. Surprise-surprise, the fish now once again preferred the enriched environment over the empty tank. If you still aren't convinced by these experiments, you are in the minority as the American Veterinary Medical *Association* had this to say in 2013:

> *"Suggestions that finfish responses to pain merely represent simple reflexes have been refuted. ... the preponderance of accumulated evidence supports the position that finfish should be accorded the same considerations as terrestrial vertebrates in regard to relief from pain."*

Just as the "debate" over the reality of man-made climate change is non-existent, so to, is the "debate" on fish pain. Scientists are in overall agreement that fish do experience pain (510). This inevitably leads us to ask what should be done. How should we now treat fish, knowing that they, like land mammals, who at least have some very weak protections, also feel pain, and suffer? Should we continue to rip them from their habitats in gigantic nets or catch them on long lines where they fight against their painful death for hours on end? Or should we introduce similar protections that end their pain as quickly as possible? Does our perception of the killing of trillions of fish every year change in any way when we know that we are not just ending their experience but causing pain and suffering to an unimaginably vast number of sentient beings? Would we request our more intelligent alien visitors to take mercy upon us? If the answer is yes, maybe we should show similar clemency to our distant cousins who are suffering in incredible numbers, largely to satiate our taste buds.

Ghosts at Tsukiji Fishmarket, Tokyo by laszlo-photo is licensed under CC BY 2.0.

A Boulder falls into the English Channel from MY Esperanza © Suzanne Plunkett / Greenpeace

Dear Indy,

When I was a child, probably about your age, Johnny used to take me fishing on his friend's boat. One day, as they were putting fish in the bucket, I was taking them out and placing them back in the sea. This went on for a while until they realized there were no fishes in the bucket. I will never forget that day.

Fishing is one of the most popular past times for fathers and their kids. I'm not sure whether you would like it or not, but you won't know because I never took you. Some of my friends take their kids fishing and don't understand why I don't take you. The main reason is that we don't need to eat fish, but other sea-based creatures do. When we remove fish, we take away their food. The other reason is that I don't think it's a good lesson to teach our children. Most people wouldn't take their child hunting for rabbits or badgers, so what is the essential difference between hunting fish? Of course, fish don't live on land, and they might not be cute, but is cuteness really a good way to decide whether an animal's life has value or not? Surely, the animal values its own life.

I've had people argue that we can't know if animals want to live or die. This is a somewhat silly argument to make. All you have to do is watch any animal try to escape when they are about to die, and you get your answer. Whether the animal is a cat, cow, pig, or fish, they will do all they can to live. They don't offer themselves to humans as a sacrifice, although this is a comfortable thought for many. Unfortunately, each individual is killed against their will. This is why you've never been fishing.

We saw in the last section that fish feel pain and suffer. Why then, do we teach children that their pain and suffering has no meaning? A much better lesson for children might be that all species are worthy of our compassion. Wouldn't this be preferable to teaching kids to discriminate between species? Maybe then we will live up to our tag of 'wise one' and move a step closer to the peaceful world we all dream of.

Whales

While many might still contend that fish feel no pain, against all evidence provided, few would argue against the fact that whales are capable of suffering. Whales have long been abused at the hands of humans, whether them being slaughtered for oil in street-lamps, being used for bombing practice in WWII or incarcerated in tiny tanks for human entertainment; some of the most intelligent animals on our planet continue to suffer at the hands of human caprice and greed.

Humans have been hunting whales for food for thousands of years with the Inuit in the Arctic Ocean relying on whales for their survival. Others like the Basques in the Atlantic and Japanese in the Pacific relied on the trade for economic growth. Beginning in the 16th century, whales began to be hunted in larger numbers by Europeans and Americans, predominantly to provide oil for street-lamps and soap; but also as a lubricant for instruments and machinery. By the 20th century, the uses for whale oil had expanded to include margarine, carriage springs, corsets, fishing poles, umbrella ribs, typewriter ribbons and the explosive material: nitro-glycerine. Right up until the 1960s, oil from the liver was used as a source of vitamin D, and blubber, bones and flesh were used to produce cosmetics and detergents (511).

As whaling spread, in order to aid the industrialization of our societies, whale populations began to decline, and it was harder to locate them near the coasts. Rather than slow down the hunting to allow populations to recover, our ancestors did the opposite and expanded the search to include the South Atlantic, Pacific and Indian Oceans. It is difficult to get accurate numbers for whale populations prior to large scale hunting, but is estimated that in the 21st century, humans devastated whale populations to the tune of around three million creatures. An estimated 276,442 whales were harpooned in the North Atlantic, 563,696 in the North Pacific and the Southern Hemisphere accounted for 2,053,956. Due to the enormous size of whales, this is considered the largest extermination of any wild animal by humans, when calculated by total biomass. Around two thirds of sperm whales, and a staggering 90% of the largest animal to share the planet with us: the blue whale, were killed. There are estimated to be a few thousand blue whales left in our oceans. The number killed may strike some readers as rather large, but they are only an estimate of those taken from the oceans; the number is likely higher with many hunts going untraced. This annihilation was all done in the name of "development". While the use of quotation marks might appear strange, they are used to highlight our odd, continued belief in human development, even as human development looks likely to actually mean the sixth extinction and our own eventual demise. As one whale species became harder to locate, whalers would simply switch to another, and then another, until commercial hunting was outlawed temporarily in the 1980s (512).

Whaling is back in the news recently as Japan abandoned its pretence of hunting for scientific purposes and withdrew from the International Whaling Commission (IWC) to hunt commercially. After its withdrawal, Japan must now abandon the Antarctic and only hunt in its exclusive economic zone. A strict quota of 383 whales has been set. This is only 60% of the number it hunted under the auspice of "science" in 2018 (513). It seems when it is on its own doorstep, the Japanese government cares more about the damage it inflicts on these elephants of the sea. Some Japanese justify this barbaric behaviour as the continuation of tradition, but as in any tradition that involves inflicting mass suffering; this is no justification. If it were, women would never be able to enter the workforce, slavery would be commonplace, and "witches" would continue to be drowned to death. It is not only Japanese who continue to slaughter these incredibly intelligent and sophisticated animals. The Faroe Islands, who have been whaling for around 1,000 years continue to turn the ocean red with an estimated 838 pilot whales killed annually for the past 300 years (514). Again, tradition is given as justification, but whaling was a tradition of many nations who have nevertheless moved on. While Iceland announced it will be ending their whale hunt in 2024, Norway has steadfastly refused to end the practice. They have now taken the mantle of largest hunter of whales on the planet with 1,278 being killed in 2018 (515). In the same year, Iceland killed a protected blue whale off its coast which on closer inspection was a rare hybrid blue and fin whale. Blue whales have been protected since the 1960s after being hunted to the brink of extinction (516).

Whales have been known to explore the depths of our oceans with the deepest known dive recorded at 2,992 metres. To put things in a little context, that is a rough equivalent of three Mount Snowdons, the highest peak in the author's birthplace; coincidently also called Wales. Their diving skills are not the only thing about whales that are deep, so it seems is their thinking. Whales, like humans have large brains and most animals with large brains have certain characteristics, namely they live long lives, are sociable, have complicated behaviours and they take special care of the few children that they produce. They teach their children life skills and raise them slowly until they become independent of their mothers and reach sexual maturity. Whale brains have spindle neurons which are associated with abilities like recognition, memory retention, communication, perception, adaptation, problem-solving, understanding and even reasoning. Toothed whales even have some special powers that humans would consider superhuman. They use echolocation to help them navigate murky waters, hunt, and even it seems, check on each other's pregnancies (517).

Scientists now believe that human intelligence may have developed as a way of coping with existing in large and complex social groups. This theory is known as Cultural Brain Hypothesis and might also apply to whales and dolphins. Sperm whales, who have the largest brain on the planet communicate in local dialects; orcas each have individual names which other pod members use to call them; and it is even possible that some dolphins practice the art of gossiping about absent pod members

(518). In 2018, the world was gripped by images of a female killer whale known as J-35. The mother carried her calf, who had died shortly after birth, for a full seventeen days in what is thought likely to be a show of grief. It is common for southern resident killer whales, but these displays of grief usually last less than a day. These are the same whales who are struggling to feed themselves due to human overfishing of their food source: chinook salmon (519). The largest creature on Earth, the blue whale, has developed such incredible communication skills that it is able to communicate with other blue whales across thousands of kilometres of ocean. These mostly solitary animals use a frequency well below the ability of a human ear, but nevertheless make it both the largest and loudest animal to share our planet (520).

Understanding what we do about the intelligence and social nature of these majestic animals should be enough for a species that calls itself the 'wise one' to want to leave them in peace, but we will see that it isn't only the people of Japan, Iceland, Norway, and The Faroe Islands that are abusing these deep thinkers of the deep blue.

Starting in the autumn of 1961, humans began a perverse experiment with whales, in this particular case, Orcas or killer whales as they are known for their incredibly effective team hunting skills. A single female whale was noticed swimming in Newport Harbour, south of Los Angeles. She was corralled and eventually hoisted onto a flatbed truck and transported to the nearby Marineland of the Pacific "amusement" park. The next day, she went berserk and swam at full speed into the tank repeatedly until she convulsed and died. In the wild orcas would be expected to travel around one hundred kilometers per day, so is it any wonder that orcas in captivity suffer enormous amounts of stress and often succumb to violence against their trainers; who we should really call their oppressors. You might think that after this original debacle, the owners at the 'amusement' park would have realised the folly of their calamitous decision. Of course, the answer would be no; the next autumn, September 1962, they took their forty feet collecting boat: the Geronimo, to Puget Sound in Washington State to look for another wild animal to entertain the masses of civilized animals back on shore.

Once again, things did not go to plan. The A-Team this was not. After a month of searching, two orcas, a male and a female appeared out of the mist and appeared to be chasing after a porpoise. After the orcas chased the porpoise round the boat a few times, the men on board threw a lasso around the female who duly dove down deep in a fit of panic. She ended up getting caught in the heavy nylon line and wound it around the propeller shaft which in turn immobilized the boat. She had around 250 feet of rope around her and as she got to the end of her tether both figuratively and literally, her male companion appeared, and they both started to swim at full speed towards the boat. They charged several times, always turning at the last opportunity but managing to thump the boat with their flukes as they did so. At this point, Frank Brocato, who was chief animal collector at Marineland grabbed his 375-magnum rifle and shot the male who then disappeared. The female was shot ten times before she died. The head animal collector at Marineland then towed the body to shore where she was weighed and measured before being rendered into dog food. Brocato then took the teeth as souvenirs (521). These are people who will pretend to care about wild animals and argue that their facilities are necessary for education and conservancy. People often argue that wild animals also kill other animals, like in this case the orcas chasing a porpoise, and this somehow gives humans the justification to also do the same. The simple truth though is that no other animal considers itself to be civilized like we do. If we were truly civilized, then killing animals for any other reason than survival would be seen as abhorrent behaviour. According to Merriam Webster's on-line dictionary, the definition of civilized is *especially characterized by taste, refinement, or restraint*. If we cannot restrain our impulses, then are we truly civilized?

One would have thought that the disastrous events in Puget Sound might have ended our plans to hold orcas captive, but just two years later, the first live orca was exhibited in Vancouver Aquarium. He was harpooned in Vancouver harbour and died three months later. He was named Moby Doll (521).

A year later, another orca was caught accidently in a fishing net in British Columbia. This wonderfully complex creature was sold to Seattle Public Aquarium for $8,000 in cash and became the first killer whale forced to perform tricks for clapping, seal like humans. The male orca was named Namu and he endured thirty-one years in captivity before his death from infected water. A year after Namu was caught in a fishing net, the owner of Seattle Public Aquarium, probably using the successful, albeit accidental capture as inspiration, devised a technique to capture wild orcas using netting. This method was used to capture more than two-hundred orcas by the early 1970s. Most of whom were sent to Sea World (521).

In 1970, there was public outcry in Washington State as eighty whales were corralled, and several died in the process. The dead whales had their bellies slit before being weighted down with chains. The outcry, unfortunately, wasn't enough to turn the tide against the inhumane practice of keeping wild animals in captivity for the entertainment of the great 'wise-one'. In 1976, an assistant to Washington State Governor, Dan Evans was out sailing when he witnessed Sea World captors using aircraft and explosives to herd and net whales. After receiving this information, Dan Evans sued Sea World and the whales were eventually released. Sea World lost its permit to capture wild orcas in Washington State and to date Washington State remains a sanctuary for them. Unfortunately, it isn't a sanctuary for their wild prey: the chinook salmon (521).

Not to be deterred, in 1976, Sea World worked a little more clandestinely than dropping bombs from airplanes and hatched a plan to steal whales from Iceland. Whales were airlifted from Reykjavik to Holland and then forwarded on to Sea World in San Diego where they were turned into performing clowns. Nine whales were acquired for Sea World in this way. Between the years 1976 and 1979, twenty-one Icelandic killer whales were captured for the amusement of humans in the U.S., France, Canada, Japan, Hong Kong, and Switzerland. A further four orcas were caught in Iceland in 1989 (521).

Not to be outdone by their U.S. and Icelandic competitors, the Japanese started to get in on the depravation. Beginning in the 1980s, the Japanese started to hunt false killer whales for food; with the most beautiful sold to aquariums and the U.S. Navy. These are members of the dolphin family, but they are large and resemble killer whales in the shape of their skulls and dark colourings. Animals are supposed to be caught in seine nets, but the Japanese practice drive fishing, where fishermen make noise under water, which confuses the animals, and they are driven into small coves where they are brutally murdered, often in front of their offspring who are then sold to aquariums throughout Japan (521). The murders continue in Japan up to this day and were highlighted in the Oscar award winning documentary movie: The Cove. To date, 2,000 dolphins, 227 beluga whales and forty-three orcas are still kept in captivity worldwide where they continue to be used as profitable entertainment. The longest tank in the world is a mere seventy meters long and twelve metres deep. In the wild, orcas have been recorded travelling more than 9,400 km in forty-two days and reaching speeds of thirty miles per hour (522).

The massive ocean warming that is being driven by our "progress" is yet another obstacle to the whales' survival. Add to this, the impact of overfishing, and the future does not look bright. The keystone species in the Southern Ocean is krill and they support penguins, orcas, whales, seals, seabirds, and many fish species. As waters in the Antarctic warm, the krill may either migrate or decline in numbers (523). By 2100, it is estimated that they may lose between 20% and 55% of their habitat (524). This is already helping push the Southern right whale to the edge of oblivion as they are struggling to adapt their behaviour as fast as the waters are warming. They are hunting different species further north, but the number of infants they produce is declining. Instead of birthing every three years, they are now breeding every four or five years (525). While this is partly due to warming, it is also being driven by our demand for farmed fish. This is where the vast majority of Antarctic krill end up, and the whales end up losing out to human demand (526). At the northern tip of the planet, bowhead, narwhal, and beluga whales are equally challenged by migrating prey due to warming waters (527). There, they face a hostile future that is sure to include a mass extermination event of the Arctic's inhabitants. Once the Arctic summer sea ice has fully retreated, the world's fishing fleet will descend upon the pristine seas and one of the last remaining refuges for ocean wildlife will be lost forever.

If the absolute tragedy of enslaving these resplendent wild animals for our own entertainment and stealing their food isn't enough to convince you of the need to leave our wild cousins in peace, then maybe a more pragmatic reason will be. Whales alive today act as giant carbon sinks: literally. Each whale stores tonnes of carbon each and when they die, they sink to the bottom of the ocean where the carbon is taken out of the atmospheric cycle for hundreds of thousands of years. It is estimated that around 30,000 tonnes of carbon is safely stored in this way each year and if we can help whale populations to rebound to their pre-commercial whaling days, then this figure could increase to 160,000 tonnes. It is even possible they do more to help our fight while they are alive. As whales feed on tiny marine organisms like krill in the depths of the ocean before heading to the surface where they poop and pee, these nutrients then stimulate the growth of phytoplankton which act as huge carbon sinks just like land-based vegetation. Most of this carbon gets recycled as it is eaten, but some that die also sink to the bottom where they are taken out of the system. It is estimated that sperm whales in the Southern Ocean draw 200,000 tonnes of carbon out of the atmosphere each year in this way. To be as pragmatic as humans can possibly be, the service each whale provides in this manner is estimated by the International Monetary Fund (IMF) to be worth $2 million. If whale populations could be brought back to their pre-commercial whaling numbers, then scientists estimate that 1.7 billion tonnes of carbon could be captured. This is more than the carbon dioxide emissions of Brazil, home to another huge carbon sink: the Amazon rainforest (528).

It seems that whether the reason is humaneness or carbon capture, it is vital that we reverse the damage we have inflicted on these giants of the sea. If we don't, then it could be us sinking into the dark abyss.

Dear Indy,

I hope after reading the last few sections that you will understand why we have never taken you to an aquarium or zoo. We even pulled you out of a kindergarten visit to Miyajima Aquarium when we were living in Tenno, Hiroshima. As an educator myself, I see absolutely no useful lesson in taking young human children to these prisons that display children who were stolen from their families and often witnessed their family members being murdered. The only lesson one can take from them is that humans are callous entities and lack empathy for any being that cannot communicate in human tongue.

I have my own shame to confront as I was one of these clapping seal-like zombies on a trip to Sea World, Florida sometime in the late 1980s.

The name of the orca that was forced to entertain me then was Kalina. She was the first surviving baby born in captivity. She was part of a show called Shamu even though the original Shamu had died in 1971. The name was trademarked like a can of cola.

I remember my excitement waiting in that stadium as thousands of people cheered. I had no idea what I was doing as she jumped out of the water and came crashing down with a thunderous plop, sending huge flumes of water into the crowd who cheered louder and louder. The terror and confusion this poor creature must have endured on a daily basis as she was deprived of any natural stimulants or the company of one of her own kind was unfortunately beyond my comprehension. My parents even bought me an inflatable 'Shamu' for me to play with in the hotel pool.

Kalina went on to give birth to Keet in 2004 and the cycle of abuse continued. In March 2016, Sea World finally decided to end its breeding programs meaning that this generation will be the last to suffer in their 'amusement' parks. Whether they survive in the wild as overfishing continues, and the seas warm, is uncertain. At least, we as a family are no longer contributing to their demise through our diets or entertainment. Hopefully, others will begin to make the connection too.

Sharks

While whales and dolphins enjoy mass public support for their conservation, albeit with a few outliers in Norway, Japan, Iceland and the Faroe Islands, the largest fish in our oceans enjoy no such backing. Sharks are destroyed in such huge numbers that it is difficult to know the true number who lose their lives at the hands of humans. The best estimates range from 63 million to 273 million each and every year (529). These incredibly important species are being hunted to extinction in a vain attempt to make our beaches safer for human entertainment purposes, as bycatch in fishing nets, on long lines intended for carnivorous fish like tuna or they are increasingly caught as trophies or for shark fin soup.

In 2021, in total, there were 137 cases of shark attacks on humans. Thirty-nine of these attacks were provoked, meaning humans initiated the interaction in shark habitat, so less than 100 were malicious. Of these 'malicious' attacks, four were attacks on boats, five were doubtfully carried out by sharks, and one was on a person who was already dead (530). While some may contend that this is a large enough number to warrant the mass culling of sharks, let's take a look at how many of these attacks resulted in death. The answer is nine. Therefore, if we were to take the approximate number of sharks killed by humans each year of 150 million and divide that by the number of humans killed by sharks, we arrive at a figure of 16.7 million. For every fatal shark attack on humans, humans exterminate 16.7 million sharks. In 2019, the number of fatalities was just two (531). To put this into context, two people were killed by cows while out for a walk in the U.K. in 2020 (532).

This does not seem to be a fair fight, and maybe we need to reappraise our relationship with these predators from our ancestral home.

Our opinion of sharks has long been dominated by Hollywood. Since Steven Spielberg's Jaws hit the big screen in 1975, sharks have become public enemy number one. The movie clouded the judgement of an entire generation and while its director, Steven Spielberg, seems to show little in the way of remorse, the writer of the book the movie is based on, certainly did. Peter Benchley based his 1974 book on the town of Montauk in Long Island, where the pier was decorated with shark fins, heads, extracted jaws and the full carcasses of unfortunate sharks caught by local fishermen. It was the 1964 killing of a 2,064 kg great white that originally piqued his curiosity and the former white house speech writer turned his attention from one set of predators to another (533).

The impact that the book, and more importantly the movie, had on great white sharks and sharks in general, was to portray them as vengeful animals, lurking just beneath the dark surface, waiting to rip you limb from limb. This encouraged thousands of regular people to take to the water in an attempt to rid humanity of these vindictive beasts. According to the Jaws author, this led to half of large sharks along the east coast of North America being wiped out. Further research estimated an 89% decline in hammerhead numbers, 79% in great whites and 65% of tiger sharks between 1986 and 2000 (534).

Until his death in 2006, Peter Benchley spent the latter half of his life fiercely defending sharks. On a dive in Costa Rica in the early 1980s, Peter's eyes were opened to the true savagery in our oceans. While diving in paradise conditions, Peter came across a section of seafloor littered with the remains of sharks who had had their fins removed before being unceremoniously dumped back into the ocean to suffocate slowly and painfully to death on the sea floor. After returning from this trip, Peter's eyes were opened to the real victims in our unequal relationship. This inspired the writer and bringer of shark death to use his book earnings to finance a shark conservation movement. Speaking shortly before his death, he told the London Daily Express that he *"could never write that book today because sharks don't target human beings and they certainly don't hold grudges (533)."*

Unfortunately for Peter, and more importantly for sharks, humanity didn't really get the memo. Today, it isn't just scared beach goers hunting them from pleasure boats that sharks have to contend with. Today, they are up against an armada of death never before seen on planet Earth.

In 2018, the Hong Kong Shark Foundation reported that over 100 million sharks were being killed each year for the multi-billion-dollar shark fin industry. Hong Kong is at the centre of this barbaric trade with around half of all fins passing through its ports on their way to the soup bowl. A staggering 85% of Hong Kong restaurants continue to offer shark fin on their menus and at just over $6,800 HK ($866) for 604.8g of fin, the industry is extremely profitable. Around seventy-six different species of shark end up in dinner bowls in Hong Kong, but while it may seem easy to point the finger of blame at China, it is worth noting that over 100 countries are complicit in this oceanic savagery (535). China has also witnessed an 80% drop in demand since 2011 after a long conservation campaign (536).

Japan is another Asian country with a large finning industry. In the town of Kesen-numa, on a single day, seventy-five tonnes of blue shark, ten tonnes of salmon shark and three tonnes of short fin mako were on show. This trade occurs six days a week all year round (537). Other Asian countries including Taiwan, Macau, Thailand, Vietnam, and Indonesia also have flourishing fin markets. Fishermen as far away as Wales are now profiting from this ruthless exchange. In May 2019, a video was shared online of around 100 dogfish washed ashore in West Wales with many missing their dorsal fins (538). While this is almost certainly illegal in Welsh waters (any sharks caught are supposed to be landed with their fins intact), these sharks represent just a fraction of those caught and discarded as bycatch.

While it may be easy for some to blame Asia for the sorry state of sharks in our oceans, the truth is that most sharks killed by humans are killed unintentionally by the fishing industry. That is, the largest cause of shark declines is not the Asian shark fin soup customer, but the regular fish consumer. Bycatch is the term used for the sea life removed from the ocean by accident while targeting a specific species. This is the most common cause of shark declines, accounting for 66.9% of shark species reported by the International Union for Conservation of Nature (IUCN) to be facing conservation threats (539). Indeed, sharks are the biggest victim of bycatch. Due to the opaque nature of the fishing industry's bycatch, accurate numbers of sharks caught as by-catch are difficult to find and many areas of the planet carry out little research. What is known is that large industrial fishing operations and smaller artisanal fishers are both complicit. The largest bycatch come from longlines, deep-sea and coastal trawl fisheries.

In the Pacific Ocean, around 3.3 million sharks are caught as bycatch on long lines intended for tuna and other large fish, while in the north-east Atlantic Ocean, bycatch has reduced hammerhead populations by 89% and thresher and white sharks by 80% (540). Around the island of Madagascar, foreign fishing vessels take around 4,300 tonnes of shark each year with a further 3,800 ending up on locally owned boats. Indonesia is a key player in the overfishing of sharks with over 100,000 tonnes being dragged from their oceans every year. Around 15% of this amount is due to bycatch from the tuna industry. Canada and the United States also perform well in the unintentional killing of sharks' league with between 71,000 and 93,000 tonnes of blue shark being caught on long lines intended for swordfish between 2000 and 2006. In the southern hemisphere, the small nation of New Zealand punches above its weight with 2,700 tonnes of shark being caught by the tuna industry. South Africa reports its bycatch by individuals killed and be-tween 1995–2005, between 39,000 and 45,000 sharks were killed as by-catch each and every year (541). The list could go on and on. The fishing industry in each and every nation is contributing to the destruction of sharks through wasteful practices and most bycatch is not declared so the numbers quoted above are certain to be hugely underestimated.

In total, 62% of shark species are facing major conservation threats and due to their late maturity and few offspring, unless fish consumption starts to decrease very soon, they face a very bleak future (539). There are many who would like to see the end of sharks, but such undeveloped thinking could have disastrous consequences.

As apex predators, sharks are crucial in the maintenance of oceanic ecosystems. As an example, in areas without the presence of sharks, sea turtles tend to overgraze seagrass leading to the destruction of seagrass meadows, but when sharks are present, the sea turtles spend less time in the same area and the seagrass, which is an important habitat for other fish, shellfish and birds, remains. This seagrass also acts as a huge carbon sink so sharks, like whales, are critical in our fight against climate change. Sharks also play a vital role in coral reef systems. By removing sharks from coral reefs, smaller predators like grouper fish that prey on herbivorous fish flourish and the herbivores are wiped out. This leads to the spread of algae which can quickly overgrow a coral reef. This in turn leads to a loss of biodiversity in the reef system (542).

To make matters worse, in our battle against viruses, including COVID-19, squalene, a natural oil from shark livers, is used to produce some vaccines. Scientists say the oil helps to create a stronger immune response and half a million sharks may need to be murdered in order to save human lives (543). It's worth mentioning that these are the same human lives at danger in the first place from our addiction to destroying nature and abusing non-human animals. Humanity seems so short sighted that we are about to exacerbate the problem initially caused by destroying nature by destroying more of nature.

The future of humans and sharks are intertwined, but our indifference to their suffering could have severe consequences for the survival of both species. Unless drastic action is taken soon, it might be too late to reverse the fate of both.

Ocean Water

Imagine exploding a Hiroshima-sized atomic bomb in our oceans every second of every day for 150 years. Metaphorically, that is exactly what we have done. Our oceans have absorbed over 90% of the heat that we have trapped in our atmosphere due to our burning of fossil fuels. They are also responsible for soaking up a quarter of our carbon dioxide emissions. As if this wasn't enough, we are ramping up emissions and heating has now increased to the equivalent of between three and six atomic bombs per second. Let that sink in for a while. Between three and six atomic bombs going off every second since 1990 (544). In the time it has taken you to read this paragraph, we have added the equivalent energy of between 60–120 atomic bombs to our oceans.

Our oceans are our lifeblood. We are reliant on them for the air we breathe, our water cycles, the food many of us eat, many medicines originate here, and they are home to more than 300,000 species of animal, some of which we are yet to even discover.

The ocean is also our ancestral home. Life first crawled out of the waves around 530 million years ago and today many of us enjoy diving back in for the calm soothing quality we find. There is a term for what happens when we dive into water, the mammalian dive reflex, and it alerts our vital organs immediately by sending blood rushing to them and our body begins to conserve oxygen (545). It happens unconsciously. It's as if we are reconnecting with our history. Whether it's through surfing, diving, swimming, or just frolicking in the waves on a summer's day, the ocean has always been there, waiting to accept us back, often with a not so warm embrace.

It is because of the ocean's constant acceptance that our reciprocal destruction of the oceans is so difficult to understand. Today, we are witnessing the complete annihilation of our ancient home and in the coming decades, there may be nothing left to keep us company when we dive back in. Now, some may think that this is hyperbole, but we've already seen that, due to overfishing, some scientists are predicting fishless oceans by 2050. Now, when you add into this mix the fact that our oceans are sucking 90% of our added planetary heat beneath their sanguine waves, it is easier to understand what led the U.N. Secretary General António Guterres to say "Sadly, we have taken the ocean for granted and today we face what I would call an ocean emergency. We must turn the tide." He was speaking at the U.N. Ocean Conference in Portugal in June 2022. He added that "We cannot have a healthy planet without a healthy ocean." Sadly, no hyperbole is at play here.

In 2014, scientists identified a huge swathe of the North American Pacific Ocean where the temperatures were more than 2.8°C (5°F) above the norm. This area of unusually high temperatures was called *"the Blob"*. The warm conditions continued unabated for two years. This, unfortunately, was not an anomaly. There have been tens of thousands of these events over the past forty years with many remaining small. The larger and longer lasting oceanic heat waves are now becoming much more likely (546).

Marine heatwaves are defined as areas of ocean that are significantly warmer than the average thirty year baseline temperature for at least five days. Scientists are sure that these heatwaves would not be occurring were it not for human caused climate change. The scientific community even took the unusual step of carrying out a formal attribution study which is used to prove the link between the cause and effect of our burning of fossil fuels and the heat waves. The research published in Science Brief found that 84-90% of all global marine heatwaves occurring between 1986–2005 were attributed to human-caused climate change. Scientists believe that if we allow temperatures to rise by 3°C (5.4°F) then parts of our oceans will never escape these heat waves. A study in Nature Journals found that between 1925 and 2016, marine heatwaves were 34% more likely and lasted, on average, seventeen days longer. They found that the average year in the 1980s had about twenty-five days of heatwaves while the average in the 2010s was fifty-five days. They went even further by stating that whether humanity changes course or not, the oceans will be in a near-permanent heat-wave state (546).

Now, what does this mean for us, some will ask. As we know on land, the heat affects everything from plants to the largest herbivores. As temperatures rise, plants begin to move away from the equator towards the poles and if they want to survive, the animals move with them. This is, in effect what is happening in our oceans, but unlike on land, the problem with ocean heating is not the amount it heats, but the speed with which it rises. Many organisms just cannot cope with the speed of change, and this leads to reduced growth rates, increased chance of disease and higher mortality. Many sources of food simply cease to exist which leads to further problems up the food chain.

The mass of warm water stretching from Mexico to Alaska and known as "the blob" caused sea nettle jellyfish to all but disappear and young salmon to starve to death. It was also blamed for killing half of foraging fish populations, starving seabirds, and initiating a collapse in cod. Tuna moved as far north as Alaska and the whales followed into territory where crab fishing lines and shipping were dangerous (546). A heatwave in 2015 in the Tasman Sea killed abalone and oysters in an eight-month period while an earlier record-breaking heatwave off the east coast of Canada and the U.S. in 2012 pushed lobsters northward. In 2011, a heat wave off the western coast of Australia uprooted sea, fish, and sharks. The blue swimmer crab industry in Australia's Shark Bay was brought to a halt for eighteen months to allow the crabs to stabilize. Stretching back to 2003, a heat wave in the Mediterranean Sea desolated marine life. The waters in a heat wave warm so quickly that tropical fish were seen in water off Tasmania in what is normally subpolar (547).

The heatwaves are continuing to get hotter, larger, and longer and in August 2020, there was a 9.8 million square kilometer stretch of Pacific Ocean off the coast of California that was 4°C (7.2°F) warmer than usual. This is already above the level of warming that scientists warn could irreversibly heat the ocean for good and it was the size of Canada. Nep2ob, as it is known, is the largest known oceanic heatwave recorded since satellite records began in the early 1980s. In fact, 2019 was the hottest year ever recorded in the oceans with average temperatures now 1°C warmer that when records began.

According to the Guardian, The King Salmon Company in New Zealand was closing 3/4 of its operation in May 2022 as warm waters led to the deaths of 42% of its fish, compared to 17% in 2018. The operators tried towing the pens out to cooler waters but mortality rates were still considerably higher at 37%, compared to just 10% in 2018. The company's chief executive Grant Rosewarne said that he had never heard of the term marine heatwave when he joined the company, but recently there had been three of them.

Particularly worrying to scientists are the regions with high levels of biodiversity. The coral reefs of the Caribbean, seagrass in Australia and kelp forests off the Californian coast are hit hard by marine warming. When these areas face biodiversity die offs, it is likely they will have a cascading effect. One impact of the warming is that fish populations decline. Global fish catch is down 4% and in some places as much as 30% (548). One such place that is impacted by rising sea temperatures is the Caribbean where per capita, people consume thirty-five kg of fish annually. According to the Food & Agriculture Organisation (FAO), the Caribbean fisheries sector is the most vulnerable in the world and with 44.42 million people living in the area and mostly reliant on food from the sea, it is easy to see why (549).

India is also vulnerable to the warming oceans that surround the sub-continent. Since 2017, there has been a 54% decline in Indian oil sardine which traditionally accounted for 30-40% of fish caught in the region. It now accounts for around 5%. While there are several reasons for this decline in sardine population, including overfishing, bottom trawling and illegal fishing, ocean warming is likely a strong contributing factor. As the oceans warm, the phytoplankton that sardines feed on is reduced, leaving them little alternative but to migrate or starve (550).

The amount of heat being absorbed by our oceans is so fantastical that even the mammoth size and depth of them is no obstacle to the ocean heating. In their complete enormity, they have a surface area of 361.1 million km², but it is their depth that makes them so elusive to humans. Their average depth is 3,682 meters but at their deepest, in the Mariana Trench near Guam, they reach almost 11,000 meters and in total our oceans contain 1,337,986,366 cubic kilometres of water. Unfortunately, this vast size is not enough to resist the onslaught of human caused oceanic heating with water between 1,360m and 4,757m being warmed by 0.02-0.04°C in the decade up to 2019 (551).

This heating is negatively impacting the marine food chain. The source of all foods in the ocean is plankton, and plankton in cooler waters are usually bigger, richer in fats and contain more calories than plankton in warmer waters. Essentially, the warming of the oceans, as we saw on land, will cause plankton to deteriorate in quality. This will affect the entire food chain. These warming waters can also impact humans living thousands of kilometers away. Heatwaves can cause toxic algae blooms like "red tide" which kills fish and makes eating shellfish dangerous to us. They also make the surrounding air dangerous to breathe (552).

A further way in which warming oceans impact humans is through more frequent super charged storms. Whilst rising atmospheric temperatures make the atmosphere less stable, the impact in the oceans seems to be inverse. With warmer waters, scientists have found that less cold oxygen and nutrient dense water from the ocean depths moves upwards and mixes with the warmer water on the surface. Additionally, surface water is less able to absorb carbon dioxide that then mixes with cold water and drops to the bottom, in turn dissipating heat. This is known as stratification and globally there was a 5.3% decline between 1960 and 2018. The Pennsylvania State University climate science professor Michael Mann, who co-authored the research, stated these warming oceans might help to drive more *intense, destructive hurricanes* (553).

As if ocean heatwaves, man-made plastics, and sea-life die-offs weren't enough for our oceans to deal with, there is an additional threat facing them right now - acidification. As we have seen, the oceans are absorbing the majority of excess heat being produced by humans, but they also absorb around 30% of the carbon dioxide we release into our atmosphere. The carbon dioxide enters the oceans and is absorbed by the seawater, which is usually alkaline, and creates carbonic acid. This acid isn't particularly strong, but it has lowered the pH of the upper ocean layer from 8.2 to 8.1. Pure water has a pH of 7 and is considered neutral. This doesn't sound alarming because of the logarithmic scale used, but it actually accounts for an almost 30% increase in acidity, a figure conspicuously close to the amount of CO_2 being absorbed by our oceans (554). Rather more ominously is that reef development is believed to stop at pH 7.8 (555).

This increase in acidity is already wreaking havoc and if left to continue it will help push our oceanic systems to the point of complete collapse. Acidification is of special concern for any form of sea-life that uses calcium carbonate to form shells or exoskeletons, such as oysters, clams, urchins, and crabs. More acidic water holds less calcium carbonate so there is less available for these organisms to calcify and the ones that are formed already are in danger of dissolving. Scientists carried out research on the Great Barrier Reef to understand the impact that estimated acidification levels there would have by 2100. They found that calcification dropped by a third (554).

It is hardly necessary for scientists to carry out such research as the effects of acidification can already be seen on our reef systems around the world. Coral reefs are the largest carbonate producer (73%) and account for a quarter of calcium carbonate buried in marine sediments (556). Coral ecosystems are essential to our survival in many ways, from providing food for millions of humans to protecting coastlines from storms and erosion. They are also the spawning grounds and nurseries for many species of fish that support countless fishermen. From an even more pragmatic stance, they provide novel medicines and provide jobs and income in many coastal regions. From a less humanistic perspective, they are also hotbeds of biodiversity with more than 800 types of hard coral species and 4,000 species of fish. Although coral reefs make up less than 2% of the seafloor, they sustain an estimated 25% of marine life (557).

The largest and most famous reef of all is the Great Barrier Reef that stretches down along the east coast of Australia. It is the only living thing visible from space and at 2,300 km in length and covering 344,000 sq. km, it is understandable why. The reef formed 7,000 years ago, and since then, it has offered its bounty to aboriginal Australian populations. Unfortunately, the reef is in a crisis of our making. Just in the past twenty-five years, its corals have more than halved and scientists are warning that without massive cuts in greenhouse gas emissions, she will soon be unrecognizable (558). Whilst much of this destruction is being caused by ocean warming, as we have seen, it is much harder for corals to calcify when there is less calcium carbonate in the ocean due to increasing acidification.

The Great Barrier Reef is one of many that is on life support with around a third of the Caribbean's sixty-five reef-building species facing a disease outbreak also believed to be caused by ocean warming. In the same process, known as bleaching, the staggeringly colourful corals are turning white almost overnight and dying completely within months (557). These bleaching events had been occurring once every twenty-five to thirty years. While they can recover after a bleaching event, it takes around ten years. These events now happen every five to six years and there is simply insufficient time for the reefs to reform, even if ocean temperatures temporarily drop significantly enough. Along the Florida coast, only 2% of original coral cover remains and scientists are not hopeful for the survival of the remainder (559). By mid-century, around 90% of the oceans' coral reefs are expected to suffer from bleaching and according to Mark Eakin, coordinator of the National Oceanic and Atmospheric Administration's Coral Reef Watch program, "It is clear already that we're going to lose most of the world's coral reefs" (560). Additional research from the American Geophysical Union collaborates these research findings by projecting that 70-90% of coral reefs will disappear within the next twenty years and by the end of the century, most current reef sites will be uninhabitable. They identified both ocean warming and acidification as the main threats to the survival of our reef systems (561).

To heap even more pressure on our oceans, human behaviour is leading to deoxygenation. Whilst Earth's magnificent forests are rightly renowned as the origin of much of the oxygen we breathe, another natural oxygen factory gets slightly less attention. At least half, but possibly 80%, of our O_2 is produced by photosynthesis in our oceans (562). While, many are aghast at the destruction of our forests, especially the Amazon, fewer seem to concern themselves with the absolute devastation being waged in our summer playground.

Just as trees absorb the Sun's light energy and then use it to convert water, minerals and CO_2 into oxygen, oceanic organisms carry out the same process of photosynthesis beneath the waves. Some photosynthesis is carried out by kelp and other sea grasses, but the majority of this process is undertaken by microscopic single-celled organisms called phytoplankton. These tiny creatures live at the surface of our seas, lakes, and rivers and as they absorb the Sun's heat, they remove carbon dioxide from the ocean and release oxygen. Forget dogs, phytoplankton is our true best friend. They not only allow us to take nice deep breaths, but they act as huge carbon sinks as their bodies absorb the carbon in the ocean. They are then eaten by zooplankton who in turn are eaten by larger animals and in this way, phytoplankton are not only to be celebrated for allowing us to breathe, but they are also responsible for the food that many poorer humans rely on for their survival (563).

Unfortunately, just as we are destroying the forests that allow us to live, we are doing our best to destroy our life-giving oceans as well. Phytoplankton are diminishing around the world, largely due to our warming oceans, mass acidification and to the complete decimation of sea life in order for humans to enjoy the momentary taste of fish flesh on their tongues. Just as trees need to be fertilized to carry out photosynthesis, so do phytoplankton. The main source of this fertilization is whales. One blue whale can produce around 200 liters of nitrogen and iron-rich feces in one go, and as we have seen, whale populations are in a perilous state (564).

To put things into context, an area the size of the European Union has become starved of oxygen since the early 1960s. Some areas have lost 40% of their oxygen whilst the number of areas with zero oxygen have quadrupled. Whereas the overall oxygen content of the oceans has dropped by just 2%, most of this drop has occurred in parts of the ocean with the most abundant wildlife. At current rates, the decline in oxygen will triple by 2100 (565). Due to intensive animal farming on land, we are creating ocean dead zones all around the world and due to the damage we are inflicting on marine life, we are further depleting our oxygen stores that were first formed 2.5 billion years ago at the end of the Archaean Era. Add to this the absolute decimation of land-based forests to graze cows and grow soybeans for farmed animals and it is becoming clearer than our diets must change if we are to continue with the 200,000-year-old human experiment.

Mangroves

Mangroves are one of the lesser heralded ecosystems on planet Earth. Along with forests and peatlands, they have been cooling our planet for thousands of years. They cover around 140,000 sq. km and provide services worth approximately $2.7 trillion a year. Buried in their soil is around 6.4 billion tonnes of carbon. While this may seem small, pound for pound, they sequester four times more carbon that rainforests. They also protect against flooding, prevent erosion and are breeding grounds for many fish species, as well as home to lizards, snakes, and nesting birds. They also help to filter pollutants from the water. They lead the way in multitasking, and they work for free. Unfortunately, as with most ecosystems on Earth, they are under attack from human interference. Between 2000 and 2015, around 122 million tonnes of carbon was released from mangroves, with 75% of it arising in Indonesia, Malaysia and Myanmar. This is equal, roughly, to the total emissions of the 11th largest emitter, Brazil. In the two decades leading up to 2000, it is estimated that we lost 35% of the planet's hard-working mangroves. The building of dams upstream can impact the delivery of vital fresh water; but aquaculture, especially shrimp farming, is the largest cause of their demise. Southeast Asia is home to 85% of these shrimp farms. The author of a 2018 study, published in Environmental Research Letters says that while protecting mangroves will not solve the climate crisis alone, *"for many nations, including most small island nations, mangrove protection and restoration represent one of the most viable climate mitigation options (566)."*

It is for this reason that eating a kilo of shrimp is estimated to be four times worse for the planet than eating a kilo of beef. While cows are belching methane and they cause huge rainforest deforestation for grazing and growing feed; the clearing of these concentrated mangrove carbon sinks is far worse. Eating a surf-and-turf dinner of prawn cocktail and steak can cause as much emissions as driving across the U.S. in a gasoline powered car. For fans of the tasty crustaceans, the alternative is wild shrimp, but their numbers are dwindling due to overfishing and for every kg of shrimp caught in the wild, around twenty kg of by catch is taken. The industry is hugely damaging to the environment, aquatic populations and indeed to humans too. As was highlighted in the Netflix documentary, Seaspiracy, child labour and human trafficking are in practice throughout Southeast Asia where 74% of wild shrimping takes place. Of those trafficked from Cambodia, a staggering 59% had seen fellow crew members murdered by the captain, according to interviews conducted by the U.N. It is easy to see why this hugely harmful trade is attractive to the most unscrupulous as global shrimp sales amounted to $45 billion in 2018 and are growing by 5% a year (567).

The shrimp industry has an abysmal environmental record, there is a flagrant disregard for human rights, and its treatment of sentient beings is also slightly debased. Female shrimps in the wild will only breed when conditions are favourable, and on crowded and dirty shrimp farms, those conditions are rarely met. Farmers have found that by removing the hormonal gland, they can overcome this problem. The hormonal gland is behind one of the female's eyes; so farmers simply slice off the eye of the females and this forces them into rapid sexual maturity. Scientific research has found that this practice of eyestalk ablation is painful for shrimp, and they are given no pain relief before, during or after the procedure. In studies, those given pain relief rub their eyes less than those not given relief. The process is banned for organic shrimp in Europe but allowed for non-organic; and it is practiced everywhere around the world to maximise profits (568). While peeling the shell off a shrimp and throwing it on the BBQ may seem like an inconsequential act, it in fact has an outsized impact that is devastating to our environment, sea life, climate change and human rights.

This chapter has helped to illustrate how interconnected the web of life is. Humans have attempted to distance themselves from the rest of Earth's creations and with terrible effect. When Paul Watson stated that *"if the oceans die, we die,"* he wasn't being hyperbolic, he was giving a measured analysis of our reality. If we are to continue living on this planet, then we will need to largely leave the oceans alone to do what they do best: nurture the life on which we depend.

Dear Indy,

I became aware of the damage being inflicted on our coral reefs just before you were born. I read a BBC article that said around half the planets' reefs would experience bleaching events by 2030 and by 2050, that would rise to 95%.

I remember that gut feeling that you may never get to experience the same wonder that I have when scuba diving or snorkeling. The oceans are incredible, and you, like your mother and I, really enjoy our time at the beach. You are eleven now and starting to enjoy looking under the surface after finding it too scary for a while.

I have been trying to push you as much as possible without putting you off for the very reason that these ecosystems that have been here for so long are starting to disappear. Already, when we visit Southeast Asia, there is plastic everywhere in the seas. This has not always been the case. When I first arrived in Asia in 2000, I don't remember seeing any plastic in the oceans, although I'm sure there was some.

I took my open water diving course in the Philippines in 2001 and loved every minute of my time under the waves, swimming with turtles and sharks. The solitude and sheer beauty keep you going back for more.

To think that we might have no reefs left by the time you are my age is unbelievably sad. And it is all so preventable. We know what we need to do. We need to get serious with plastic; and force manufacturers to use biodegradable packaging or materials that are actually recyclable.

Next, we need to give the oceans time to heal. We can do this by paying fishermen to collect all the fishing gear that they have dumped in our oceans. People worry about the economic impacts of stopping fishing, but we are going to need as many people as possible to get out there and collect the trash we have discarded. Most governments subsidize the industry more than they contribute to GDP so we can continue to pay the subsidies for the industry to clean up their mess. We can also completely ban the killing of all sea mammals and bring an end to industrial fishing.

Finally, we need to move immediately to renewable sources of energy to reduce temperatures and turn the tide of acidification. If we can do these things, scientists say the oceans can rebound fairly quickly. The race is on!

Fresh Water

While our forests and oceans receive deserved attention as bio-diversity hotspots, our inland water bodies, such as rivers, streams, ponds, wetlands, and lakes, are a little overlooked. Globally, around 300 million natural lakes add up to about 4.2 million sq. km of Earth's surface with a further 335,000 sq. km being man-made. In addition, our global network of interconnected waterways amounts to a further 500,000 sq. km and between 12.8-15.8 million sq. km of our Earth is made up of wetlands. Combined, these ecosystems cover less than 1% of Earth's surface but are home to 10% of all known animals and a third of vertebrates (569). They are natural habitat for seventy species of freshwater-adapted mammals, 5,700 dragonflies, 250 kinds of turtles, 700 birds, 17,800 fish species and 1,600 types of crab. These ecosystems also provide services to billions of people, many of whom are living a hand to mouth existence (570).

Unfortunately, these ecosystems are under continuous attack from a variety of human activities which we will look at in this chapter. Human societies have grown up along rivers for the obvious benefits that access to freshwater brings. Many of our largest cities are built on and around river ecosystems: from Paris to London, and Ho Chi Minh to Beijing. At the basic level, since we've lived alongside rivers, humans have been impacting water quality from our simple day to day needs. While in most wealthy countries today, our waste is treated before entering the waterways, around the world, and historically in rich countries, this is and was, simply not the case. Globally, around 80% of our wastewater is dumped into our rivers without being treated (569).

With the human population rising by hundreds of thousands each and every day, and most of this growth coming in poorer countries lacking adequate sewage systems, the severity of the threat from untreated human waste on river and lake biodiversity is fairly salient. The impact on human health is equally pernicious. This is not to lay blame on developing nations because the vast majority of damage being done to our freshwater ecosystems stems from the industrialised world.

Since the industrial revolution, our inland waterways have increasingly suffered. During its coal mining heyday, rivers in Wales infamously ran black with so much coal dust, sewage, and industrial waste, that it was impossible for fish to survive. In the mid to late 20th century, as industry was shifted to areas with cheaper labour, the pollution followed closely behind. Wales was able to clean its rivers up and the fish eventually came back, but the problem simply shifted from Wales and other coal producing nations of the nineteenth and twentieth centuries to China, Australia and elsewhere.

Now, of course, it isn't only coal production that reduces water quality and biodiversity levels. Industry in general impacts life in our blue arteries. Today, in order to quench our energy demands, two-thirds of the world's longest rivers no longer flow freely to the sea, and more recently, the huge global demand for animal products such as meat, dairy and eggs is also helping to slowly kill life in and around our inland water ecosystems. Our wetlands, which hold twice as much carbon as all the world's forests, are also under attack. Add to this chemical cocktail that now constitute our rivers, the pesticides, herbicides and synthetic fertilizers from industrial farming and our freshwater ecosystems are in extremely poor condition.

Untreated Sewage

As we have seen, freshwater habitats around the world face a number of threats that include: pollution, habitat degradation, and a fast-rising human population. While the majority of damage being done to freshwater ecosystems is occurring through industrialisation in one form or another, the rising human population cannot escape scrutiny. As we have mentioned, the vast majority of humans being added to the planet are entering existence in developing nations. If the opposite were the case and most new babies were being born into wealth, then we would truly be facing a catastrophe because human population multiplied by the expendable income enjoyed in many wastefully affluent societies is the real creator of nightmares. As it stands, if all 7.96 billion humans were to live the life of the poorest on our planet, then our current population could live sustainably. Unfortunately, many of us in the richer global north are now so entrenched in our consumerist lifestyles that we will not accept any drop in our standard of living. This leads many rich consumers to lay blame at the feet of the poorest and most vulnerable in our global society. In this chapter, as well as a following section, evidence will be provided to support the claim that the real problem is not simply with too many people, but more to do with the consumption of the richest.

Let us begin in one of the countries that will be most affected by the climate crisis: Bangladesh. In a country considerably smaller in size than the United Kingdom, Bangladesh has a population of 171,047,328. More than ten million Bangladeshis are considered to live in extreme poverty - less that $1.90 a day. For these people, flushing toilets connected to sewage systems are not on the radar. Raw human waste is usually emptied into canals connected to larger streams and rivers. The impact this has on freshwater ecosystems downstream is significant. In an international research project conducted by universities in Bangladesh, the U.K. and Australia, scientists analysed biodiversity by using a before-after-control impact methodology. The project began in 2006 before untreated wastewater was released into the Barnoi River. After eight years of untreated wastewater discharge, the abundance of fish downstream, compared with those further upstream from the discharge, dropped by 47% and species biodiversity dropped by 35%. When compared with biodiversity levels prior to the project, fish abundance had dropped by 51% and species biodiversity had dropped to 41%. The pH level, dissolved oxygen and water transparency were all lower in the areas downstream from the discharge. More than 41.3% of the species recorded were threatened with extinction in Bangladesh; and all but one had suffered significant declines (571). These results will be replicated across the globe as the bacteria in human waste use up the oxygen in the water to decompose the waste. This causes a lack of oxygen and river species die. These bacteria, such as E. coli can also infect the water which in turn causes disease that can spread through the human population. With one of the biggest threats to people in developing nations being access to clean drinking water, this is of specific concern.

The impacts of human waste in affluent countries creates a slightly different problem; but nevertheless a problem that beautifully illustrates the difference between the haves and the have nots. In 2016, in Seattle's Puget Sound, when sewage-treatment wastewater was analysed, it was found to contain a concoction that included Prozac, caffeine, cholesterol medicine, OxyContin, ibuprofen, bug-spray and even cocaine. The levels were high enough that these drugs were present in the tissue of juvenile chinook salmon. In total, eighty-one drugs and personal-care products were found in water and tissue samples. Even areas further afield from the discharge points, which were considered pristine, showed elevated levels of many chemicals. While the impact on humans has not been studied, it is believed to be low. The impact on fish, however, is noteworthy. Juvenile Chinook Salmon migrating through contaminated estuaries in Puget Sound die at twice the rate of fish elsewhere. It is believed that the drugs impact fish growth, behaviour, reproduction, immune function, and antibiotic resistance (572). A similar problem has arisen in the Florida panhandle where populations of bonefish have dropped by 50% in four decades. Researchers at Florida International University (FIU) found that of ninety-three bonefish sampled, all of them tested positive for at least one pharmaceutical: including heart medications, opioids, antifungals and antidepressants. A single bone fish tested positive for seventeen different drugs; eight of which were antidepressants that were as much as 300 times above the therapeutic level for humans. Even the food bonefish feed on: like crabs, shrimp and small fish were found to contain on average - eleven contaminants. While researchers are trying to identify the impacts fully, they believe the pharmaceuticals affect the bonefish's ability to reproduce, possibly by making them more antisocial. Scientists say the problem is global and are urging improvements to wastewater infrastructure (573).

In England, only 14% of English rivers were considered of good ecological standard in 2020 and 36% of damaged waterways were due to water companies discharging sewage wastewater. In 2019 alone, water companies discharged raw sewage 200,000 times (574). This had risen to 400,000 times in 2020 and the worsening climate crisis is expected to exacerbate the problem as water flow in the sewage system is projected to rise by 55% by 2050 (575)

We can see, that while the problem of untreated human wastewater in the developing world is leading to biodiversity loss; the affluent nations' consumption habits and incompetent corporate behaviour are leading to marine losses even after wastewater has been treated. Area by area, the problems are different, but the problem is global in scale, and caused by humans.

Pollutants and Contaminants

In her seminal 1962 book Silent Spring, Rachel Carson warned us of a world with chemical soups that constitute streams, rivers filled with field runoff containing pesticides and chemical fertilizers. She wrote at length of the affect these chemicals would have on plant life, fish, and animals.

Her prophetic message was brought closer to home in 2019 when scientists identified a nosedive in insect and plankton numbers in a Japanese lake after neonicotinoid pesticides were introduced to rice paddies there. Soon afterwards, scientists monitoring Lake Shinji identified the sudden collapse of smelt and eel populations which relied on the plankton and insects for food. Until 1982, the midge Chironomus plumosus and the zooplankton species Sinocalanus tenellus were abundant in the lake, but after Neonicotinoids were introduced to the area in 1993, the later decreased by 83% and the former had vanished completely from all thirty-nine locations sampled in 2016. This in turn led to the annual smelt fish catch dropping by 90% and eels by 74% in the ten years after neonicotinoids' introduction. Scientists tested several alternative causes for the sudden die offs but with the icefish catch remaining unchanged due to feeding on different species, they have ruled them out. With the neonicotinoid industry being worth $3 billion a year and neonicotinoids being the most widely used insecticide globally, perhaps further studies should be carried out as a priority, but according to Professor Olaf Jensen at Rutgers University, *"There is the issue of not seeing a problem if we don't look for it"* (576).

Another notable contributor to the destruction of our freshwater wildlife is the automobile industry, particularly the producers of tires. Tires use chemicals to prevent the rubber from breaking down, but when it rains, run-off carries fragments of old car tires into streams and rivers where they are affecting Coho salmon on the U.S. west coast. The salmon, which can reach 60cm in length are born in freshwater streams before they embark on a blockbuster expedition out to sea before returning in later life to complete the cycle. When they return from their incredible journey to reach the pinnacle of their life, they are dying in rivers before they reach their life's goal of spawning. The chemical that is causing these deaths is called 6PPD and it is leading to acute cardio-respiratory problems with fish being unable to swim upright (577). We will see later that these leading tire companies conspired to get us out of public transport and into private cars so they could profit. Their conspiracy comes with a huge cost attached.

In Florida, the state's iconic manatee is struggling to survive. Manatees are protected by the oldest wildlife protection laws in North America. Unfortunately, this protection is not sufficient to prevent close to two and a half million pounds of nitrogen, and phosphorus flowing into the manatees' waterways every year. This run-off comes from agricultural chemicals, garden fertilizers and leaky septic tanks, and causes algae blooms to form which in turn kill sea grass which is the manatees' food source. In 2021, 1,001 manatees died, many from starvation. The future of this wonderful species depends on our willingness to prevent further water pollution (578) (579).

While the pollution caused by insecticides, pesticides and tire chemicals is hard to see, the pollution from the fashion industry, as ever, is right in your face. Streams of deep magenta gush into waterways in the developing world as the fashion industry not only exploits cheap labour, but also poisons the surrounding environment in its quest to push down prices so low than exploited workers in developed countries can afford to buy disposable items for a night on the town. The fashion industry in China, as an example, produces around 9.4 billion litres of contaminated wastewater each year. Fast fashion in Asia is largely unregulated and waterways throughout China, India, Indonesia, and

Bangladesh are being devastated by a toxic cocktail of chemicals from the denim and leather industries. These waterways are the lifeblood of communities reliant on them for drinking and bathing, but they are completely lifeless due to these chemicals which are causing cancers and gastric and skin issues for those living nearby. The rivers are so obviously polluted, there is a joke in China that they can see which colour is the new 'it' by looking at the rivers. Leather tanneries dump chromium into the rivers, and this is evident in cow's milk and agricultural products. According to Sunita Narain, director general of Centre for Science and the Environment in India:

"We are committing hydrocide by deliberately murdering our rivers (580)."

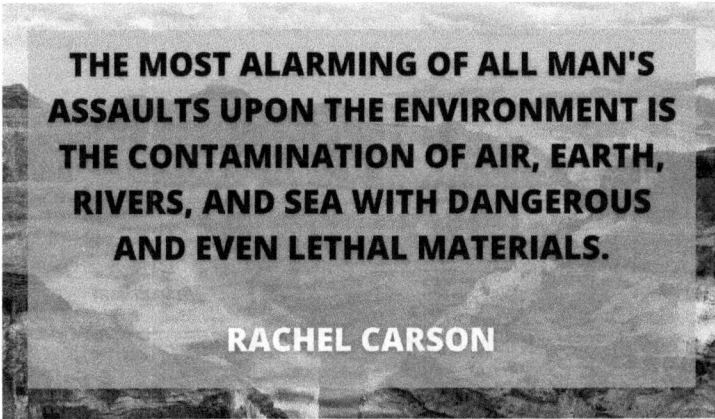

THE MOST ALARMING OF ALL MAN'S ASSAULTS UPON THE ENVIRONMENT IS THE CONTAMINATION OF AIR, EARTH, RIVERS, AND SEA WITH DANGEROUS AND EVEN LETHAL MATERIALS.

RACHEL CARSON

Peatlands

Like mangroves, peatlands are given little credit for the priceless services they provide. They cover around 3% of the Earth's surface and are present in at least 175 countries. Stretching across four million sq. km, they provide clean water, regulate climate, and store an incredible 30% of global carbon. They store twice the carbon than all the world's forests combined. Historically, peatlands have been seen as barren wastelands and humans have taken it upon themselves to drain them of water and then "improve" them by making them more productive for human use. They seem barren because the large amount of water in their soil excludes oxygen which makes it difficult for many useful (to humans) plants to live. This though, is the reason they store so much carbon. As the soil is deprived of oxygen, the breakdown of microbes is slowed down dramatically and while this decomposition without oxygen makes peatlands a methane producer, the amount they produce is less than the carbon stored through plant photosynthesis. This makes them a significant carbon sink. Around 15% of global peatlands are now degraded, with 80% degraded in the U.K. While most mangrove destruction is occurring in a few countries, peatlands are being destroyed from the tropics to the poles and they are now emitting around two GtC, or 6% of global greenhouse gas emissions every year (581) (582). In the tropics, especially Malaysia and Indonesia, the main driver of peatland destruction is palm oil production. Here, the peatlands are drained in order to grow palm oil for customers in countries around the world. As the peatlands are drained of water, greenhouse gasses equal to around seventy coal plants, or the annual emissions of Vietnam, are released into the atmosphere (583).

In the northern hemisphere, drainage for agriculture is largely to blame for peatland destruction, but another more insidious practice needs to be called out. This is the act of burning peatlands in order to increase the number of grouse that British, and foreign elites, can then shoot for sport. Across the U.K., peatlands are intentionally set alight during winter in order to provide more nutritious shoots of heather for grouse in spring. In 2018, more than 660 fires were started on grouse moorlands in just the county of Yorkshire. As the U.K. hosted the COP26 climate conference in 2021, the fact that the British prime minister is himself an avid grouse shooter, made a mockery of their aim to achieve net-zero emissions by 2050. In 2018, the government banned the burning of peatland that is more than 40cm deep, but it is the shallower peatland that needs to be restored so it can absorb carbon more efficiently (584). In 2021, the U.K. took a step towards protecting its peatlands by announcing a ban on selling peatmoss to gardeners from 2024 (585). Perhaps, they could prove their seriousness by also announcing a complete ban on the burning of peatland for grouse shooting. This may never come as shooting grouse is a favourite pastime of many of those privileged enough to be making laws in the House of Lords.

Peatlands in high latitudes act as a negative feedback loop in that the warmer the planet gets, the more carbon they will sequester. Add to this, the fact that the restoration of negatively affected peatlands promises to provide sequestration equal to that of all agricultural soils, and it becomes essential that whatever the cause of peatland destruction, it must stop immediately if we are serious about halting warming at 1.5°C (2.7°F) (581).

Dams and other Alterations

In response to rising fossil fuel emissions, it is satisfying to believe we can just change the way we produce energy and then we will be fine. It's easy to see why many people jump on this bandwagon, rather than accept that our current lifestyles are simply not sustainable with a population of many billions. As governments pass legislation banning the sale of gasoline powered cars, many quickly turn their attention to electric cars and think these are the answer. Unfortunately, they really aren't. The batteries these cars will run on will need to be mined and manufactured and fossil fuels will be used in the production. There is also the unanswered problem of battery disposal. The more logically sound answer would be to connect people and places with fast affordable public transportation; but of course this is not what the stock markets want to hear. The reason this is discussed here is that for many years we have believed that damming rivers to harvest their energy to be an amazingly green solution to providing low emissions energy. Again, as with most technological solutions to human caused problems, they simply create new problems downstream, in this case, both figuratively and literally.

The first dam is believed to have been made by ancient Egyptians between 2750-2950 B.C, but it failed within a few years. Since then, humans have made bigger and stronger dams all around the world. Today, there are around 58,000 large dams, higher than fifteen meters, with a storage capacity of around 7,000-8,300 sq. km. A further 3,700 are being planned or currently under construction. No reliable figures exist for smaller dams but an estimate of around sixteen million has been suggested. These dams combined have increased the amount of freshwater available to humans by over 7%. They also supply around 70% of the planet's renewable electricity in addition to supporting irrigation services (586). With a rising population meaning food production will need to rise in tandem with energy; more and more dams will be needed if we are to continue with our current lifestyles. Already, we have dammed about two thirds of the world's large rivers; and the impacts they have had on humans and wildlife are far reaching.

The most troubling aspect of damming rivers is the loss of life that occurs as a result. They are often sold as a green alternative to fossil fuels and whilst they are renewable sources of energy, they are not green according to aquatic ecologist Herman Wanningen. The creative director of the World Fish Migration Foundation describes the volume of water behind the dam as a stagnant reservoir with no natural habitat and no fish. As the river is no longer free flowing, the water is also warmer which exacerbates the problem. Additionally, dams stand in the way of migratory fish like salmon who can no longer reach their spawning grounds. The Columbia River in the United States saw its salmon population completely crash after it was dammed. It had been home to one of the largest salmon runs on the planet (587).

The damming of the Yangtze River in China is believed to have helped cause the extinction of the Chinese paddlefish which was one of the largest freshwater species. The species had survived the extinction event that killed the dinosaurs; but they could not survive the onslaught of overfishing and dams. Down under, the large predatory fish biomass in Australia stands at around 10% of its preindustrial level when rivers were originally modified (588). In Mexico, a third of all freshwater fish face extinction and in Japan 50% of species are threatened. The loss of free-flowing rivers is largely to blame for these losses (589).

In Brazil and Paraguay, the Itaipu Dam which was constructed in the 1970s and 1980s resulted in a 70% decline in biodiversity while the Tucrui Dam in the Amazon resulted in a 60% drop in fish numbers. In Southeast Asia, where many dam projects are planned for the mighty Mekong River: fish numbers are expected to plummet by 40%. It isn't only fish who are affected as many people rely on the Mekong for their food and the Tapanuli orangutan which is the Earth's rarest ape is also expected to go extinct if a planned hydroelectric project in Sumatra, Indonesia is completed. There are only 500 of our distant cousins left (590).

In Cambodia, on the Tonle Sap Lake which links Phnom Penh in the south with Siem Reap in the north, the water level is dropping precipitously, along with fish populations. The lake is home to the planet's largest inland fishery and one million people. While the lake was once famous for its abundance of fish, the water level is now five meters lower than usual, and the fishing nets are empty. Villagers on the five floating communities are turning from fishing to growing chilies and other crops. The cause is a mixture of climate change and the dams that have proliferated upstream on the mighty Mekong River (591).

Dams not only stop water flowing downstream, but sediment as well. This leads to less fertile land downstream and can lead to erosion of riverbeds. Dams are believed to reduce sediment by 30-40% and this sediment feeds fish and vegetation along the river. Rivers without sediment are essentially dead (590).

On the great Zambezi River in Africa, the construction of two major dams has reduced water flow on Africa's fourth largest river and caused a 30% reduction in agricultural productivity downstream. Ninety-five percent of the rural labour force rely on the river and animals and humans have been forced to migrate as a result (592).

Dams also have the potential to cause conflict between those upstream and those downstream. The Grand Ethiopian Renaissance Dam (GERD) is about to become Africa's largest source of hydroelectric power; however, Egypt is worried the dam will affect river flow downstream where it is essential for agriculture (590). Likewise, tensions may rise along the Mekong River which snakes its way through China, Myanmar, Laos, Thailand, and Cambodia before emptying into the Mekong Delta in Vietnam. The delta is one of the world's largest bread baskets and is already threatened by rising sea levels.

China, which has been on a massive dam building project, is planning to build a large dam across the Brahmaputra River which flows from Tibet into India and Bangladesh. The Chinese government argue the dam is necessary to meet their climate goals and to increase water security. Many people are rightly concerned that water supply will not increase but decrease (593). They only have to look at Cambodia to see the potential harm that will likely arise. There is a further concern for those living downstream of dams. During torrential monsoon rains in July 2019, a dam in India collapsed and more than half a dozen villages were swept away by the raging torrent. Eight people died and a further seventeen were missing after the disaster in Maharashtra state (594). In February 2021, dozens of people were killed when a dam collapsed in the Himalayan area of Uttarakhand, in India. A further dam was damaged on the Dhauliganga River. These dam collapses have become more frequent in recent years as storms become stronger and release more water in condensed time (595). With storms expected to get stronger and become more frequent as our planet warms, it is easy to see why those living downstream are reluctant to allow dams to be built on their rivers, regardless of the clean electricity they are promised.

Finally, while the energy that dams produce is technically green, they are also the source of greenhouse gas emissions, especially those built in heavily forested areas. When the land behind the dam is flooded, the forests go from being carbon sinks to carbon sources as the vegetation decays and methane is released. It is estimated that dams produce around a billion tonnes of greenhouse gas emissions each year. In some neat symmetry, as was stated, dams have increased the amount of freshwater available to humans by 7% and exactly 7% of the freshwater needed for humans is evaporated from the world's reservoirs each year as the water is warmer due to there being no natural flow (590).

Dams have enabled us to produce renewable energy, but they are strongly impacting water quality, reducing biodiversity levels, and causing large amounts of greenhouse gas emissions. They also pose a problem for humans living downstream. The climate crisis will almost certainly reduce food production; and dams have the potential to cause conflict between neighboring countries. Is it time we reconsider our options when it comes to this form of renewable energy?

Panorama of the Hoover Dam
Jorge Lascar is licensed under CC BY 2.0.

Water Overuse

A further way in which humans are impacting freshwater ecosystems is through overuse. Humans have affected river systems to such an extent that many rivers don't reach the sea at all. The Colorado River which flows from the Rocky Mountains has not reached the sea since 1998. The Colorado used to empty into the Sea of Cortez in Mexico but due to the upstream use of 9.3 billion cubic meters of water every year, the amount of water that reaches its delta is now just a trickle. It's not just the Colorado River that dries up before finding the coast, the Yellow River in China, the cradle of Chinese civilisation, no longer reaches its destination. Neither do the Amu Darya or Syr Darya in Central Asia, the Euphrates and the Tigris in the Middle East and the Rio Grande on the Texas-Mexico border. The Indus River in Pakistan is also at risk of drying up before emptying into the ocean (596).

The Murray River, Australia's longest, fails to reach the ocean 40% of the time. While drought plays a part in this, irrigation upstream is largely to blame. In the middle of a drought in 2019, farmers planted an astonishing 100,000 hectares of cotton while fish all along the river were dying en masse. These plantations will use 1000 gigalitres and a further 1000 gigalitres will be lost to evaporation in private dams as temperatures hit 40°C. To put this into context, only forty gigalitres flowed further than Bourke on the way to the sea in 2018 (597).

While the cotton industry is to blame for water overuse in the Murray River; around the world, there is an even more problematic industry that uses 29% of all available freshwater and poisons our rivers in exchange.

Animal Agriculture

As our planet warms, rainfall patterns will become more erratic, and the amount of available freshwater is likely to decrease in many areas. In warmer areas more water will evaporate to compound matters. We have already looked at some of the areas that will be impacted by water stress within the next decade. According to the U.N., there are 700 million people suffering from water scarcity and this number will rise to 1.8 billion by 2025 and by 2030, almost half of our population could be water stressed (598). Fast forward thirty years and the United Nations World Water Development Report says six billion people will face water scarcity. They further state that this number may be an underestimate. According to Professor Benjamin Sovacool from Aarhus University in Denmark, the world will run out of water by 2040 (599). With this likelihood now understood, could this moment be a good time to start managing our water more efficiently?

In total, around 92% of our total global water footprint is used for agriculture. This makes sense as food production is our most important need. What doesn't make sense is that as 700 million humans struggle to find water, seventy billion farmed animals on Earth face no such struggle. As we hurtle toward this future, it is worth looking at this simple fact to see how we in the rich world might react to the news of billions potentially dying of thirst or hunger. Will we even care?

On our planet, as hundreds of millions cope with a life of thirst, 83% of our farmland and 29% of agricultural water is used for farmed animals yet they provide just 18% of our calories and 37% of our protein

(196) (600). This is simply an inefficient use of land and water and one which might push us all to the brink.

In addition to this overuse of water, to make matters worse, the industry is responsible for 58% of agricultural greenhouse gas emissions, 56% of polluted air and 57% of overall water pollution (600). Let's take a look at the impacts the industry is having due to both the overuse of water and water pollution.

In some parts of the world, water use for animal agriculture is even higher than the 29% stated above. One such place is the Colorado River basin in the western United States where the amount rises to 50%. Water here is used to produce burgers, steaks and ice cream that keep urban city dwellers extremely well fed. Unfortunately, it causes massive problems for wildlife and humans who depend on the rivers. According to research in Nature Sustainability, this overuse is threatening around or 50%, or 700 different fish species with extinction.

Since the Gold Rush of the 1850s alfalfa has been transported west to feed cows, horses and other grazing animals used by settlers. Today, the alfalfa is grown to feed animals kept in Concentrated Animal Feeding Operations (CAFOs) or factory farms as they are more commonly known. In some U.S. western river basins, around half of all water goes to farmed cows, and their feed. Due to the combining effects of overuse for growing alfalfa to feed livestock and climate change, the average Colorado River flow is down 17% annually compared to the 20th century average. Around half of this is attributed to water over-use. By 2050, the flow is projected to decrease by a further 20-30%. The Colorado River supplies water to around forty million people (601). As if to rub salt into the wounds of people and animals in the region, the United Arab Emirates (UAE) and Saudi Arabia are now importing this alfalfa to feed to their own animals which are in turn fed to their citizens. This is placing further pressure on the already threatened wildlife in the area (602).

Lakes are also at risk with water levels in Lake Mead in Arizona and Nevada falling by almost two-thirds since the turn of the century.

Animal agriculture is responsible for 75% of this depletion and the lake hasn't been full since 1983. In total, across the western U.S., around sixty fish species are at an increased risk of extinction from depleted water tables (603).

Lakes and rivers only supply part of the water we use to drink and grow crops. The vast majority of available fresh water (97%) is found under our feet in aquifers. These aquifers supply water to around 1.5 billion people worldwide. In rural U.S., this number is 95%. Unfortunately, these aquifers are also running dry. The Ogallala aquifer is the most heavily depleted in the U.S.. It has an average depth of sixty-one meters (200 ft) and is losing between 0.91 to three meters (three-ten ft) per year. This water was formed twelve million years ago; and recharges at less than thirteen cm per year. The vast amount of depletion has been to feed cattle in Nebraska, Iowa, Kansas, and Texas which produce around half of all cattle in the U.S. (602).

During the almost decade long drought in California in 2014, it was pointed out that Californians were now tapping so deep into the ground to extract water in order to maintain their high consumption lives that they were drinking the same water as the dinosaurs once slurped. Now, whilst that is an incredibly tantalising thought, what it also means is that Californians have now drunk their way through millions of years of water which included the ice age. When we say Californians, we should say Californian agriculture as around 80% of water use in California is for agriculture. And when we say agriculture, we really should say animal agriculture as this takes the vast majority of this 80%. It's not difficult to understand why, when we consider that it takes around 125 gallons (567 litres) of water to clean a cow's stall each and every day and the average farm uses 3.4 million gallons (15,456,706 litres) of water daily. Producing one gallon (3.78L) of milk uses the same amount of water as a month's worth of showers. The amount of water used to produce one pound of beef (453g) is equal to a staggering six months of showers (604). As Californians were urged to conserve water by turning off taps or not watering their lawns, McDonald's and other fast-food outlets spent millions of dollars urging Californians to buy beef.

The situation is mirrored around the world with 1/3 of the world's largest aquifers now being sucked dry. Out of thirty-seven mega aquifers, eight are now overstressed; while a further five are classed as highly stressed. The research from NASA satellite data shows that the Arabian Aquifer System and the Indus Basin aquifer under India and Pakistan are the planet's most vulnerable. These aquifers supply water for hundreds of millions of people, and two nuclear armed nations (605).

Freshwater withdrawals per kilogram (Poore & Nemecek, 2018)

Freshwater withdrawals are measured in liters per kilogram of food product.

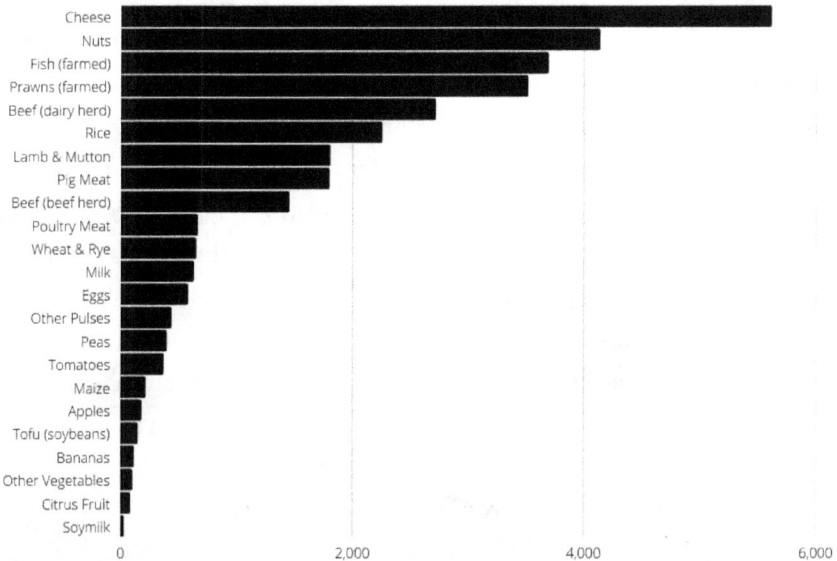

Reducing food's environmental impacts through producers and consumers.
Poore, J., & Nemecek, T. (2018). Science, 360(6392), 987-992.

As demand for animal products continues to grow around the world, pressure on our rivers, lakes and aquifers will rise accordingly. When the day finally comes when our aquifers run dry, our only choice will be to either drain our rivers or reduce our water usage by eating in a more sustainable manner. When it comes to water usage, animal products use a disproportionate amount of water to get the same amount of calories or protein than plant-based foods such as beans and pulses. The above graph highlights the discrepancy between water use for animal and plant products.

We have looked at the way animal agriculture disproportionately sucks up our freshwater supplies around the world, so now let's take a look at how the industry pays us back for this huge use of freshwater. Let's start in New Zealand where agriculture contributes around 6% of GDP (606) and employs around 84,700 people. Just over 70% of farm-land is classified as sheep and beef, dairy or mixed livestock and they use just over 40% of the land. Only a fraction of New Zealand's land is used for horticulture, orchards, and crop farming (607). This doesn't seem like a good deal, and it isn't, it gets worse.

In New Zealand, dairy farms not only use as much water as sixty million people, but they also have the environmental footprint of thirty million more (608). To get into details, due to animal agriculture, an astonishing two-thirds of rivers and lakes in the country are now un-swimmable; and three quarters of native freshwater fish species are at risk of extinction (609). The majestic landscape used to bring Tolkien's Lord of The Rings books to life is now dominated, not by Orcs and Ents, but by 6.6 million cows. Dairy contributes around 3.5% of GDP and in return, the effluence and fertilizer run-off from these cows causes eutrophication, where ecosystems are asphyxiated and toxic algae blooms as a consequence. A single outbreak of gastroenteritis in 2016, caused by sheep faeces resulted in 5,000 people becoming sick and four dying. Tourism provides almost double the GDP at 6.1%; how are tourists going to react to being told that they can only safely swim in 40% of the waterways that entice them there (610).

The example of New Zealand serves to show how much water is used in grass fed meat and dairy production and it also shows the impacts that grass fed agriculture has on our waterways. New Zealan-ders shouldn't be surprised, as Australia, where 98% of cows are grass fed, has already faced the same problem. In 2011, 80% of Victoria's rivers were classed as in very poor condition. This was again down to the waste from cattle causing toxic algae blooms (611).

Sadly, when it comes to water pollution, it isn't just cows and sheep to blame. In Wales, the River Wye, which was voted England and Wales' favourite river in 2010, and is a site of specific scientific interest; is facing an equally concerning threat, intensive chicken farms. The excessive phosphate, nitrogen, and ammonia from these 'farms' is polluting the river and turning it into a 'pea soup' where the nutrients cause plants to proliferate and the water in turn is starved of oxygen. Insects, birds, water shrews and fish are then affected. The amount of phosphate in the lower Wye has doubled since 2014; and coincides with planning permission being granted to 134 large broiler and free-range egg farms since 2015 (612).

The problem doesn't stop with cows, sheep, and chickens. Pig farming is also destroying life in our waterways around the world. Let's skip across the Atlantic to take a look at the impact intensive pig farming is having on water quality in the U.S. In North Carolina alone, where there are 2,300 pig farms, around ten billion gallons of waste are produced from livestock, with the majority coming from swine. This waste is sitting in around 3,300 waste lagoons of bright pink sludge. These lagoons regularly overflow their contents into river systems, especially during hurricanes. Alas, you don't have to wait for a hurricane to be able to see the spread of this toxic mix. In order to stop the waste overflowing, it is routinely sprayed on fields as a fertilizer. This causes health issues for people living nearby and in 2018 and 2019, $473.5 million was awarded to 500 local residents by Smithfield Foods who eventually only had to pay $94 million because of limits on punitive damages (613). Studies have linked emissions from pig farm emissions to increased stress, anxiety, fatigue, mucous membrane irritation, respiratory conditions, reduced lung function and elevated blood pressure (614). A further study in 2018 found that people living near CAFOs had higher rates of infant mortality, deaths from anaemia, kidney disease and tuberculosis (615).

Ground and surface water are also contaminated by the runoff that contains high quantities of nitrates and phosphates. The Neuse River in North Carolina witnessed a fish kill that resulted in the deaths of hundreds of thousands of menhaden, *"as far as the eye could see in both directions*". The small fish are vital in the food chain and seen as a canary in the coal mine for river ecology. CAFOs are again to blame for these deaths. This example can be extrapolated out across the nation and in Iroquois County, Illinois, more than 492,000 fish were killed between 2005 and 2014 after waste from a single farm seeped into Beaver Creek. Little, if any aquatic life remained. In total, 108 km of the states' rivers, creeks and waterways were polluted in that period with some farmers simply dumping thousands of gallons of manure into the imperiled waterways (616).

We have looked at some specific examples but let's look at the broader picture. Animals in CAFOs are fed excessive amounts of heavy metals like copper and zinc to promote growth and prevent disease in chickens and pigs. Incredibly, other metals in animal waste can include cadmium, lead, mercury, and arsenic. These metals then accumulate in the soil before contaminating water supplies which can lead to gastro-intestinal and liver disorders amongst other health problems. Copper is also toxic to fish and aquatic life in waterways. Additionally, zinc can also kill fish, damage algae and other sea life (617). When it comes to nitrogen emissions, the planetary boundary for emissions sits between sixty-two and eighty-two terragrams (Tg). The total nitrogen emissions from meat and dairy production alone contribute around sixty-five Tg and therefore, this single industry is responsible for pushing our planet past its lower limit safe boundary. The largest cause of this nitrogen production is from growing crops for animals (68%) followed by the build-up and management of manure (618).

In response to growing concerns about its burgeoning mountains of manure, pork and dairy producers are attempting to create a fresh market for their waste. The project between Smithfield Foods, Dominion Energy and Align RNG aims to capture methane from animals and then convert it into a biogas that will be piped into the living rooms of U.S. homes and businesses. Other alliances are emerging between energy companies and meat producers across the U.S. The two most destructive industries on the planet are joining forces, in what many are calling a classic example of green washing; to continue polluting our environment and avoid any meaningful change. Calling the biogas renewable is a stretch as the animals will need to be fed vast amounts of energy that will come from elsewhere. Capturing the animal's waste also does nothing to address the much larger problem of cow burps. Creating a new profitable market for animal waste will spur larger investment in animal production and any methane stored from waste will pale in significance to the additional methane released through cow burps and the diminished ability of forests and soil to sequester carbon due to deforestation (619). This is yet one more example of technology coming to the rescue and simply creating more problems.

When it comes to what we can do to limit our freshwater use and reduce water pollution, the finding is clear according to Joseph Poore from Oxford University, as reported in the Guardian:

> *"A vegan diet is probably the single biggest way to reduce your impact on planet Earth, not just greenhouse gasses, but global acidification, eutrophication, land use and water use."*

"hog confinement system" by friendsoffamilyfarmers is licensed under CC BY-ND 2.0.

Birds

"We ate the birds. We ate them. We wanted their songs to flow up through our throats and burst out of our mouths, and so we ate them. We wanted their feathers to bud from our flesh. We wanted their wings, we wanted to fly as they did, soar freely among the treetops and the clouds, and so we ate them. We speared them, we clubbed them, we tangled their feet in glue, we netted them, we spitted them, we threw them onto hot coals, and all for love, because we loved them. We wanted to be one with them. We wanted to hatch out of clean, smooth, beautiful eggs, as they did, back when we were young and agile and innocent of cause and effect, we did not want the mess of being born, and so we crammed the birds into our gullets, feathers and all, but it was no use, we couldn't sing, not effortlessly as they do, we can't fly, not without smoke and metal, and as for the eggs we don't stand a chance. We're mired in gravity, we're earthbound. We're ankle-deep in blood, and all because we ate the birds, we ate them a long time ago, when we still had the power to say no."

This striking poem from Canadian author Margaret Atwood, in her book The Tent, is a fair, but arguably tame description of the horrors that humans have inflicted on our avian cousins. The lone survivors of the last extinction event that wiped out non-avian dinosaurs sixty-six million years ago can trace their ancestry back to a group of meat-eating dinosaurs called theropods. This group contains the infamous T-rex immortalized in the Jurassic Park franchise. When we look into the eyes of birds, we are looking at 150 million years of evolution. You would think this fact alone would demand an instant respect; alas our species is new money, our arrogance abounds and while we like to preach about respecting our elders; this sadly applies to just living, breathing humans (620).

In the pages that follow, we will examine some uncomfortable truths surrounding the atrocities we have committed and continue to commit against our Earthly elders. These atrocities will be split into two sections that focus on our disrespect for birds in the wild and the barbaric behaviour we show to birds we breed for food. It is important to look at both as they are intricately intwined for the more domesticated birds we breed, the less wild birds there are left. Today, a staggering 70% of birds on our planet are bred for human use and just 30% are free to soar like an eagle. In a few decades' time, the term 'birds eye view' may have completely lost its meaning.

Domesticated birds

Wild birds

70%

30%

Avian biomass distribution on Earth
https://www.pnas.org/content/115/25/6506

Domesticated Birds

Chickens are the most populous bird to have lived in human history. The descendent of Southeast Asia's red jungle fowl were domesticated around 8,000 years ago and today, at any one time, there are approximately twenty-three billion chickens on our planet. That's around three for every human. In 2019, more than seventy-two billion chickens were slaughtered worldwide. In the United States, 99.9% of broiler chickens (chickens bred for their flesh) are factory farmed in giant sheds where they struggle for survival for six weeks with up to 125,000 others. In the U.K., the number is 95% and in Japan 90%, mostly imported. The global average for factory farmed chickens is 70% (621) (622) (623). Wherever people consume a lot of chicken, the majority will be from factory farms. These birds' first mistreatment stems from their birth where they are hatched in giant incubators without their mother's protection. Then they will be crammed into shipping crates and sent to factory farms where these fluffy yellow chicks will have the top part of their beaks cut off with a hot blade. This has been commonplace on farms in the United States since 1940 when a farmer in San Diego realized that his chickens were unable to pick and pull at each other's feathers if he used a blowtorch to burn away the upper part of the beak. Over the years, we have evolved the practice to using a hot blade instead of a blowtorch and this practice is now widespread around the world (624).

Unfortunately, we have devolved as a species as any empathy we might have spared for birds has been replaced by complete indifference. You might argue that we are being cruel to be kind by removing these baby chicks' beaks; but the reason we do so is because we cram so many of them into these monstrous sheds that they suffer from such stress that they take it out on each other, much like humans would under such conditions. The industry likes to claim that it's like having your fingernails cut, but according to a U.K. animal welfare committee set up by the government, the reality is that the beak is not simply a fingernail but a complex mesh of bone, horn, and sensitive tissue. It would be more akin to having the top of your finger cut off than the fingernail. This is all done without anaesthesia, and it causes great and prolonged pain (624).

After being fed a concoction of growth hormones and antibiotics, they will balloon in weight by around 50g a day. By just day nine, the legs can barely keep the oversized breast off the floor and by day eleven, they will be twice the size of a traditionally bred chicken. By the time they are thirty-five days old, they look like Buzz Lightyear on extremely potent steroids. By this time, their organs and legs cannot keep up with the unnaturally rapid growth and heart attacks and organ failure are common. They will then be left to die in agony on a sea of ammonia that causes their feathers to fall out and burns their skin. Many others die because their legs cannot support them as far as the water nozzles. In a complete reversal of the meaning of life, the unlucky ones who make it to six weeks will be once again crammed into shipping containers and driven to their slaughter. They will be hung upside down in full view of others and at a speed of 150-175 chickens a minute, they will be attached to a conveyer belt that drags their necks past a sharp blade that is supposed to cut their throats. Unfortunately, the kill speeds are now so fast that many birds are not killed, and are then submerged in boiling hot water in order to allow their feathers to be removed more easily. These conscious birds are literally boiled alive. This is the six-week life of a broiler chicken. In the wild, chickens can live to be eight years of age, so these descendants of the dinosaurs are still just babies when they are slaughtered (625) (626).

If the last few paragraphs were uncomfortable to read, you might want to skip the next few because the atrocities are ordered from least to most debased. Using the word 'life' to describe what broiler chickens are forced to endure makes a mockery of what should be a positive experience; but other birds experience far more pain and suffering. Every nauseating experience broiler chickens experience is shared with chickens we raise for eggs; but these chickens have a use that far exceeds six-weeks, so they are forced to endure a whole lot more until they are unceremoniously chopped into pieces to feed to our pets, who we treat like royalty.

There are more than seven billion egg laying hens on our planet so almost enough for each and every woman, man and child. In the U.K., 48% of these hens are caged, in the U.S. the number is 74% and in Japan, 96.1% of egg laying hens live out their days in cages (627) (628) (629). The lives of these chicks begin much like the broiler chicks but with one major difference. Whereas broilers can be both male and female, eggs are only laid by females. What happens to all the males? On day one, shortly after the chicks are hatched, they will be carelessly tossed onto a conveyor belt where they are sexed by humans. The females continue on to their two-year journey of suffering while the males take a slightly shorter path to the end of a completely unnecessary experience. Their conveyor belt takes them to a macerator which shreds their fluffy yellow innocence to pieces until they are no more. That's the entire experience of these fortunate male survivors of the fifth mass extinction.

Such is life, it's the females who are forced to withstand the greatest torment. As all women will attest, producing an egg requires a great deal of energy, so much in fact, that like humans, chickens naturally only produce between ten and fifteen eggs per year. Unfortunately, there is nothing natural about the misery of these egg laying hens. Hens' bodies in the modern age have been manipulated to such an extent that they now lay an egg every twenty-two hours. As the hens grow older, the number of eggs they lay decreases. Some farmers counter this with a method known in the business as 'forced molting. Hens usually replace their feathers prior to winter. At this stage, they stop laying eggs in order to focus their energy on producing feathers to keep them warm. Forced molting is used when a hen's egg production declines due to such high production. Rather than slaughter the hen, farmers have discovered a way in which to get an extra few hundred eggs from her before ending her life. During a forced molt, food and often water are removed from the hens for between seven and fourteen days. The hens then stop laying eggs and produce feathers instead. Their reproductive system is thus given time to rejuvenate, and she will be expected to produce another cycle of eggs. She will do this the whole time spent in a small cage where she cannot even stretch her wings. Is it possible for most humans to try to place themselves in the shoes of these gallant would be mothers? Their entire life spent in a cage or a cramped factory, laying painful egg after painful egg that will never hatch and then being starved in order to force her to produce more and more when her body is saying "please, no more". After around two years, she will be completely spent and become uneconomical. This will trigger her delivery to the slaughterhouse; where in thanks for her years of obedient service, she will have her throat cut and be chopped into pieces. This is the life of an egg laying hen (630) (631) (632). In the time it has taken you to read this section, approximately 688,000 chickens have been slaughtered in the manner explained above, for food (633).

As promised, each new section describes more suffering than the previous. Finally, we arrive at what is potentially the most depraved action of all and something that no homo sapien should consider acceptable behaviour. This time, the victims are not flightless chickens, but air worthy ducks and geese, and the product is not flesh or eggs, but liver. A diseased and massively enlarged liver to be precise. The process to be described is the production of the French delicacy: Foie Gras.

Fois Gras is an expensive, but nutritionally valueless meal of diseased and fatty liver. To produce this diseased liver, a tube is slammed thirteen centimeters down the throat of the bird and a daily total of 1 kg of corn is forced down the tube five or six times a day for between two and three weeks until the liver increases from an original weight of 50g to a bare minimum of 300g. Ducks are typically killed at 100 days old and geese at 112 days. Ducks can live to be twenty in the wild while geese can live to be twenty-four. This practice of forcing tubes down the throats of birds in order to produce something of no nutritional value should shame us all. These are sentient animals that wish only to grace our skies with their splendour, and we treat them as mere objects to bend at our will. As if the sheer cruelty involved isn't enough to make us instantly outlaw this barbaric practice, the absolute waste should be the final straw. To produce 1 kg of liver that has no nutritional value, twenty kg of nutritional corn is forced into each bird's mouth. This waste, at a time when almost a billion humans go hungry at night is a further stain on our species (634) (635).

The feed we grow for broiler chickens and egg laying hens is further removing forests from our planet and this is where wild and domesticated birds interconnect. The U.K. alone imports three million tonnes of soybeans annually from Brazil, Paraguay and Argentina and the vast majority of it is fed to factory farmed chickens (636). In Japan, numbers are similar to the U.K., and these will be replicated everywhere that factory farms are prevalent (637). The more domesticated birds we raise for food leads to the removal of wild bird habitat and if we don't act soon, the only birds with whom we share the planet will be chickens, ducks, and geese.

The main justification for our lack of compassion and respect for birds seems to stem from their reputation as simple, stupid creatures. We call unintelligent humans birdbrain, people lacking courage chicken, and useless footballers are described as headless chickens. These depictions of the most commonly eaten and abused animals on planet Earth serve an excellent purpose. It has always been expedient to describe others, from whom we wish to steal or manipulate, as stupid, savage or barbarian. White Europeans used this tactic to their benefit while plundering Africa, the Americas and Asia. Humans at large use it to justify their continued exploitation of all non-human animals, and especially domesticated birds.

The truth, again rather inconveniently, is that birds are far more intelligent that we give them credit for. As the South African writer Frans de Waal points out in his book of the same name, Are we Smart

Enough to Know How Smart Animals Are? While historically, we may have been forgiven for passing birds off as lacking intelligence; only the ignorant and misinformed can enjoy such an excuse today.

There is not a week that goes by without more evidence of bird intelligence being delivered directly into our hands via glowing screens. The following description of birds is from the Smithsonian Magazine in 2016:

> *Birds are smart, good at solving problems, learning, remembering, way-finding, even conversing in a meaningful way (638).*

In his book, Bird Brain: An Exploration of Avian Intelligence, Nathan Emery of Queen Mary University of London states that a number of bird species should be considered more as 'feathered apes' than 'birdbrained'. Scientists' initial assertion that birds were incapable of higher thinking stems from their research on bird brains. They found no cerebral cortex and thus decided birds were unintelligent, but research has found that a different part in bird brains, the pallium, carries out similar functions to the cerebral cortex in mammals. Additionally, they found that different parts of bird brains had shorter connections than in mammals and this may allow birds to make more rapid decisions (639).

But, what of the humble chicken, surely, they must lack any abilities other than pecking for food? Not so, according to research published in 2015. Researchers at the University of Adelaide found that chickens experience boredom, frustration, and happiness. In addition, they can count, show a level of self-awareness, and can even manipulate others by Machiavellian means. Students working with chickens to understand psychology and cognition were shocked to find that chickens were a lot more complex than their preconceptions had allowed. The same tests were carried out with poultry workers and their preconceptions were similarly obliterated. In other research, chickens have shown empathy and self-control (640).

The evidence is mounting, but whether or not this will affect our consumption habits is still to be seen. The jury is out, but time is running out for the chickens' wild cousins.

Battery hens -Bastos, Sao Paulo, Brazil
Secretaria de Agricultura e Abastecimento do Estado de São Paulo Agriculturasp is licensed under CC BY 2.0.

Wild Birds

Our behaviour towards birds in captivity is mirrored by both the present and historical indifference we have displayed to our feathered friends in the wild. Whilst we may be able to claim we were unaware of their plight and suffering in the past, this is certainly not true in our present times. As Ms. Atwood stated in her poem, we have the power to say no, but we can barely even bring ourselves to consider the word, yet alone display the courage of our convictions.

Due to human activities, wild bird populations are in serious decline. Out of 10,994 species, 48% (5,245) are thought to be in population decline, and 13.5% (1,481) are threatened with global extinction (641). Be it from our warming planet, industrial fishing, plastic, deforestation, pesticides, or the illegal wildlife trade, we are hugely endangering the existence of our modern-day T-rex friends. In the following pages we will look in more detail at these activities, starting with the climate crisis and the impact it is having on birds.

As was discussed in an earlier chapter, as the temperature warms, fauna slowly but surely make their way north to find cooler climes. The same is true of avifauna. In Europe, breeding bird populations were found to have moved an average of one kilometre per year towards the pole, according to data from 120,000 birdwatchers in forty-eight countries. Since 1997, the 539 native bird species documented were found twenty-eight km further north. Generally speaking, this could be seen as a positive as the larger the area a species inhabits, the less likely it is to go extinct; however, the reason for this geographical spread is likely not from a sudden population increase, but because of habitat deterioration. While 35% of the birds studied had increased their breeding range, 25% had gone in the opposite direction and the rest had remained stable. The problem for migratory birds is that they have evolved to be able to carry out their migrations to perfection and as migrations become longer, they are put under enormous stress. There has only been one bird extinction in Europe since 1990: the common button quail, but Dr Iván Ramírez, senior head of conservation at BirdLife Europe and Central Asia believes that as birds are forced into areas that are suboptimal for their survival, their populations will crash, and extinctions will escalate (642).

To make matters worse, the positive feedback loop of warming atmosphere and increased fire risk is resulting in larger, more intense fire seasons like the one at the end of 2019 in Australia. The fire killed an estimated 180 million birds. In North America, where huge forest fires enveloped California, Oregon, and Colorado in the summer of 2020, hundreds of thousands of migratory birds were found dead in New Mexico as the smoke from the fires forced them to migrate earlier than they were prepared for and according to bird watchers, they fell from the sky on route to Central America. Dead migratory birds, including warblers, bluebirds, sparrows, blackbirds, and flycatchers were also found in Colorado, Texas, and Mexico. It is possible the number of deaths will reach into the millions. Birds have been known to be sensitive to toxins in the air and it is for this reason, canaries were the figurative and literal canary in the coal mine. Scientists remain unsure whether it was toxic air that caused the birds to perish or simply starvation (643) (644).

As the oceans have absorbed most of our man-made warming, it stands to reason that sea going birds will be affected accordingly. As plankton levels drop due to higher temperatures, so do small fish populations which seabirds rely on for sustenance. This leads to birds starving. The result of this was highlighted in 2015-2016 when 62,000 common murres (guillemots) were washed up on the coast of the Pacific Ocean due to the 1,600 km long body of warm water which affected the north-eastern Pacific between 2013-2016. The number of birds washed ashore is believed to be just a fraction of the total number of deaths which are estimated at around one million.

In a similar situation, from late June to early August 2019, thousands of Short-tailed Shearwaters washed up on Alaskan and Russian beaches. After investigation, the birds were found to have no food in their stomachs. Since 2015, puffins, and auklets have been added to the list of birds, including shearwaters and guillemots, suffering starvation in Alaska. Each wave of deaths has been associated with unusually warm ocean temperatures. Months later, thousands of shearwaters were washed up on Sydney's famous beaches. The birds were making the 14,000 km trip back from Alaska to breed; but krill and other fish species that they rely on to make the trip at full strength were in short supply due to rising sea temperatures (645). This single species is facing problems at both ends of our planet and their struggle provides even more evidence of the interconnectedness of life on Earth.

Scientists predict much more warming in the coming decades and with it, more and more sea bird deaths as a result. Whether at sea, or on land, the changing climate is going to test avian species' ability to survive. We shall soon find out whether their fortune in the fifth extinction will be mimicked in the sixth: the first to be caused by a single species (646).

While warming waters are a growing concern and can quite rightly be blamed for the deaths of these seabirds, there are two much larger threats causing them to starve, those of industrial fishing and plastic pollution. Seabirds have been traversing our oceans in various shapes and sizes for sixty million years, but in the past sixty years, their numbers have declined by 70%. These birds of the extreme can dive 200 meters and fly around the globe without rest; but they are being pushed to their limits by our love of fish and all things plastic.

There are almost 350 species of seabird who have evolved to survive on every continent on Earth. Unfortunately, they have not yet evolved to cope with a fast-rising human population placing ever-growing demands on the oceans. As more and more humans decide that fish is for dinner, less and less food is available to seabirds who are in direct competition for fish with land living humans. Our desire for the flesh of fish has helped drive 28% of seabird species to the edge of extinction with 47% in decline. They are twice as likely to face extinction as land-based birds. Some seabirds have fared much worse with the above-mentioned shearwaters and petrels declining by 79%, albatrosses by 69%, frigate-birds by 81% and terns by 85%. In Japan, seabird populations have dropped to between 3-35% of their baseline levels in 1990. There are only between 3-35 seabirds depending on species, for every hundred that existed before the Berlin Wall fell (647).

The situation for the African black-footed penguin is even worse. Africa's only penguin species is facing extinction within five to fifteen years. The cause of this extinction is down to over-fishing of sardines, anchovies, and pilchards which the penguins need to survive. These small to medium sized fish account for 37% of all fish caught and a staggering 90% of them are ground up and fed to farmed animals on land (648). We are essentially driving these penguins to extinction in order to satisfy our love of animal flesh. The lack of food is pushing penguins to extend their swim for food from twenty km to thirty km and as a result, they have insufficient energy to feed themselves and their young. Many eggs and chicks are being abandoned as a result (649). Antarctica's iconic King Penguin does not have much longer than their African cousin according to a 2018 study; with the flightless bird expected to vanish by the end of this century due to a mixture of rising sea temperatures and over-fishing in Antarctic waters (650). Here, overfishing of krill to feed to farmed salmon and shrimp is removing the penguins' food source (651). The main catchers of krill are South Korea, Norway, Poland, and Japan. In order for humans to satiate their tastebuds, we risk losing one of the most recognisable animals on our planet and quite what the knock-on effect will be, we are yet to find out.

Industrial fishing is at the heart of these problems. Currently, only one percent of our oceans are off limits to the fishing industry. A study in 2012 found that whenever fish abundance dropped below one third of maximum levels, seabird populations also fell. Co-author of the study Ian Boyd believes we need to protect one third of the ocean from fishing if seabirds are going to survive the human onslaught (652).

While humans are hurriedly removing fish from the sea, we are swiftly replacing the fish with plastic, and it is unsurprisingly finding its way into the stomachs of seabirds. It's now estimated that ninety percent of seabirds' stomachs contain plastic. Birds often mistake plastic for fish eggs and ingest it like it were. Plastic that has been found in seabirds includes bottle tops, toys, model cars, and lighters. This plastic is not only toxic but can also lead to starvation as it obstructs their bowels. Birds are even feeding bits of plastic to their babies resulting in large numbers of deaths (652).

As if a lack of edible food coupled with an abundance of inedible food wasn't enough for seabirds to contend with, they also have to contend with the fishing industry's indiscriminate killing, or bycatch as it is commonly known. Birds are increasingly being caught on long lines intended for carnivorous fish or in massive nets that stretch out for kilometres. Longlines, intended to catch tuna can measure 128 kilometres in length and are being blamed for pushing 75% of all albatross species to extinction (653). Every year, at least 160,000 and potentially in excess of 320,000 seabirds are caught on long lines and drown along with turtles, dolphins and other marine life not being targeted (654). The birds follow the baited hooks and when they dive into the frigid waters they become caught on the hooks. In more disturbing news in the south-west Atlantic Ocean, many fishermen have taken to slicing the top part of the beak off in order to free the bird from the hook before tossing the birds into the ocean to die. There are ways to free the birds from the hooks without killing them; but due to expediency, they are being ignored. Many of the birds thrown overboard to die are endangered such as the northern royal albatross and the spectacled petrel. Seabirds can take up to ten years to reach maturity, so these losses are completely unsustainable (655).

Unless humanity changes course very quickly, it is likely we will no longer share the planet with these extreme-loving birds and as with all extinctions, there will be a ripple effect. An example of such an effect can be found in coral reefs. Seabirds fly hundreds of kilometres out into the ocean to feed and when they return, they deposit their droppings, known as guano, at their colonies. This guano then leeches into the ocean where it fertilizes reefs. Compared to areas with no seabirds due to rats and other predators, reefs have been found to flourish when seabirds are present, and fish grow quicker and larger as a result (656).

As human activities drive the extinction of ocean-going birds, so it is on land. The main culprit is deforestation. With the rate of forest loss now estimated at one football pitch of forest every second, it is easy to understand why (657). By replacing biodiverse forest cover with kilometre after kilometre of uniform crops or grazing land for cattle as far as the eye can see, we are removing any vestige of hope for many bird species and unless we stop the rate of loss very quickly, it is likely we will lose species after species until there is no more bird song to remind us of what once was.

As was mentioned previously, not only loss of habitat is to blame for population declines, the degradation of available land is also a root cause. In North America, land degradation is leading to rapid declines in bird populations. While extinction is not on the immediate horizon for these species, populations have been declining steeply since 1970. There are now estimated to be 2.9 billion fewer birds on the continent than just fifty years ago. In total, that is a 29% reduction across the spectrum from threatened species to the more ubiquitous birds like sparrows, warblers, finches, and swallows. These birds are an integral part of the ecosystems as they help control insects, and they also help to plant seeds through their droppings (658).

The situation is similar in France where one third of all birds have disappeared from French farmlands since 2001. Some familiar farmland species like skylarks and grey partridges have seen even more alarming declines of 50% and 90% respectively since the mid 1990s (659). The U.K. saw a 9% decline in farmland birds in just five years between 2010-2015. Since 1970, there has been an overall decline in farmland bird populations of 56%. Some species like the grey partridge, turtle doves and tree sparrows have all declined by more than 90% in the past fifty years (660). Germany has witnessed the loss of 421 million birds since it was reunified in 1990 (659). A quick Ecosia search for the reader's country will no doubt provide similar findings. The causes of these farmland bird losses are diverse, but in the U.K., as an example, they include the removal of hedgerows in order to maximise productivity, the change from spring sown to autumn sown cereal varieties, land drainage and grassland management, increased density of livestock like sheep and cows who remove food available to birds, and the subject of the following paragraphs, the use of pesticides and insecticides to control weeds and insects (661).

As we saw in the insects chapter, in our lifetime, we have instigated a global insect genocide and there may be none left by the end of the century if current trends continue. The largest reason for these losses is insecticides like neonicotinoids which contaminate entire ecosystems. The majority of land-based birds rely on insects for at least part of their diet. In North America the figure is 96%. Insect eating birds peck their way through 400-500 million metric tonnes of beetles, flies, ants, moths, aphids, crickets, caterpillars and other anthropods every year on our boundlessly fascinating planet. Unfortunately, as insect numbers plummet due to the use of these chemicals of mass destruction, so to do bird populations. Even if birds do not directly rely on insects for their food, they may be indebted to them for pollinating the flowering plants that provide their seeds or fruit. The makers of these chemicals will argue their products are not the cause of bird declines but in areas where insecticide use is limited, bird populations rebound. This would appear to be a straightforward case (662) (663). The president of the American Bird Conservancy, Michael Parr stops short of directly blaming the companies but says:

"We don't have categorical proof, but it'd be very surprising if the billions of pounds of insect poison being spread across the landscape wasn't connected to the decline of insect-eating birds." (658)

While causation may be difficult to ascertain in the case of insecticides, the case of vulture deaths in India and Africa is firmly closed. In early 2020, almost 1,000 hooded vultures were poisoned to death in Guinea-Bissau. The mass poisoning has pushed this already endangered species to the brink of extinction. The birds were feeding on feral dogs who had been poisoned near rubbish dumps. Other vultures are the victims of inadvertent poisoning as they scavenge poisoned bait intended to kill lions, hyenas, and other wildlife. In the Indian subcontinent, vulture populations have plummeted by 99% after ingesting a toxic veterinary drug fed to cattle (664). It is estimated that in Canada alone: 2.7 million birds are killed from pesticide ingestion (641).

Add to the chemical mix, herbicides like Roundup, which is the most widely used herbicide for preventing weed growth, but which also reduces available plant and insect food for birds, and we start to see the connection between human activity and the decimation of entire ecosystems. It seems clear that urgent, radical changes are needed to our farming and fishing industries if we are to avoid the complete collapse of these ecosystems in the coming decades.

Birds, as we have seen are under a pervasive human caused onslaught, but there is still one more destructive human activity that has been overlooked thus far. Tropical rainforests are some of the most biodiverse regions on our planet, and when intact, they should be home to thriving ecosystems with the human ear experiencing a chorus of diverse sounds, much of which produced by our two-legged friends. There are forests however, which are in otherwise pristine condition and void of chemical use or human agriculture which are disconcertingly quiet and strangely empty. What could be the cause of this discombobulation? The answer is the illegal wildlife trade.

Forests across Southeast Asia are waking up to birdless mornings due to the rampant use of snares which are used to indiscriminately catch birds that are then sold on the open market as pets. From 2010-2015, in the Southern Cardamom National Park in Cambodia, the number of these snares removed by forest rangers from the Wildlife Alliance almost doubled from 14,364 to 27,714 (665). In the Indonesian archipelago, white cockatoos and parrots that are caught in this manner are forced inside plastic drinking bottles while alive and smuggled out of the country and sold for as much as $1,500 each. White cockatoos are listed as endangered by the International Union for Conservation of Nature (IUCN), but this doesn't stop them being sold in avian markets or crammed into drainpipes and smuggled out of the country. According to the IUCN, there are as few as 43,000 left in the world, and their numbers are being decimated by the illegal wildlife trade (666) (667).

The problem is global in scale with the U.S. State department estimating the trade worth $10 billion annually. This makes birds the most common contraband with between two and five million birds being traded each year (668). In the Amazon, some captured birds are killed for their prized feathers, skins, or other body parts while others are sold to foreign collectors who prefer the birds alive to admire their beauty from within a cage. These practices are helping to push some bird species to the brink of extinction and their survival perches in the balance (669).

We have arrived at the end of part two which has contained some uncomfortable truths about the perilous state of the ecosystems on which we rely for our existence. A paper published in Nature in 2020 describes our problem in simple numbers. At the beginning of the 20th century, the study reports, human produced objects like steel, concrete, bricks, and plastic were equal to about 3% of the world's total biomass (plants, organisms, shrubs, and animals) but by 2020, the weight of our objects now outweighs nature and since the Second World War, it has been growing by 5% a year and in effect doubling every two decades. Just the plastic we have produced now outweighs all land and marine creatures combined (670). As we decimate the natural world, we are replacing it with plastic, concrete, and metal. We are swapping breathing fish for plastic talking fish and breathing trees for concrete skyscrapers.

As the COVID-19 pandemic continues, many of the problems concerning deforestation and our indifferent attitudes to non-human animals have certainly come home to roost. This is a time of great opportunity. It could be the time our species changes course and heads in a more sustainable and compassionate direction, or it could be the time we push the pedal to the floor and accelerate over the cliff edge. The following quote from Charles Dickens in a Tale of Two Cities sums up our predicament succinctly.

"It was the best of times, it was the worst of times,
it was the age of wisdom, it was the age of foolishness,
it was the epoch of belief, it was the epoch of incredulity,
it was the season of Light, it was the season of Darkness,
it was the spring of hope, it was the winter of despair,
we had everything before us, we had nothing before us."

Banksy: Detail
eddiedangerous is licensed under CC BY 2.0.

Dear Indy,

After reading the past pages, I would be an awful father if I finished on such a bad note. As I mentioned, this is an amazing opportunity for our species to change direction. The past two years have been a mixed bag for our family. We moved from renting a small suburban house in Kumamoto to buying a 7,000 sq.m former mandarin farm on an island in the Seto Sea. Our lives certainly changed. It was also the time of COVID-19. A time in which we wore masks every single day and sacrificed friendships and after school play times. I am also writing this the day after being woken in the middle of the night to hear that Grandpa Jonny had tested positive for the virus. It has certainly been interesting.

Everything in Part Two of this book is interconnected, the virus, deforestation, and our diets. We have caused this disease by our inhumane treatment of wild and domesticated animals. 75% of the new and emerging viruses are zoonotic, which means they originate in animals and are passed to humans. It is likely that this virus originated in a bat in a wild animal market in Wuhan, China, but since the outbreak, there have been additional outbreaks of swine flu in China, Poland, and the U.S. There have also been outbreaks around the world of avian flu and these viruses have the potential to kill many millions of people if we continue to keep animals in the filthy dirty conditions we currently do.

Deforestation also increases the risk of these diseases spreading to humans as we come into increased contact with wild animals. There have been many reports this year urging us to prevent further deforestation in order to prevent future pandemics.

The best way to do this is to change our diets. Dietary change could eliminate factory farms and wet markets, and with them 75% of potential diseases. According to researchers at Oxford University, removing meat and dairy from our diets could reduce the land used for agriculture by 75%. All that area could be reforested, and plants and animals could once again thrive as they have for hundreds of millions of years. The forests would also suck carbon dioxide out of the atmosphere. The opportunity is there for us to grab but unfortunately, I worry we might not take it.

Lluis Barba - The Garden of Earthly Delights by Bosch (2007)
Cea. is licensed with CC BY 2.0.

THREE

How we Got Into
This Mess

M any believe that our current predicament is shared by all tech-
nologically advanced species in the Universe. The Great Filter
theory claims there are an estimated forty billion Earth sized planets
in the habitable zone in solar systems throughout the galaxy, many of
which are much older than our Milky Way, and any species living on
them would face existential challenges like our own. An example of
such a challenge is research published in the journal Science, which
claims that the switching of magnetic poles around 42,000 years ago led
to a weakening of the Earth's protective shield. The research proposes
that this weakening increased the solar radiation which had dramatic
effects on the environment. It may have been what forced humans and
other Neanderthals to live in caves to escape the heat and also to us
using red pigment to offer protection from the Sun. This would have
led to increased competition that possibly spelled the end of Neanderthals. Should this prove to be accurate then a modern day switching of
poles could devastate power grids and satellite networks, and provide
evidence for a limited time frame in which technological development
can occur (671). Consequently, as soon as a species begins to develop

technology, it will be faced with a choice of risking its extinction to get off the planet or opting to live in a sustainable manner that conserves ecosystems. If the first path is chosen, the Fermi paradox claims the challenge must be exceptionally difficult and leads the species to extinction. If the second option is chosen, enormous challenges like the one just described may lead to severe disruption which limits population size and at some point, the star will eventually die, and all life will perish. This theory attempts to explain why we have found no real evidence for extra-terrestrial life thus far (672).

We are currently at the latter stages of the first option in the Great Filter and the route we are on seems increasingly likely to cause the majority of our species to become extinct in the foreseeable future. Even if, against all odds, we were to succeed in colonizing other planets, the ensuing result will be that the 99.9% will be left on a largely uninhabitable planet and the 0.01%, in its current form, would likely exterminate everything in its path in a rush for ever more riches.

Nevertheless, it is not the belief of the author that our predicament was inevitable, and the following chapter will lay out the key events and decisions that have led us to the planetary collapse we are witnessing today in chronological order. While we may all disagree on the causes, a case will be built to demonstrate that deep and profound systemic change is necessary to avoid the worst possible outcome for the inhabitants of planet Earth. The list is not intended to be exhaustive but simply to highlight major influences that have shaped our destiny. Changing lightbulbs and switching to electric cars, while reducing CO_2 emissions, are unlikely to help us ultimately avoid the extinction of life as we know it on our planet of limitless wonder.

While the agricultural revolution 10,000 years ago certainly got the ball rolling when it comes to freeing up humanity to start producing non-food items and allowed some humans to accumulate wealth, it wasn't until the bronze age in 3,500 BC that archaeologists began finding differences in the way people were buried; some with bronze armour and others without. The diet of those able to produce bronze also shifted from mostly grains to large numbers of fish. Wherever bronze spread so too did the earliest form of inequality, until the first city states were developed in Mesopotamia: modern day Iraq. Here, according to Fabian Scheidler in his book The End of The MegaMachine: A Brief History of a Failing Civilization, the first authoritarian states with despotic leaders emerged. These states were able to distribute goods and supply a military and frequently fought wars with other city states to steal their accumulated wealth. They also mark the beginning of the human acceptance of limited freedom in return for state security. Here, within the walls of the city, as long as the rules of the leader were adhered to, the leader would offer protection from other city states. If the rules were broken however, the leader would give a sample of the physical violence on offer elsewhere. This was an early form of what we refer to as a protection racket (673). It is also similar to the Faustian bargain we have made with our governments today who spy on everything we do in order to provide us with "security".

This was the first time in human history that man had been coerced by man and Scheidler explains this coercion was down to three factors that he calls the Three Tyrannies. The first of these tyrannies is physical power, something he attributes especially to militarized states. Here, the state is able to "encourage" others to do as they are told or risk the threat of violence. We have moved from bronze axes to nuclear weapons and those with the most destructive weaponry are able to suppress those without.

The second of the tyrannies is structural violence, a tyranny much more subtle than the first and something many hardly even notice. The greatest example is that of wage work. On the face of it we often feel like work is a choice but when we break it down, failure to comply with instructions in a job eventually leads to the loss of employment which then has knock on effects that eventually lead to physical violence if for instance you fail to move out of a property that you can no longer afford to pay the rent on after losing your job. This structural violence is often seen as legitimate by the majority of citizens today with the help of an acquiescent media owned in large part by the ruling class. The structural violence is dependent on inequality to function effectively as a tyranny.

Our love of land ownership dates back to Mesopotamia around 3,000 BC. Here, as private property, in the form of family or clan land, occurred out of formerly egalitarian villages, there were lottery winners and losers. Some were awarded fertile land while others not so fertile. Add in irrigation which benefited some areas more than others and soon differences in harvest arose (673).

Today we would say that the men with the largest harvest were the smartest, hardest workers while laziness must be the cause of those with smaller harvest, when in reality luck was, and is a massive factor. As harvest inequality grew, wealthy men who owned productive land could allow less wealthy men to till their land in return for food. The rich could then dedicate their time to trade which further exacerbated inequality and also led to the buying of slaves who had been taken as prisoners of war. This further led to the introduction of non-personal credit where rich landowners would provide part of their harvest to poorer farmers who experienced a poor season. The borrower would then have to pay back the lender the next harvest but if this was not possible, the poor farmer would lose his land and possibly even place his family in the servitude of the lender, or in the case of women, they were often forced into prostitution to help pay off debts. At this point, the land began to be held by small numbers of men who used structural violence to control others. Slavery and prostitution have been part of our civilization ever since and today in many regions of the world, people without are traded by those with. Whether it is wage slaves working on Qatari World Cup stadiums having their passports removed until they fulfill a quota of work or Vietnamese workers smuggled into the U.K. illegally to grow cannabis or forced to sell their bodies, inequality is of the upmost importance. Likewise, debt has been a constant form of structural violence throughout history and failure to pay your pound of flesh is often backed up by the first tyranny, physical violence.

The third tyranny is ideological power. The first recorded use of writing was not to impress the ladies but to record goods bought and sold in Sumer, largely cattle, grain, and human slaves. Scribes became an important asset for rulers and firstly clay tablets but then papyrus and parchment were used to record the debts owed in the form of promissory notes. This was the earliest form of money and replaced the personal relationships that had usually allowed borrowing and lending. Now, those with excess could lend to those without even if they were unacquainted, safe in the knowledge that the debt was recorded and failure to repay would lead to an additional slave. Beginning around 1800 BC writing was also used to systemize laws which prevented subjective decision making while often legitimizing unjust power. Until modern times, the ability to read and write had been the luxury of those with time on their hands and money available to purchase books and this further led to the written word being used to keep the illiterate in their place. Most existing religions owe their survival to the written word of God and those able to read the word of God were able to exert power over those who could not. Today, literacy is taken for granted in most countries, but as we saw in the last chapter, the media who disseminate information are mainly controlled by the richest in society or the state itself and are frequently used to legitimize tyrannies one and two, physical and structural violence (380).

Money, Military, Complex

During his farewell speech on January 17, 1961, Dwight D. Eisenhower chose to warn U.S. citizens of the dangers of the Military Industrial Complex. He warned that the close cooperation between defence contractors, the U.S. Congress and the U.S. military could allow the rise of misplaced power with disastrous results. While the warning seems to have fallen on deaf ears, the danger was not novel and had existed since the sixth century BC.

In Greece, the "Cradle of Democracy", the first market economy arose, and this economy was bound tightly to the military. Until this time, farmers had been content to grow their own food. They provided a portion of their harvest to the state who then paid civil servants with the collected harvest. If this situation had endured, our modern system would have remained impossible as soldiers were part-time and always had to return home to sow their fields. Additionally, transporting supplies over long distances was impractical and looting only provided temporary provisions. What was needed was a system of money so that a standing army could be paid, and goods could be purchased wherever the soldiers were.

Unfortunately for the state, most farmers saw no advantage in this as they could provide for themselves given favourable enough weather conditions. To alter the situation to their benefit, Athenian elites introduced a law stating that all taxes needed to be paid in silver coins produced in state owned mines. Farmers were then forced to sell part of their harvest at market in order to earn coins to pay their taxes and the monetary system was born. The state then used these coins to build

a standing army which conquered more and more land. The main bene-
ficiaries were the private entrepreneurs who leased the silver mines
from the Athenian state and were massively enriched. These mines
were worked by 20,000 slaves working ten-hour days. The silver they
produced was used to pay the navy who then expanded the empire and
captured more and more slaves who produced more and more silver to
fund more and more soldiers to capture more and more land and thus
a perfect cycle of growth was developed (380). Our "democracy" and
market system from the outset were tied to slavery, war, and theft. If,
for any reason a farmer could not pay his taxes, there was now an army
ready to apply force. This cycle has endured, but today, the Athenian
state has been replaced by western corporations who use the military
might of the state to open up new markets, forcing the poor into sweat-
shops and sucking the material wealth out from underneath them with
the help of rich kleptocrats and tax haven secrecy. Today, Scheidler's
three tyrannies are on full display. Structural violence is essential to
persuade nations to adhere to the capitalist modus operandi. Examples
include Cuba, the Soviet Union and China. If nations are stubborn,
then physical violence is used to pressure them to conform. Think Iraq,
Afghanistan, Libya, or Syria. From a survivalist perspective, it's therefore understandable that the leaders of North Korea and Iran consider
it important to develop nuclear capability to ensure they are not
overthrown. Of course, they are practicing their own three tyrannies
on their own people in much less subtle ways. Structural violence is
additionally used to force low wage employees to work long hours in
dangerous conditions to produce cheap goods to supply to Western
consumers where they are vital to maintain the mirage of perpetual
growth and prosperity as wages stagnate or fall and the ecosystems collapse. Credit companies are on hand to lend to those in the West who
cannot afford that new TV, car, hot tub or even house. Failure to keep
up with the repayments will lead to legal action and possible seizure/
eviction. The bank, just like the rich landowner 5,500 years ago, gets to
keep your property. Then, the ideological power of the media and state
education are used to create the conditions that allow the first two

tyrannies to reign supreme, largely unquestioned. In the modern era, as many in the world have abandoned their belief in an all controlling God, their belief has been placed instead in the ability of man. Just as God was deemed to control all nature including that of humans, today, technology is claimed to be all powerful and many in the world have placed their bets on human engineering to win the day over the natural world, of which we are part (380). The fruits of winning this bet will be enjoyed by the few, but the stakes for losing will be felt by all.

The system of silver, war and slavery helped to fund the empire of Alexander the Great whose 120,000 mercenaries were paid half a ton of silver a day. The silver mine at Laurion couldn't cope with this enormous demand so Alexander forced local populations to mine their own silver and once again, former subsistence and communal economies were required to pay their taxes with coins. Soldiers were then able to pay for goods with their silver salary and the military industrial complex and market system spread further and further (380). Today, corporations gently persuade nations to participate in global capitalism through incentives like International Monetary Fund or World Bank loans. These loans are not being offered out of the goodness of the financial institutions' hearts. They are linked to the government opening the financial system to international banks and the privatization of services like energy and water and mineral resources. Global corporations are only too happy to provide their services which undercut those on offer locally and the nation becomes trapped in an attempt to escape the crushing debt it has accrued. Country after country in the global south exports its rich natural resources to the growth addicted nations in the global north. Should a country decide it doesn't want to play this game then structural violence is used to soothingly encourage them it's in their best interest. Among the tools on hand is an embargo, such as those placed on the Soviet Union, Cuba, Iran, Iraq, North Korea, and Venezuela. Should an embargo not have the desired effect then regime change is another popular option. The CIA has built up an impressive resume in the field of physical violence with coup d'états and uprisings organized in Cuba, Iran, Grenada, El Salvador, Venezuela, Panama,

Honduras, and Nicaragua. Of course, if the leader is particularly strong then Western military intervention is another possible tool. In recent times, Colonel Gadhafi, the once best friend of former British Prime Minister, Tony Blair, was overthrown and murdered in the Libyan capital, Tripoli. In 2003, the leader of Iraq, Saddam Hussein, was captured and hung in his hometown for the 1982 murder of 148 Iraqi Shi'ites in the town of Dujail. President Bush and Tony Blair, who were responsible for the deaths of hundreds of thousands in Iraq in order to overthrow Mr. Hussein were not hung for their crimes against humanity.

Thousands of years ago, as leaders accumulated ever greater wealth and built-up larger armies, the power was centralised like never before. It was difficult for citizens to resist the will of the state as the army was on hand to enforce rules with physical violence. Centralised systems continued to grow as more lands were conquered, more slaves acquired, and more silver was mined. The Roman empire, which stretched from the Mediterranean Sea to the North Sea, spread the use of silver money everywhere it marched. Fresh mines were dug in conquered Spain, France, and Tunisia, with slaves toiling day and night until their deaths to supply the empire's insatiable demand, but after centuries of depletion, the mines were eventually abandoned by the end of the second century AD. The fate of the mines may have been sealed by soldiers returning from the eastern Mediterranean harbouring the deadly Antonine Plague which killed approximately half the Roman population. What is thought to have been smallpox or measles also killed miners and the silver supply dried up meaning soldiers could not be paid, and the system was weakened enough that external attacks ended the Roman empire by the fifth century AD (674).

Today, money is no longer made from silver and many transactions are made digitally. The silver mines have been replaced by cobalt for our cell phones, uranium for our power, and lithium for electric cars, but the system is still largely the same as it was in Roman times. Countries are "encouraged" to join the global game of capitalism and their people are subjected to slave like conditions in order to extract their resources which fuel the global game of greed.

306 ~ SIMON WHALLEY

The case of lithium is a perfect example of extractivism at its best. Large amounts of lithium are being mined in South America. In Argentina and Chile, indigenous people are protesting against the mines which they say use huge amounts of already scarce water and create toxic pollution which is making their lands uninhabitable. In neighbouring Bolivia, the socialist leader, Evo Morales, who was already unpopular with the West, after unsuccessfully trying to take back control of the nations' water, gas, and oil resources from multinational corporations, further angered the global north when he asked for something that was unheard of. What was his request? In exchange for granting corporations access to the largest lithium supplies on the planet, he wanted to actually produce the batteries in Bolivia; rather than merely supplying the raw lithium to be exported to the global north where the batteries were to be produced. This would have allowed Bolivia to become an industrial nation but global capital, Tesla and the World Bank had different ideas (675). In 2019, a coup was instigated to remove Evo Morales from power and install a Tesla friendly capitalist regime so that Tesla had a steady supply of lithium for its electric cars. The world's richest man, Elon Musk, who owns Tesla, joked about the coup on Twitter saying *"We will coup whoever we want! Deal with it."* This brazen attitude is straight out of Roman times where the rich could, and did, take it all. There was widespread anger on the streets of Bolivia and a new election was called. A socialist government was duly ushered back into power but without Evo Morales at the helm (676).

Monotheistic Faith

Up until social hierarchies arose in Mesopotamia, humans had revered their fellow animals and the female form as is evidenced in cave paintings from the time. From animism in Japan to druidism in Western Europe, nature was honoured. However, shortly after man began controlling man through the money military system, religions started to focus on individual, universal salvation. Was it a coincidence that as life on Earth became more and more of a grind that religions began to offer moral life plans that provided relief from Earthly suffering? (673) Additionally, the money-military-slavery complex caused the traumatisation of entire populaces. As the Greeks and then the Romans subjugated populations, raped, and murdered siblings, parents, and children, those left behind must have suffered enormously. This was a time of incredible suffering and religion was there to offer much needed solace and also manipulation. In the modern era, as economic injustice has shredded the social fabric within communities, the environment is again rife for manipulation but this time instead of an all-powerful God, enter the demagogues who will bring back the good old days, AKA, Donald Trump, Viktor Orbán, Jair Bolsonaro and Nigel Farage.

It was in present day Iraq that the first religions occurred and just as most of us find Tom Cruise and his Scientology nonsensical, what must those hearing for the first time of all knowing, all creating Gods in the sky have thought? It seems they were equally unimpressed as the religious epics of the time were enacted for the King's court, rather than the masses who continued with older traditions (673). Nevertheless, as time went by, a growing normalization occurred and by 1800

BC, monotheistic religions had begun to take hold. They were to have a drastic effect on the lives of every species on our planet. The first of the Abrahamic religions to take root was Judaism and in the words of The God Delusion author, Richard Dawkins, it was:

> *"a tribal cult of a single fiercely unpleasant God, morbidly obsessed with sexual restrictions, with the smell of charred flesh, with his own superiority over rival gods and with the exclusiveness of his own chosen desert tribe."*

In Palestine, under Roman occupation, Christianity, a slightly less aggressive version of Judaism was founded by Paul of Tarsus, and after the collapse of the Roman empire, Islam emerged with a new holy text that was, like the Jewish Torah, uncompromising. The Koran also explicitly encouraged military conquest to spread the faith. Today, more than half the world's population believe that an Abrahamic God is in the sky watching over them. Many will argue that religion has a positive influence on humans so let's get the positives out of the way first. Some religious organisations provide charity in times of hardship, and it is certainly comforting to believe that when we die, we will float up into God's cloud house to drink tea with all our friends and family. For many, that certainly provides consolation, especially at the end of life. So too though does believing that Santa Claus delivers presents to all the children on the planet in just one reindeer methane burping adventure the night before Christmas. It doesn't, however, make it true, and should anyone past adolescence still believe in Santa, we would be concerned about their development. This is not a book about belief in monotheistic spaghetti sky monsters though, so I will leave that to Richard Dawkins, Christopher Hitchens and Sam Harris, whose books on the subject are must-reads. Here, we will focus on how our certain belief in non-Earthly fictional characters has helped lead us to the Earthly destruction we are causing today.

Anyone harbouring serious beliefs that any of the Abrahamic religions are peaceful is potentially displaying an ignorance of history. While the good books might advise you to turn a cheek or stipulate that you must not kill, all three religions at the very least have tolerated war and suffering in God's name and at worst have been the justification for conflicts around the globe. The Roman philosopher Lucretius had this to say on the subject:

"To such heights of evil are men driven by religion."

Since the Roman reign of Antiochus IV, when Palestine was conquered, and Jews burned and crucified for their membership in what Rome saw as a cult, religion has been used as a justification for torture, abuse, and theft on a colossal scale. From the Middle East to Southeast Asia, Russia to China, all over Africa and the Americas and Ireland, continental Europe and the Indian sub-continent, religion has pitted human against human in ways that have found no cure, and men, and women, are frequently ready to end their earthly existence, and that of others, to appease their all-forgiving God. This is not to say that people who believe in God are violent people. That is clearly not the case. Having said that, if the only thing stopping people from committing acts of violence is a vengeful ever-present God, then shouldn't society be concerned? Today, the majority of religious people will consider themselves peaceful, yet many frequently offer support for oppressive governments who claim the blessing of God and as we will see later, the vast majority continue to condone extreme violence in the name of ancient texts.

According to emeritus professor of sociology at Cambridge University: Michael Mann, Christianity realized early on that satisfying Rome and her empire's need for salvation was imperative for the fledgling religion to spread far and wide. Christians could see that something in the Romans' lives was leaving them with a yearning for redemption but what could this need be when their empire was at its peak? Mann believes they had the greatest need for salvation precisely because of Rome's imperial contradictions, often masquerading as achievements. Rome was a civilized society and yet the behaviour of her military was anything but. The citizens could enjoy some material wealth, yet it was founded, as we saw, on slavery, theft, and violence (673). Compare this today with the United States of America where material wealth has reached its zenith, of all the industrial nations, religious belief is especially strong. Here, 53% of Americans say religion is very important in their lives whereas in France the number is 11%, Japan 10%, Sweden 10% Germany 10%, the U.K. 10%, and China 3% (677). Additionally, 74% of Americans believe in a God, 68% in the divinity of Jesus Christ, 57% believe Jesus was born to a virgin, 72% consider miracles to be possible, and a staggering 58% of Americans think that the Devil lives in a place called Hell (678). Could the fact U.S. religious belief is high, and that their material wealth is also built on slavery, theft and violence be a coincidence? Perhaps not, as during the height of the British Empire, religious belief was also strong. In fact, belief in protestant Christianism was one of the three Cs of Empire, along with Commerce and Civilisation. Quite what was civilized about British imperialism is hard to fathom but believing that your religion is superior to others certainly won't hurt. The second British Empire (1750-1870) coincided with a series of religious wars against the Catholic French and Britain's imperial expansion was linked to protestant evangelism where missionaries often led the way. In their heyday in the 17th Century, the Dutch also turned to religion, this time Calvinism, which taught that God split people into groups of the chosen and the eternally damned. This belief was to come in handy as the Dutch raped and pillaged from those they deemed to have been damned.

Similar thinking was at play down under where the former Australian Prime Minister; Scott Morrison, a Pentecostal Protestant, believes in prosperity theology; where people who are rich are so because it is God's will and that those who work hard and save are God's people. Scott Morison considers himself to be sent by God with social equity and social justice not at the forefront of his agenda. Quite what his motivation would have been to help those of his citizens in need is difficult to see; likewise, his stance on the climate crisis was problematic as his religion believes strongly in the second coming of Christ. It isn't hard to see why he put so little energy into reducing Australia's carbon emissions even as the country was either burning or drowned under flood waters. Pentecostal Christians further believe that all other religions are false and that their believers are on the road to an eternity in hell (679). Mr. Morrison hardly seemed well placed to care for all Muslim, Buddhist, Hindu, Atheist and Jewish Australian citizens.

Stretching from the crusades in 1096 to India and Pakistan's grievances, the Iran-Iraq War, the troubles in Northern Ireland and the continued Israeli apartheid in Palestine, religion has been the source of the conflict. Catholicism was also spread through the sword in Central and South America where indigenous leaders were forced to acquiesce or be killed during the Spanish invasion of the Conquistadors. Hernando Cortés was a fastidiously pious man who insisted on removing his hat whenever he met a Catholic priest and used to become physically ill when seeing the idols indigenous people worshipped. However, he saw no issue with enticing Indian rulers into traps and then slaughtering them along with their men. This behaviour is at odds with the teachings of Jesus Christ and what type of all-knowing, all-seeing God would condone the killing of his creations? After witnessing the mass genocide of their people, indigenous Indians were converted to Catholicism at the sharp end of a sword, and they are among the most devout Catholics on the planet today. How is it possible that people who have witnessed the treacherous behaviour of an invading force in the name of a foreign God can then be converted into the faith of these dominant colonizers? It is possible that modern day inhabitants of the Americas are identifying with their aggressor in a mass form of Stockholm Syndrome. This escape mechanism is often used by people who have suffered trauma at the hands of a more powerful force, in this case the heavily armed, and as we will see later, a heavily funded, Spanish invasion force.

Christianity, according to Saint Paul is founded on four pillars. The first is that history is leading us to a divine moment where believers will find salvation and unbelievers will be damned. The second is that all people are part of this single story and the third states that whether people believe it or not, there is only one universal truth and no alternatives. The final, and perhaps, most dangerous pillar is that whoever knows this truth has the right and obligation to lead less informed people onto the same path (380).

These four pillars have been used to legitimize European expansion for the last 2,000 years, alternatingly under the guise of Christianity, the enlightenment, or more recently the free-market economy that must be followed..., or else. Just as former British Prime Minister Margaret Thatcher infamously said there were no alternatives, Europeans have claimed superiority in their ideology over everyone else and today, the march of Americanism is finally fulfilling the third aim of Christianity by Americanizing the entire world into a McDonald eating, jeans wearing, Disneyfied oneness. Augustine of Hippo once wrote that the gospel must be proclaimed everywhere and there should be no land without a church. Today, churches have been replaced by Western brand names like KFC, the GAP, Starbucks, and Coca-Cola.

Religious extremism has also led to non-state sponsored terrorist attacks on civilian populations in the U.S., Europe, Australia, Africa, Southeast Asia as well as the Indian subcontinent and it continues to be used as justification for Imperialism. Speaking shortly after nineteen Islamic terrorists, who incidentally believed they would receive 80,000 servants and seventy-two wives for their good deeds, flew aircraft into the World Trade Center and Pentagon in 2001, the American right-wing writer and chat-show host, Ann Coulter, opined, *"We should invade their countries, kill their leaders and convert them to Christianity."* It's certainly no coincidence that the insurrectionists who stormed the Capitol Building on January 6th, 2021, were largely Evangelical Christians, marching under Jesus's banner to implement God's will to keep Trump in the White House with signs declaring JESUS SAVES and GOD, GUNS & GUTS MADE AMERICA, LET'S KEEP ALL THREE. In exit polls after the 2016 election, a staggering 81% of white evangelical Protestants voted for Trump, with 64% of white Catholics and 57% of white mainline Protestants also voting for the shrewd, calculating bigot with the biggest ... eh... brain (680). Quite what these virtuous Americans see in Trump to place their trust in him is difficult to say when he has objectified women, slandered those "shithole" countries, locked kids in cages, created a Muslim ban, and quite frankly, encapsulates everything Jesus Christ would have objected to in humanity. Whatever it was, they certainly are following thy leader. Writing in War Is a Force That Gives Us Meaning, Chris Hedges, the son of a Baptist minister no less, warns:

"There is a danger of a growing fusion between those in the state who wage war—both for and against modern states—and those who believe they understand and can act as agents for God (681)."

The racism that has existed in the U.S. since its creation, was and continues to be, aggravated by white Christians' righteous belief in their superiority over those of darker complexion. Speaking in his first autobiography, the abolitionist and social reformer, Frederick Douglas states that through the conversion to Christianity of his former slave master, Thomas Auld, he hoped Auld would free his slaves or at the least treat them more fairly. Unfortunately, what happened was that Auld doubled down on his cruel intolerance and became more hateful. Douglas goes on to explain that Auld used his newfound belief in God as a license to inflict even more pain and suffering on his slaves and often quoted this scripture as he whipped his "possessions," *"He that knoweth his master's will, and doeth it not, shall be beaten with many stripes."* Again, this is not to say that all Christians are racist. This would be an abominable lie. Many Christians today still believe in the teachings of Jesus Christ and live their lives accordingly, by among many other examples, providing shelter for desperate refugees; it simply provides evidence for the ability of man to commit horrific violence in the name of an imaginary all-seeing God. Christianity was invented as a tool against barbaric Roman oppression and preached of agrarian love and compassion. There is no evidence that Jesus was himself a Christian but perhaps simply someone who believed that love was more powerful than hate and that enlightenment could give people the strength to endure the huge hardships that Jews were experiencing due to Roman tyranny. However, the Christian religion that was founded in his name has ended up supporting the actions of oppressive states who instigate the exact opposite of the teachings of their own savior. The Indian born writer, Deepak Chopra, divides Jesus into three interpretations. The first is historical Jesus who is discussed by St. Paul and the Roman historian Josephus. This is the man who preached of love and compassion. Then there is the theological Jesus who he claims was created by institutions later on and used for their own purposes to control populations. The third Jesus is the God-consciousness that we all have within us and exists in everything we see (682). The Jesuits in the U.S. and Canada are a good example of Jesus being used as a tool of oppression using the second interpretation. For over a century, they were involved in the sale of thousands of African slaves in order to raise funds to build churches and colleges, including Georgetown University. It took until 2021 for the problem even to begin to be addressed with the announcement of reparations possibly amounting to $1 billion (683). Concerningly, A report released by The Office of The Director of National Intelligence in March 2021 stated that the most lethal threat in the U.S. is presented by racially motivated violent extremists (684). More worryingly, the head of the Public Religion Research Institute, Robert P. Jones, claims that if you were trying to recruit for such a white supremacist cause on a Sunday morning in the U.S., you would have more success outside a Christian church than the local coffee shop. He adds that *the more racist attitudes a person holds, the more likely he or she is to identify as a white Christian" (685)*.

The trauma that emerged from the mobilization of mass despotic armies also produced another response, Apocalyptic thinking. This thinking stems from huge despair and desperation which must have been felt by generation after generation and led ultimately to the *Book of Revelations* written around 90 AD. The writer describes the complete annihilation of the universe with plagues and catastrophes raining down on Earth and an endgame scenario emerging. Today, as plagues and catastrophes rain down on us, once again caused by collective greed and indifference to the suffering of others, apocalyptic thinking engulfs us once more, though this time, with the threat of climate catastrophe, the extermination of nature, and the continued threat of nuclear war, the possibility is now acutely in our hands.

Apocolypse
Ted Van Pelt is licensed under CC BY 2.0.

Religion Versus Science

It is true to say that both religion and science have been a force of good and evil in our world. The teachings of Jesus Christ and Mohammed may have included love and compassion for our fellow beings, but they have been subverted by the powerful to aid their oppressive causes. Likewise, science has proved valuable in many fields from medicine to modern machinery, astronomy to astrophysics, and evolution to electricity. Without science, we would have no modern comforts, no TVs, no airplanes, no internet, no vaccines, and no lights at night. We would, also, however, have no guns, no weapons of mass destruction, no nuclear weapons, no climate crisis and no sixth extinction. It seems that worthwhile causes are always apt to be manipulated by those with power. The Republican party's co-opting of Christianity, and the Manhattan Project are two examples that spring to mind. But, while monotheistic religion, in its almost 4,000-year history has improved the lives of people in no discernible way (other than Islam in medieval times), science has at least offered a blueprint to understand the universe we live in. While science is far from perfect and vulnerable to politics and financial pressure, the scientific method is the best tool we have for discovering the secrets of our existence. It is clear, however, that it can be used against humanity at the literal flick of a switch, and the conveniences it has afforded humanity are in large part causing our downfall.

In The God Delusion, Richard Dawkins describes two stories of people receiving new evidence: one of a resolute scientist certain of a truth, and the other of a fundamentalist Christian scientist equally certain of a truth. The first story involves a well-respected scientist at Oxford University's Zoology Department who is certain that a microscopic feature inside a cell is not real, but an illusion. This gentleman had for years extolled his students of this fact, but when a visiting cell biologist from the U.S. gave a thoroughly convincing, evidence based, presentation at the university, the professor shook the presenter's hand at the end and said, *"I wish to thank you, I have been wrong these fifteen years".* The second story concerns a promising young scientist who received higher degrees in geology and palaeontology from Harvard University. Coming from a fundamentalist Christian upbringing, this young gentleman now had a dilemma on his mind. As a child, the Bible had convinced him that the Earth was less than 10,000 years old, but his research in radiometric dating of rocks at Harvard would have suggested the Earth is about 4,000,499,990 years older than the scriptures give God credit for. Additionally, he would have studied fossilised remains of dinosaurs that were not mentioned in the Bible but are clearly more than 10,000 years old. Faced with this quandary, the young man decided to go through the holy book and cut out all the passages that disagreed with the scientific evidence. After completing his project, the book was so scant that he could barely lift it without it falling apart. This was a crisis for the scholar who saw a clear choice between believing in the evidence he had seen with his own eyes or putting his faith in a book that was patently incorrect. Due to his early indoctrination, he chose to follow the second path and gave up on achieving his dream of teaching at a respected university. Of course, there was a third option of accepting the Bible as merely allegorical, as does Pope Francis, but fundamentalism won the day (686). In the first story, someone's long held belief had been shattered by strong new evidence that contradicted their certainty and the scientific method prevailed. In the second story, someone was presented with strong new evidence that changed their worldview completely, but this person decided to jettison reason and logic to cling to their conviction in an ancient text. These two stories concern well-read men capable of logic and reasoning, but not all are so adept and fundamentalist ideas are more easily adopted by those less educated and facing difficulties in life. This has led, and continues to lead, men and women to take up violence against one another in the name of an all-seeing sky God.

While monotheistic religions have long endorsed the killing of men in the name of thy father, they have taken a very different position when dealing with unborn foetuses. Here, the conviction of pious believers of many faiths is that all life is sacred and therefore abortion is morally wrong. The very same people who believe life is sacred and protest outside doctors' surgeries where abortions are carried out support the death penalty and celebrate when alleged terrorists are murdered in front of their families. This is contradictory beyond belief but when you consider that people believe in a sky God without even a hint of evidence, logic isn't the greatest strength of religion. This is why the conspiracy theory group QAnon is so well supported within the Evangelical Christian movement. If you believe there is an all-powerful, all-knowing man in the sky watching over everything we do and that when we die, we will meet him at the pearly gates of Saint Peter, then it's not much of a stretch of the imagination to believe that Donald. J Trump is fighting a war against a global paedophile ring led by the dastardly Tom Hanks. The anti-abortion voters are largely made up of people with misogynistic views and this is hardly strange considering they believe in a patriarchal God and worldview. Less than half of anti-abortion respondents in a U.S. survey wished there to be equal numbers of men and women in power (687). In contrast to 80% of pro-choicers who wished for more equality. Gore Vidal sums things up succinctly:

"The great unmentionable evil at the center of our culture is monotheism. From a barbaric Bronze Age text known as the Old Testament, three anti-human religions have evolved - Judaism, Christianity, and Islam. These are sky-God religions. They are, literally, patriarchal – God is the Omnipotent Father – hence the loathing of women for 2,000 years in those countries afflicted by the sky-God and his earthly male delegates."

Abortion, whether you agree with it or not, has had a hugely positive impact on our planet. Considering that there are around 125,000 abortions each day, that amounts to between 40-50 million per year (688). At present around 385,000 babies are born each day and there are approximately 150,000 deaths. Without women having the option of bringing a child into the world or not, our population would be astronomically higher and the sixth extinction that much closer. Of the millions of abortions carried out each year, around 45% are done so unsafely, predominantly because they are illegal in large part due to a belief in an all-seeing sky God, not due to logic and reason.

The Catholic Church has further muddied the waters in sub-Saharan Africa where some twenty-two million people are infected with HIV. Here, his holiness the Pope advised Catholics not to use condoms as they would likely exacerbate the epidemic. This is in direct opposition to the science that says condom use is highly effective in preventing HIV and other STDs. Speaking in 2009, Pope Benedict XVI claimed the fabric of African life was threatened by a contraception mentality. The Church's advice was to go against our natural instincts and abstain from sex altogether. In response, Rebecca Hodes, from the Treatment Action Campaign said the Pope's religious dogma was more important to him than the lives of Africans (689). The abstinence that his saintliness encourages has resulted in an epidemic of child sexual abuse by his very men of the cloth and somehow, he is arrogant enough to preach to others how to live their lives. Joseph Ratzinger, or Pope Benedict as he was known between 2005-2013, is believed to have known of the predatorial sexual behaviour of his clergy back in the late 1970s but did nothing to stop it (690). Quite why anyone on planet Earth should take moral advice from such a man is a mystery.

Another of God's missionaries was Mother Teresa. Being critical of Anjezë Gonxhe Bojaxhiu, as was her human name before she became flawless, is not advised. The Saint from Macedonia is admired around the world for the amazing work she did with the poor and destitute of Calcutta. On face value, it is hard to criticize a lady so much loved for building hospitals and preaching God's love. When you look behind the façade of moral perfection, however, things are not quite as pretty. In his book, The Missionary Position: Mother Teresa in Theory and Practice, the late Christopher Hitchens documents some extremely unsavoury acts the miraculous lady from Skopje carried out. When treating patients, often children, no pain medicine was prescribed as she believed suffering was God's will. All that was offered were comforting words and an assurance of God's love. Needles were rinsed with cold water and reused. The medical care offered in her institutions did not even match those of hospices. You may think this was due to lack of money, but her organisation had a budget of more than $29 million a year. Can anyone imagine the reaction they would have if they took their child to hospital in pain and the doctors refused pain medication on their belief that it isn't God's will? When the First Lady, Hillary Clinton, visited her New Delhi orphanage in 1995, babies usually wearing thin cotton diapers were suddenly deemed worthy of expensive Pampers and newly stitched clothing. Ms. Teresa's Missionaries of Charity in New York even refused to allow an elevator to be installed to aid less abled people on the grounds that modern conveniences were forbidden. This, from a woman that would jump on a plane at the drop of a hat to raise a few million dollars. She flew to Ireland to oppose the legalisation of divorce, then congratulated one of her main donors, Princess Diana, on her divorce from Prince Charles. Another of her donors was the U.S. fraudster Charles Keating who donated $1.25 million of other people's money to Mother Teresa. He also gave her use of his private jet. So much for abstaining from modern conveniences. Mr. Keating, who stole more than half a billion dollars from small investors, was sentenced to ten years in prison. During his trial, Mother Teresa wrote a letter to the judge asking him for clemency. She argued that Mr. Keating always tried to help the worlds' poor even as it was overtly clear that the money he gave to the worlds' poor was not his money. It is interesting that Mother Teresa is so well remembered for her love and compassion for the world's poor because in 1981, while in Washington D.C., on another flight that didn't require any modern convenience, she said this:

"I think it is very beautiful for the poor to accept their lot, to share it with

the passion of Christ. I think the world is being much helped by the suffering
of the poor people (691)."

It is evident that while religion offers consolation in our darkest individual moments, it has subsequently been responsible for many of humanities' worst events. In its obvious hypocrisy, it is worthy of little but disregard; but what of science. At first glance, science is nothing but a method of gaining an understanding through logic and experimentation, but science has also become a way for the powerful to gain advantage over the powerless, and at the macro level, for humanity to dominate nature absolutely. The Israeli historian Yuval Noah Harari describes both science and religion neatly when he says the core values of religion are about order and the main values of science concern power. He elaborates, *"Science, as an institution is interested in gaining the power, gaining control over the world, to be able to gain control over diseases, over the human body, over the environment, over rivers and animals and forests (692)."* Equally, science has always been used by the powerful to gain advantages. This is why the science of geography was so interesting to the European kings and queens who saw the financial benefit of "discovering" new territories and disposing those already there of their resources. Science, like religion, has come to inherit a universalism where indigenous knowledge has been discarded as backward and anything that cannot be neatly slid into a scientific classification is labelled as pseudo-science or quackery. The fact that this morning's knowledge could end up in the dustbin by tea-time is lost in the wilderness and we single-mindedly follow the blind path towards "progress". As the world becomes more and more secular, our faith in sky gods is being replaced by an absolute faith in technology. Like religion it seems, there is no other way. An unswerving trust in the ability of technology to fix the problems that technology caused is a form of religion in itself, this time however, we are worshipping at the altar of humanity. As the world burns, storms rage, and record temperatures are recorded on every continent, adults everywhere relax in front of brightly lit screens, safe in the "knowledge" that the tech geniuses are about to save the day with electric cars, brain implants and geoengineering. The bet is on, and nature bats last.

Science Worship
jurvetson is licensed under CC BY 2.0.

SALE ENDS TODAY
Miquel. C

Dominion

"And God said, Let us make man in our image, after our likeness; and let them have dominion over the fish of the sea, and over the fowl of the air, and over the cattle, and over all the earth, and over every creeping thing that creepeth upon the earth."

This short paragraph from Genesis 1:24-26 has caused so much pain and suffering on our planet that it deserves attention. Again, the hypocrisy is astounding. This is not to say that the scripture is at fault. It is the interpretation and actions of Christians that are exceptional. Genesis was apparently written by Moses while The Ten Commandments were apparently written by God himself. Why then do Christians follow the word of Moses and not the word of God? The Ten Commandments were crystal clear. Though shall not kill. This is unambiguously telling people it is immoral to end the existence of someone, and yet Christians point to the word of man to support their right to dominion over all the animals on planet Earth. Some will argue that killing is sometimes necessary and point to nature as justification. Lions eat animals, they will say. Unfortunately, lions have little choice but to eat other animals as they are carnivores and require taurine to survive. Lions also kill lion cubs of rivals; but we don't justify infanticide on the basis that lions do it. Additionally, lions fulfil a vital role in their ecosystem of controlling the number of herbivores. Humans patently do not need to eat animals to survive, and we certainly do not fulfil any meaningful role in our ecosystem, other than destroyer.

Today, the U.S. military machine tramples the rights of poor down-trodden people and then the President, whoever he is, says those magic words God Bless America. Why would God bless any country, let alone a country that brazenly breaks the majority of The Ten Command-ments? Thou shall not steal. Just ask any one of the peoples that the U.S. has robbed from, the indigenous Indians for a start. Thou shall not covet. Considering capitalism is essentially built on envy, this is hardly being adhered to. Remember the sabbath day, to keep it holy. Prob-ably shouldn't have shopping malls and sports events on Sunday then. Thou shalt not commit adultery. Monika Lewinsky or Stormy Daniels anyone? Thou shalt not bear false witness against thy neighbour. Pre-tending that you've been attacked by another country to justify war probably fits neatly under this one. I could go on, but the Lord only had ten rules and rape and paedophilia weren't on his list. There is no intent on the writer's part to single out the United States here as all believers of the Christian and Judean faiths are similarly guilty of such duplicity, both historically and in the present day. The U.S., however, does have the largest number of fundamental Christians, armed to the teeth, and just an inch away from violence at any given time, as was apparent on January 6th, 2021, during the capital insurrection.

While fundamental Christians have sabotaged the true teachings of Jesus Christ to support their narrow-minded intolerance, it is Chris-tianity at large which is responsible for using ancient texts to justify the mistreatment of all non-human species on planet Earth. When questioned about their right to use all non-human animals as mere commodities, the average Christian will point straight to the good book as vindication. They never stop for a second to consider the elephant in the room which is their literal word of God, the Ten Commandments, and quote straight from a passage which was written by a mere mortal: Moses. Whether it's exposing our simian cousins to radiation, forcing CO_2 down Peter Rabbit's throat, locking Peppa Pig in a small metal cage, confining baby calves in wooden boxes, shoving tubes down the throats of geese, keeping bears in cages to steal their bile, flogging horses to their deaths or battering sheep with hammers, we do so safe in the knowledge that God has given us dominion to act as we like towards our fellow Earthlings.

How could a species that harbours such noble conscience carry out the above abominations? This is where religion has been extremely useful. God made man in his image; we are told. We have souls; we are told, but animals do not; we are told. When you are contributing to abuse on an unprecedented scale, this must offer some level of comfort, but if we are honest with ourselves, we will admit internally a couple of things. One is that God did not invent man; man invented God. And we did it to escape the trauma inflicted by man on man. It had the noblest of roots. The second is that we all deep down know that the way we treat non-human animals is deeply wrong. Yet, we continue on this hugely immoral and painful path that not only diminishes what it is to be human, but threatens our very survival.

While both the Torah and Quran make the case for treating our non-human cousins with reverence, the behaviour of modern disciples is decidedly at odds with the scriptures. The Torah states clearly that *"It is forbidden to inflict pain on any living creature,"* and that *"on the contrary, it is our duty to soothe the pain of any creature"*. Likewise, the Hadith promotes reverence for all forms of life and the fair treatment of non-human animals. In practice, both religions do the exact opposite. While the majority of developed nations have scant, if any, in the case of Japan, animal welfare laws, one law they do have is that animals must be stunned unconscious when the knife is dragged across their throats, and they are shackled upside down to bleed out. This law was put in place to at least prevent a hugely painful death. It needs to be stated that even in this system, animals routinely have their throats slashed open when they are fully conscious, and their eyes are wide open as their bodies are dismembered along a conveyor belt that is a real hell on Earth. With religious slaughter, however, stunning animals is strictly prohibited. In both Kosher and Halal slaughterhouses, animals are conscious when they have their throats cut and are left on the cold blood drenched floor all alone to die an agonizing death. Religion is so powerful and animal suffering so inconsequential to human-animals that even in secular states, the laws to protect animal cruelty are bent to the will of religious beliefs rather than to protect the innocent. In allowing the already insufficient laws protecting animals to be broken because of an ancient belief in sky gods, we are placing man's belief in fictional characters over the very real suffering of non-human animals.

Christianity is certainly no better in its attitude towards the plight of non-human animals. Saint Augustine called vegetarianism absurd and a superstition. He went on to say that Christ never killed a human even if they were guilty and yet killed animals who were wholly innocent. This, to the mighty man of his time, was sufficient reason to ignore the suffering screams of God's creatures. He finished by saying that *"It is by a very just ordainment of the Creator that they [animals] have been subordinated to our use."* Pope Pius XII refused permission for a society for the prevention of cruelty to animals on the belief that doing so would confer that humans have duties and obligations to inferior species (693). Things have hardly improved, and Pope Francis refused a $1 million challenge to go without animal products for a whole month in 2019. The Million Dollar Vegan campaign, spearheaded by twelve-year-old climate change campaigner, Genesis Butler, offered to donate $1 million to the charity of his Holiness's choice should he accept the challenge. The project was created to allow the Pope an opportunity to lead by example in reducing his dietary carbon footprint, but the Pope declined the challenge for reasons unknown.

In the Orthodox Church, Matthieu Ricard notes that vegetarians were split into two categories. In his soul-searching book, A Plea for the Animals, he explains that those who abstained from eating meat as self-punishment were deemed "good vegetarians" and those that refrained from killing for compassionate reasons were called "bad vegetarians". There is a modern-day equivalent to this with those who give up eating animal products on doctor's orders being widely accepted by society while those who do so on moral grounds often being ex-communicated from the flock.

Philosophers of the cloth, so to speak, have taken dominion to the extreme like no others to commit acts of incredible cruelty. Descartes is one such man. He believed, in accordance with the scriptures than man alone possessed a soul. In his soulless eyes, he compared non-human animals to machines. His theory of animal automata led scientists of the day to nail live dogs to tables by their paws and cut open their bodies as they writhed and screamed in agony. These movements and sounds were to him but reactions of a machine and did not confirm the ability to feel pain. He ascribed the complexity of these machines to God, our creator. Emmanuel Kant also considered animals to be unconscious of themselves and unworthy of any moral obligations. Thankfully, reason was found in Voltaire who detested such barbaric practices and questioned why dogs would have all the same internal organs and apparatus of sensation as humans if they were not meant to feel anything.

As we have seen, an incongruent paragraph from a mere mortal has outshone the word of God and led to non-humans being persecuted on a scale never before witnessed. The Endowed Professor of Animal Ethics and Welfare at the University of Pennsylvania, James Serpell, summarizes the situation articulately:

> *"For more than two thousand years, European religion and philosophy have been dominated by the belief that some supernatural and omnipotent being has placed humanity on a moral pedestal, very much above the rest of creation. On the basis of this point of view, we have exercised absolute dominion over the rest of living beings, and we have even held the belief that their only reason for being was to serve our own egoistic interests. . . . The point of view of the early Christians, according to which animals were created solely for the benefit of humanity and the Cartesian idea that they were incapable of suffering are only mutually compatible variations on the same theme. Both have furnished human beings with a permit to kill, a permit to use or abuse other forms of life with complete impunity" (693).*

When it comes to the largest problem facing humanity today, the looming inhabitability of our planet, religious belief again stands in the way of us tackling the issue. While the more progressive Pope Francis published an environmental Encyclical letter in 2015, urging Catholics to reconsider their connection with Mother Earth, and talked of her crying out because of *"the harm we have inflicted on her by our irresponsible use and abuse of the goods with which God has endowed her"*, U.S. Christians of all denominations are taking a rather different approach. Research carried out in 2018 found that between 1990 and 2015, as the planet warmed and natural disasters proliferated across the U.S., concern around the climate crisis among pious Americans actually declined by a third. Compared with non-religious Americans, Christians were less likely to prioritize environmental protection over economic growth and were also more likely to believe that the threat of climate catastrophe was exaggerated. The research's author, David Konisky, of Indiana University's School of Public and Environmental Affairs, stated that many Christians believe human dominion over the Earth absolves them of any duty to protect the environment (694). Their devout Christian and Muslim cousins in the U.K. similarly display little urgency for environmental issues, especially the climate crisis, here due to their belief in an afterlife and more amazingly 'divine intervention' (695). It is perplexing to wonder why God would intervene to offer salvation to the single species responsible for breaking all his rules in order to trash the planet he spent the best part of a week working on.

As we have seen, religion certainly offers a sense of hope in an afterlife to those who need such comfort and there are millions of believers of all faiths who still believe in the moral teachings of Jesus Christ and Mohammed, and they act in compassionate ways to their fellow brethren to satisfy their ever-watching God. On the flip side, religion has been used to justify war, theft, and genocide. It stands in the way of scientific understanding and is a bastion of hypocrisy. It has also given humans a very false sense of dominion over nature and by offering people a second chance at a life in the clouds, it is also standing in the way of meaningful action on the climate crisis. Religion arose as a way of coping with oppressive regimes, but it is now being used to justify them. Only time will tell whether religion goes down in history as one of the causes of the end of human civilisation.

Dear Indy,

We have still not kept our promise of taking you to as many different houses of God as we can. You have seen your fair share of Buddhist temples and Shinto shrines, but you have yet to step inside a Church, Synagogue or Mosque. We want you to see the beauty of all the architecture, the smells and the sounds that make these places so special.

I was raised in the Church of Wales which is part of Protestant Christianity. Some of my earliest memories, good and bad, are from St Mary Magdalene's in Cymbach. Here, when I could barely walk, I would attempt to climb the altar to reach the booming voice of the Vicar, Geoffrey Gainer. It was here, on Sundays that I learned of the Good Samaritan, and I can still see the man's bloodied body as he lay on the road to Jericho, having been beaten, robbed, and left for dead. It was also here, that Johnny Whalley, drunk and disorderly, was banging on the Church door begging to be allowed in for midnight Mass one Christmas Eve. At the age of 13, I actually decided to confirm my faith, with Geoffrey Gainer carrying out the service at Saint Peter's in Swansea. My main motivation at this time was spending time with my friends who lived near the Church. I was never very religious, but I vaguely believed that there was probably a God, and his son was probably white with long blonde hair and blue eyes, just like me in the early nineties.

I had my first doubts when I moved to Taiwan in 2001. Here, for the first time, I was exposed to Buddhist and Taoist temples and learned of a God that didn't have long grey hair and a white beard and I began to wonder. I figured that either one of them was correct and all the others wrong, or that they were all wrong. For the first time, I started to feel that religion was fundamentally at odds with a peaceful life. How could people believe that their God was universal?

It wasn't until your mother, and I moved to Hahajima in the middle of the Pacific Ocean in 2009 that I thought more on the subject. I had an extremely easy job, and I was able to read a lot with my feet up under the desk. I read 'The God Delusion' first, then 'God is Not Great' and finally 'A letter to a Christian Nation' one after the other. My belief in an all-seeing Sky God ended on a small island in the middle of the ocean surrounded by nature.

Return of Silver Coins

In what were known as the Dark ages, life for the many was in some ways actually more preferable to what came before or after. During this period, peasants no longer had to finance large armies, so taxes were lower. Slavery was ended apart from in Spain and rich landowners' power was limited as they could not finance standing armies to suppress the populace. This was a time of strength for individual citizens, and landowners were forced once again to compromise. Fabian Scheidler provides one example of such compromise in 579 AD. Here, the Frankish king Chilperic I tried to collect taxes from the citizens of Limoges, but the people were not so inclined and decided to burn the tax files. The king promised meekly to never again request taxes. This would have been impossible during the Roman empire as well funded armies would have likely crucified the tax evading hordes. According to Scheidler, this period was a time where man's ability to control other men or nature was limited and this is why our elites would like us to believe it was a period of vast darkness, rather than a period of equity for the masses and sustainability for the planet. Unfortunately, this relative equality was not to last and when coins made a comeback, the rapacious desire for them would propel us faster and faster to where we are today.

Beginning in the tenth century, the Medieval Warm Period and agricultural innovations led to increased food surplus and the population rose as a result. The landlords were able to increase their wealth disproportionately and as coinage made its comeback, divisions once again began to appear within society. One such landowner was the church, and its officials began living luxurious lifestyles decidedly at odds with the humble existence of Jesus and his followers. Resistance to the elaborate riches displayed by the elites arose and were similarly stamped out in ruthless fashion, leading to the torturous reign of the Inquisition in the 13th Century. The church introduced new forms of heresy for anyone criticizing the wealth and power of the elites. The growth of the late Middle Ages continued to rise until climate and disease combined to shake Europe to its core. From 1315–1322, between 10-25% of the population died of famine and just thirty years later, the Black Death spread its way from the trading ports of Eastern Italy across the continent reducing the population by another third. The plague which killed between 75-200 million people did have a silver lining in that the resulting reduction in population and food production caused the elites to lose income and with it the power to dominate their citizens. The balance of power had shifted from there being too many people for too little land to too few people for too much land. This was a time of vast uprisings with peasants and craftsmen working together with an aim of creating an egalitarian society, free of class. Revolts occurred in Bruges, Italy, and England where the Tower of London was occupied in 1381. The English king was forced to acquiesce to many of the demands, ranging from increased wages to the end of bondage, at least until he could gather enough soldiers to crush the rebellion. This show of force did little to quell the insurgence with egalitarian communities emerging throughout Europe and requiring the full force of both the church and state to return the continent to the control of the powerful and wealthy, completely reversing the intent of Jesus and his early disciples. Those in charge had been deeply concerned, and they were desperate to hold on to the powerful positions they held. They were about to unleash on the planet the all-devouring monster that we know today as capitalism, and this wasn't driven by any particular thirst for knowledge or global conspiracy, it was a simple unplanned survival instinct.

After the collapse of the Roman military-money-slavery complex, power shifted back east to Constantinople, Aleppo, Cairo, and Baghdad who were trading luxury items with China, India, Persia, and

North Africa. The Italian city states of Venice and Genoa were at the Western edge of this trading system and home to Europe's most wealthy merchants, immortalized by Shakespeare. The fabulous wealth these merchants amassed enabled them to stay out of the powerful reach of the Vatican and other influential entities. Out of a population of 150,000, the Venetians boasted a navy of 36,000 people. With the use of over 3,000 ships, this navy escorted the traders everywhere they ventured. Soon, money would flow from these ports into the hands of explorers like Genoa born Columbus and Europe would wrap its powerful tentacles around the world, sucking up resources wherever they landed and mimicking the Roman military-money-slavery complex in egregious fashion.

Whereas the Arab merchants were separate from the state and did not resort to violence, the Merchants of Venice and Genoa were backed by massive military might. They were also supported by unswerving Catholic ideology and funded the first blood drenched crusade in 1104 in return for trade bases, monopolies, and other privileges in the eastern Mediterranean. These were the early days of modern capitalism, and the juggernaut was being positioned to become the hegemonic force it is today. During the fourth crusade to Egypt, the Catholic backed military machine got as far as Istanbul, then Constantinople. The Venetians had already cut a pre-crusade deal to receive three-eighths of any booty as a return on their investment. In 1204, they provided the ships and loans and in exchange, they received vast riches in the form of gold and silver which the Eastern Roman Empire had spent hundreds of years amassing. Buildings of all religious denomination were destroyed and thousands of people in the Byzantine capital were either raped or murdered in the process. The modern-day equivalent was the invasion of Iraq with countless killed, infrastructure destroyed and the loot going to multinational oil companies and those contracted to rebuild what the military arm of capitalism had destroyed. The only difference being that the bill wasn't footed by the corporations that profited, but by the U.K. and U.S. taxpayers.

As in the times of the Roman Empire, the princes of the Middle

Ages were faced with similar problems. Their armies were made up of farmers, so any military campaign had to be short, or the threat of food shortage would loom large. Mercenary armies were one solution to this problem, but these were problematic too. In the early fourteenth century, as in pre-Roman times, the economy was largely subsistence based with taxes paid in kind. Farmers had little interest in going off to risk life and limb in return for silver coins that were largely worthless. There was also a shortage of silver.

Once again, the elites worked tirelessly to create the ideal conditions necessary for war and capitalism to flourish. Finding fresh supplies of silver was hard enough with depleted mines being abandoned after the collapse of the Roman Empire. How though, could they get the population to embrace silver as payment when there was little to buy, other than food? Again, they settled on an insistence that taxes be paid in silver. Farmers and craftsmen were once again required to produce more than they needed to sell part of their produce in return for silver. The elites now had an abundance of silver to pay troops and they had created a market for the silver to lubricate. Now, all they needed were people desperate enough to die in distant lands for even more distant princes. This is where, according to Scheidler, a curious inheritance law common in Europe came in handy. Only first-born sons were able to inherit land with the others left to find their own path as farm labourers or craftsmen. In times of rapid population growth, there were usually a number of unemployed men looking for income. William the Conqueror used 7,000 such men from all over Europe to invade England in 1066. These men, flush with silver coins, could then purchase food, weapons, clothing, or sexual favours as the army marched. Scheidler points out that people were now contributing to the very system they had been trying to overturn in earlier agrarian uprisings. Today, we rarely consider the oddity of the military being comprised of similarly downtrodden people. It has become so force de rigueur that in the U.S., the only path to an education for many is to enlist in the armed forces and this goes largely unquestioned. Poor people of colour are "gently nudged" into fighting wars against other poor people of colour in order to be guaranteed of an education. This, in the home of the free, land of the brave.

Throughout history, wars have spawned innovation unlike any other times. When humans are faced with adversity; they often get creative. Whether it is sanitary towels, tea bags, vegetarian sausages or radar, war has been the catalyst for modernity. In the late Middle Ages, with the upswing in metallurgy that was necessary to provide the head-to-toe steel-plated armour worn by Europeans in battle, not to mention the silver coins needed to purchase them, new technologies were being created at pace. The silver mines had to go deeper and deeper, elevators and rails were invented hundreds of years before the first railways and the water-lifting wheel replaced 600 now redundant workers. Then, from the east, gunpowder arrived and soon, the elites had been sent tumbling into an arms race that has yet to end. Firearms were first used in the Hundred Years' War, but it wasn't for another hundred years that cannons were capable of smashing a hole through castle walls. In 1450, French guns destroyed an impressive sixty fortresses in just four days and drove the English out of northern France. Just a few years later, the Byzantine Empire crashed down around Constantinople as half-ton cannonballs rained down from the sky. In a battle for survival, newer and better weapons and defences had to be produced at lower prices. It was no longer enough to train and energise soldiers to win in battle. Princes and kings now had to raise as much capital as possible in order to pay armaments workers and soldiers. This encouraged a quick monetization of the economy and wage labour was enforced to build stronger fortresses and more powerful weapons. Thousands were uprooted from their lands to work in the iron industry and these people generated the later labour necessary for the industrial revolution. The iron industry also needed huge amounts of water to drive the bellows and rivers were dammed or diverted to supply it. This angered farmers whose irrigation water was depleted, and forests were slain for build-ings, equipment, and fuel. Rivers were polluted with ore and fish died as a result. The English so thoroughly destroyed their forests to build ships and weapons, that facing shortages of charcoal, Sweden, with lush intact forests, was able to become the predominant canon maker and it wasn't until coal use was adopted that England was able to exert its influence once again (380). The embracing of black gold nuggets delayed the exhaustion of materials for a few hundred years, but it also put us on the collision course with the climate that we are on today.

As we have seen, rulers felt compelled to raise investment for weapons and security and this is where the banks, especially those established in 12th Century Italy came in. Both Spanish and German Crowns were indebted to the Banco di San Giorgio in Genoa but investing in wars was a risky business, so some smart banks insisted on monopolies in return for loans. In the Sixteenth Century, Jakob Fugger began what we know of today as campaign financing, but in essence is bribery. Here, he loaned his favourite royal candidate enough money so that he could pay his way to become the next emperor of the Holy Roman Empire. In return, Fugger was rewarded with monopoly mining rights that afforded him complete control over his production chain, from mining, to smelting and sales. He became fabulously wealthy as a result and used this wealth to finance the Habsburg dynasty, and with the help of Genoese banks, the genocide that was about to be unleashed on the Americas (380).

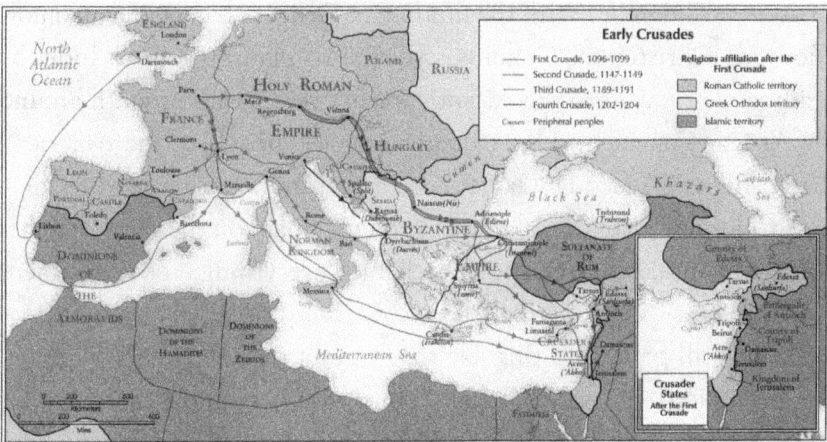

"Early crusades" by Norman B. Leventhal Map Center at the BPL is licensed under CC BY 2.0.

Genocide of the Americas

T he year 1492 will forever be etched in the minds of the indigenous peoples of the Americas. While it is celebrated in the U.S. and among her European compatriots, it is a day of infamy for those on the receiving end of what can only be described as genocide. The feverish demand for capital accumulation by those elites trapped in an arms race drove the destruction which ostensibly began forty years earlier after Constantinople fell to the Ottomans. Until this point, trade could pass through on its way to Europe and these profits could go into improving weapons. With the trade route affectively blocked by the Ottomans, the Europeans began looking for a sea route to the east. The Portuguese were first to build bases along the East African coast and then the Spanish turned to an Italian pirate who went by the name of Cristoforo Colombo. With the funding of Genoese banks, the man now known as Christopher Columbus started to plan his trip to India by heading west, out into the Atlantic, rather than heading south around the African continent. Unfortunately, for the people of the Americas, the Portuguese Vasco Di Gama didn't reach India until 1498, by which time Columbus had long planted the flag of Spain on the sandy beaches of the Bahamas, and in the process claimed everything on the continent for her majesty, Queen Isabela I.

Many still believe to this day that Columbus was a naïve explorer, but his actions and words suggest otherwise. Columbus stated in his logbook that *"With 50 men, we could overpower the lot and do with them as we please."* In fact, he not only did do with them what he pleased by enslaving 500 of them to help him sell the second trip he was already planning, but promised the Spanish Crown, *"as much gold and as many slaves as you like."* Unfortunately, for him and indigenous peoples, the money for this second trip was coming from the banks of Genoa, Augsburg, and Antwerp; and they wanted a nice little return on their investment. The pressure was now on Columbus and his men to provide just that return on investment, in the form of gold. It was this desire for gold that drove the butchery about to reign down on the Americas, as unfortunately, the continent was not home to much of the stuff. Natives over the age of fourteen on Hispaniola were ordered to deliver gold every three months or have their hands cut off and bleed to death. Faced with this demand, which could not be met as gold was so scarce, the indigenous people tried to escape to the mountains where they were hunted down and hanged or burned alive. These Arawak Indians were reduced to half their population of 250,000 within a couple of years and as it became apparent that there was no gold to be found, the Spaniards decided to force them to work on plantations instead, where they died in huge numbers. Between 1515 and 1550, the Spanish reduced the number of Arawaks from 50,000 to just 500.

As we saw in the wild animals section, wherever the Spaniards went, they were accompanied by disease which caused the deaths of many. This unseen plague, though, is highly unlikely to account for the true numbers slain across the continent. It is estimated that around forty-seven of the fifty million population died within 150 years of Columbus's "discovery". Scheidler points out that even if 90% of those deaths were caused by disease, there were still millions put to the sword in the name of the military, money, slavery complex, with the help of Catholic ideology. Additionally, the majority of those disease caused deaths arose from the inhumane conditions indigenous people were forced to endure in order to turn a profit for the financiers back in Europe. At this stage, I'm sure some will be denying that such terror was cast from Europe across the Atlantic Ocean, so the words of a Catholic Bishop present at the time might help assure you it did:

"The Spaniards do nothing save tear the natives to shreds, murder them

and inflict upon them untold misery, suffering and distress, tormenting, harrying and persecuting them mercilessly."

These were the words of Bartolomé de Las Casas, and he goes on to describe soldiers tearing babies from their mothers and smashing their skulls against rocks while others were thrown alive into rivers to drown. In case, this evidence is too distant from Columbus, perhaps the man's own words will suffice to highlight the character of the man so applauded by Europeans and their decedents. Writing to a friend of the Spanish queen, he said, *"There are plenty of dealers who go about looking for girls; those from nine to ten are now in demand, and for all ages a good price must be paid (696)."* In memory of Jesus and his twelve disciples, thirteen men would be hanged from trees and burned alive.

The true motivator for these punishments was not Christendom though, but the conquistadors' desperate need for gold to repay their debtors back in Europe. Not only had Columbus borrowed vast amounts for his invasion, but individual soldiers had to pay for their own armour with whatever valuables they could steal. An Aztec account from the time describes Europeans holding gold up in the air as if they were monkeys with a vacuous look of joy on their faces. Even the Spanish conquistador Hernán Cortés is believed to have expressed the "disease of heart" which him and his men suffered from and that could only be cured by the accumulation of gold. It was not gold, the Spaniards found though, but silver, high in the mountains of Bolivia. After torturing and killing the indigenous Indian population, the conquistadors finally found the precious metals they so craved and they duly turned the small town of Potosi into both nightmare and dreamland. It was a dreamland for the Europeans who paved many of the streets in silver extracted from the mines and a living hell for the tens of thousands of slaves forced to work deep underground, at any given time, digging and smelting. It is estimated that around eight million people died in the mining operations in Potosi and the surrounding vegetation was obliterated by poisonous gasses from thousands of smelting fires burning on the slopes. When the Spaniards ran low on local slaves, they tried to replace them with slaves from Africa and these people too, shortly died. The mountain was emptied of 40,000 tonnes of pure silver, which resulted in the mountain collapsing, and shipped it off firstly to Spain where it was counted before the banks' pound of flesh was sent on to the expedition creditors in Genoa and elsewhere. Unfortunately, for the town of Potosi, it isn't only silver that sits under its expansive mountains. The area is also home to the world's largest lithium deposits, and with electric cars about to replace the combustion engine, it is about to be torn to pieces once again, this time to satisfy the profiteering car companies, battery makers and their financial creditors (697).

There was perhaps another reason behind the Spaniards' sadistic cruelty towards the Indians. In addition to their debt to European bankers, they were also suffering after fighting a centuries-long war against the Muslim Moors. Throughout the Dark Ages, Islam had been experiencing a flourishing that resulted in prosperity and progress, but with the help of the Catholic Church and again, the banks of Italy, they defeated the Islamic reign in Spain by 1492, and forced both Muslims and Jews to leave the Iberian Peninsula. This long war had created a population that was hugely traumatised and accustomed to the kind of violence the Conquistadors exhibited in their quest for riches. After arriving in the Americas, the Spanish could not understand why the indigenous people refused to physically punish their children. This had become so ingrained in their psyche (380).

In today's hyper partisan world, the argument supporting this violence and theft is that the perpetrators were simply a product of their time and therefore cannot be judged by today's standards. Whilst it is always problematic to do so, it is also often too simplistic to blame past crimes merely on our improved moral standards. While Europe was raping and pillaging across the Americas, Africa and Asia, the traders of both the Islamic and Chinese Empires were doing so without the threat of violence. The Chinese were sailing as far as the Persian Gulf and African continent with as many as 27,800 men and 1,500 vessels by 1405, but they displayed no imperialistic ambitions and wished solely to create trading relationships (698). Likewise, during the Crusades, it was Christians ransacking Muslim cities and screaming Deus Vult, God wills it, and not the other way around. Muslims never rode across Europe on horseback to burn French and English cities to the ground. Today, once again, it is the Imperial West that invades Muslim countries and redirects their vast resources to satiate their desire for constant growth, and it is the U.S. that encircles China with 400 bases in the Philippines, Okinawa, Japan, and South Korea. There are no such Chinese bases in Mexico, Canada, or Cuba. While the U.S. Navy sits just outside of Chinese territorial waters and points missiles at the Middle Kingdom, the Chinese Navy poses no such threat to U.S. cities (699). Yet, through the billionaire media, we are led to believe that China is the aggressor, not the other way around. It appears the difference between these cultures was the belief in a single universal truth that the Europeans and their heavenly descendants held firm. Indeed, Columbus himself was sure that his actions were part of a plan to save the world from an apocalypse brought on by a fight between holy goodness and heathen evil.

As interesting as the motivations were, as the vast supply of stolen silver found its way from the Americas into the hands of European elites, inflation peaked at the same time as the elite's power to suppress the people. Peasants were removed from their land to allow landowners to graze sheep for the textile industry and many people became homeless just at the time food prices were soaring. Across Europe, wages were kept low and didn't return to their pre-genocide levels until the end of the 19th century. Scared of the egalitarian movements which had come close to removing their power, the elites used tactics involving the persecution of witches to suppress any thoughts of equality. These times, which are considered enlightened in comparison with the Dark ages, culminated in the torture of innocent women between 1550 and 1700. According to Sheidler, this persecution was not by chance, but a way for the elites to sow division in the peasant movements opposed to the enclosures of the commons, and especially at the heart of each and every peasant family. The media of the time, as they are today, were used as propaganda tools to create paranoia throughout communities and deflect attention away from the real crime that was occurring, the removal of peoples' ability to live self-sufficiently. It was here that the powerful created the myth of peasants as lazy, while the elites were hardworking, when in reality the peasants had been made homeless due to the elite's theft of silver from indigenous people in the Americas. This myth still surrounds us today with billionaires declaring themselves the rightful masters of the universe due to their sheer hard work and intelligence while those less inclined to steal, beg or borrow, decried as lazy and stupid.

By the reign of Henry VIII in 1530, the criminality of poverty, caused by the enclosure of the commons which was itself a result of theft and genocide, was being written into law with vagabonds whipped for their first offence, having part of their ear sliced off for the second, and finally killed if they refused to become a wage earner a third time. Edward VI gave powers to those who denounced someone as an "idler" for refusing to work. The 1547 laws included being forced into the slavery of the denouncer and allowing the master to request anything of the slave, no matter how demeaning. If the newly acquired slave decided slavery was not for him, he was to be branded with an "S" on the forehead and forced into slavery till the end of his days. Should he continue to protest that slavery was not becoming of his time, he was to be put to death.

As you can imagine, this treatment didn't go down to well with peasants across Europe and movements such as the Anabaptists surfaced with the intention of creating independent non-violent and self-determined communities. These movements were none too popular with the powers that be, including ironically the Church, who were not keen on the teachings of their own redeemer, Jesus Christ, whose ideals the Anabaptists had based their vision on. The peaceful insurrectionists were systematically tortured and executed with the assistance of the financed armies of Jakob Fugger, the man who helped finance the genocide of the Americas which had led to the appalling conditions the Anabaptists were rebelling against.

The Greedy Leviathan

Until the creation of the Dutch East India Company in 1602, companies were limited in life span, usually just to the end of a single voyage. The newly developed joint stock company radically changed this by existing in perpetuity. No longer were the earnings of individuals tied to their lifespans, now the company existed across generations, and it had just one aim, to increase its value. The immortal corporations of today, many of which enjoy the same protections as mere mortals, must constantly increase their capital through any means necessary. For the last 500 years, these corporations have been plundering the planet's finite resources in order to grow infinitely. This simple mathematical impossibility, foisted on us by a lifeless and limbless creature, has created the conflict we find ourselves in with Mother Earth. While the accelerator and brake are operated by humans, the path must remain the same, never ending capital accumulation. As forests are felled to graze cattle or oceans are emptied of sentient life, regardless of the damage that is inflicted, the drivers must press on. Should the capital accumulation slow down, they will simply be replaced by another driver who knows how to maximize the returns to the distant investors who could also now, not be held accountable for any damage they inflicted, other than a fluctuating dividend. This is why we are still subsidizing the very same industries, fossil fuels, animal agriculture and so forth, that are destroying our right to life. Not only was this new braggadocious beast endlessly insatiable but it also had no need to display a conscience as investors' liability was protected from the actions of the Greedy Leviathan (380).

The Dutch model, complete with its own private army, was later copied by the English who had eyed the riches of the East since they defeated the Spanish Armada in 1588. After, the voyages of Francis Drake on Golden Hind returned investment of 5,000 percent, English elites were keen to exploit the rich resources of even more countries. The English East India Company, also referred to in oxymoronic terms as the Honorable English East India Company hired its own private army who were then used to subjugate populations in India, Sri Lanka, Southeast Asia, and Hong Kong. The English joint stock company took things even further by producing their own coins and ensuring that all employees and subjugated peoples were under their criminal and civil jurisdiction.

While the dominated people of these corporate colonies suffered at the hands of armed forces intent on guaranteeing access to profitable resources, the investors back in Europe were free to gallivant around town and fraternize with kings and queens. The poor of Europe were forced into often humiliating wage slavery to produce weapons and supplies, and the poor of Asia were robbed of their resources, and in many cases killed, all the while the elites were partying like it was 1999. The descendants of these simultaneously barbaric but *civilized* thieves are today treated as Gods, with lavish weddings and front pages dedicated to newborn babies born into stolen wealth.

North American Genocide and the Birth of a Religious State

As Europe began to reap the rewards of its violent foreign conquests, a new form of religion arose. In what would come to be known as Calvinism, its followers believed that God had ordained people from birth as either chosen or eternally damned. Conveniently for those born into wealth, there was no way of changing your position and to try was against God's will. Understandably, Calvinism took hold in places where income inequality was greatest as it offered a legitimacy to those who had and delegitimised those who did not (380). In England, where Protestantism had replaced Catholicism in 1530, in an attempt to rid the Church of corruption, Calvinism attempted to continue the purification and from 1560 onwards, these virtuous Christians became known as Puritans. On the one hand they wished to return the Church to its original simplicity and on the other hand they were legitimizing extreme wealth, which was declared a sin by Jesus. With Europeans' firmly held belief that there can only be one God and that each country must have a uniform religion, there was no room for two in the bed (700). By 1620, with a rise in persecution, 102 Puritans set sail from England and after a sixty-six-day voyage on the Mayflower, they landed at Cape Cod on November 11th. After a brief stay, they sailed up the coast to Plymouth, New England, where they built a settlement (701).

While half of them died before the end of their first New England winter, twenty-four males survived to keep their lineage alive. They would not be forgotten (702). Beginning almost a decade later, a further advance party left England to prepare a colony in Massachusetts Bay. Harvard University was set up in 1636 and the English printing press was started in Cambridge three years later. Faced with increasing persecution under King Charles I, a total of 20,000 Puritans left Europe and set out to join the colonies in New England by 1640 (703). The Puritans were funded by public limited companies and poor English colonialists served as serfs for up to seven years in order to make profits for shareholders in London (380). As the settlements grew out into Connecticut, the English came into conflict with the indigenous Pequot people who intermittently killed Puritan men. On May 26th, 1637, the Puritan God loving people decided enough was enough and safe in the knowledge that God had chosen them, they began systematically wiping out the Pequots, beginning in Mystic River. The Puritans liked to arrive before dawn when the Pequots were still sleeping. By July 28th, the vast majority of Pequots had been killed and any left were sold as slaves or escaped to join other indigenous tribes (704). To justify their slaughter, the Puritans cited Romans 13:2:

> "*Whosoever therefore resisteth the power, resisteth the ordinance of God: and they that resist shall receive to themselves damnation (705).*"

This was not the first time the English had attacked native Americans, and it certainly would not be the last. The English had begun the Indian genocide as early as 1609 in Jamestown, the very first English settlement in North America. They had only been on American soil for two years before they began sending out raiding parties to demand food. The English soon moved from demands to burning down the Native American houses and stealing what wasn't theirs. For the first few years, the Indians and English had traded amicably, but with the increasing English demands, the Powhattan people finally snapped (706). At the end of 1609, the Powhatan warriors laid siege to Jamestown and starved the English. By the end of the winter, their population had been cut from five hundred to sixty-one. The Winter is known as the Starving Time, and some English were reduced to cannibalism to survive (707). The English kept on sending ships and many continued to die, but they also continued to expand, and this alarmed the Powhatans, who in 1622 decided that they needed to wipe them out for good. They attacked Jamestown on March 22nd and massacred 347 men, women, and children. This event persuaded the English that it would not be possible to live with the Indians so they decided they would simply exterminate them instead. When Columbus landed on the Americas in 1492, there were an estimated ten million native Indians north of Mexico but by 1900, their population had dropped to just 237,000 (705). The Europeans, and their diseases, killed so many men, women, and children in the Americas that the amount of CO_2 in the atmosphere dropped and the planet entered the Little Ice Age (708).

The founding fathers had envisaged a nation where religion and state were separated, but by July 30th, 1956, the evangelicals of the country had pushed for *"In God We Trust"* to be the new motto of the United States. It was soon on every coin and note in every pocket. Historians of 18th century Evangelicalism: *"readily acknowledge continuity,"* from Puritanism (709). More incredibly, an astonishing thirty-five million U.S. citizens claim to be descendants of the original Puritans. With the rise of Trump and his Evangelical base, the original Mayflowers would be proud of their hard-won fight (710).

On August 20, 1619, European sadism continued apace when the first African slaves were sold by the Portuguese to English colonists in Jamestown: Virginia. The sale of around twenty or so Angolans who had been kidnapped and transported by ship across the Atlantic marked the start of 241 years of trade in human beings that destroyed entire communities, resulted in vicious sexual abuse and torture, and caused ongoing psychological trauma. Of the approximately twelve million human beings torn from their lives, 50% died on forced marches through the African interior and a further 20% perished on the ships where they were kept in their own excrement for weeks at a time. While the suffering of those traded can hardly be contemplated, the trade also enriched many European elites and made vast profits for banking institutions like Barclays and HSBC, as well as the Church of England. The joint stock model was again the preferred model for selling human beings who were called "black ivory", and investors in these companies included the "enlightened" John Locke and Isaac Newton (380). The economies of the U.S., and Europe were driven by the free labour that was forced at gunpoint and today, modern cities like Liverpool, Lisbon, Bordeaux, Bristol, Nantes, and Glasgow can thank their city status on the forced labour of the African continent. The elites in Europe additionally benefited through the trade providing access to cheap calories in the form of sugar cane which could keep wage workers nourished on minimal wages that enabled them to maximise their profiteering.

By the 1700s, the elites were no longer killing people for failing to agree to wage work, which was in essence slavery as you earned so little as to be only able to feed and clothe yourself. The new stick they were using was the workhouse where many people who had been made homeless after being deprived of their right to self-sufficiency ended up. Life in these workhouses was intentionally appalling in order to encourage people into the alternative which was wage slavery. The soldiers of fortune had been the first wage workers who had no choice but to continue fighting for their masters or face a life of destitution, which many eventually did. Slowly, as life on their own land became increasingly impossible due to high taxes and land enclosures, people were forced into the sphere of wage labour and as the industrial revolution approached, factory systems of work became more and more common. By the time coal had replaced wood as fuel, craftsmen had already been largely removed from the complete production of goods and were now completing simple parts of production in the way that fast food and factory workers do today. In this way, peoples' ability to provide for themselves was further reduced and they were evolving from people into workers, all the time guided by a drive for maximum productivity (380).

Industrial Revolution

It would have been tempting to begin Part Three in the eighteenth century, such the impact the events of this period have had on our current problems. An argument could be made that this was the beginning of the Anthropocene with energy usage in Europe rising from just 13 gigawatts to over 1500 gigawatts today. It was the black rocks of million-year-old sunlight that were the physical catalyst, and they went hand in hand with the ideology of constant capital growth to propel us to the brink of the sixth extinction. With the invention of the steam engine, mines were able to go deeper and deeper and in turn more and more coal could be extracted. This allowed for more and more iron and steel to be manufactured which was used to build additional steam engines, interconnecting railways and canals and roads which allowed for more and more coal to be transported. This cycle of increased coal extraction and growth enabled humanity to escape the physical limits that burning trees for energy placed on growth. The escape though, was only temporary, and we have now passed through the boundary of safe levels of CO_2 in the atmosphere. Many of the techno fantasists would have us believe that all we need to do is switch from burning coal, oil, and gas to renewable sources of energy, but this would only deal with one of the problems that faces humanity. While, we have no choice but to make the switch to non-fossil fuel-based energy sources, should we make the leap by Midnight tonight, we would still be cutting down the forests to satiate our system's need for eternal growth; we would still be invading other countries to extract their resources; and we would still be destroying the oceans and causing the sixth extinction. At some stage in the near future, we will need to address the elephant in the room which is today's pervasive ideological dogma of eternal growth on a planet with fast dwindling resources and ever encroaching planetary boundaries.

Back in the eighteenth century, getting people to work in the new factories that opened to mass produce goods was not an easy task, even after the people had been removed from their land and their right to self-sufficiency. Many people preferred to be homeless than work in the satanic mills of Northern England or the perilous mines deep underground. By 1795, with many destitute, the British state belatedly intervened by introducing a minimum wage, largely due to worries over further social strife. The captains of industry were aghast at this intervention as in their eyes, workers were lazy ungrateful good for nothings. A British doctor, Joseph Townsend, commented that *"Hunger will tame the fiercest animals, it will teach decency and civility, obedience, and subjugation. It is only hunger which can spur them on to labor."* Clearly, this sentiment was not lost on others of his ilk as in 1834, the already feeble support system for the nation's poor was dismantled and workers were forced to compete head-to-head in what we now call the free market. What came next was literally Dickensian. In the following years, slums appeared throughout British cities and people lived in conditions that *"no person would stable his horse in,"* according to a government commissioner of the time. Women were hit hardest with many forced into prostitution and London being home to upwards of 100,000 desperate women. Girls as young as ten sold their bodies to satisfy the powerful men of the day. Just like people in developing countries today are forced to migrate to cities where they live in absolute squalor in order to fulfil their role as a mere cog in the capitalist machinery, so too were Europeans during the industrial revolution. They were torn away from their families, landed in cramped and squalid conditions with no safety net and no protection from the diseases that were prevalent. All the cards were once again held by those with power, and they were not frightened to play them. This is the free market system so popular today and which encompasses much of the globe, and it all occurred at a time of great economic flourishing (380).

What happened in England soon spread to other European nations and also the U.S., but the industrial revolution also impacted countries further afield and unlike the Europeans' desire to industrialize, many of these countries were not so inclined. One such country was India which boasted 27% of the planet's GDP in 1700. According to Yorkshire born American Unitarian Minister, J.T. Sunderland, India already possessed *"every kind of manufacture or product known to the civilized world,"* and *"was a far greater industrial and manufacturing nation than any in Europe or any other in Asia."* Unfortunately for India, this was not part of Great Britain's plan, and they went on to completely dismantle Indian industry. The thriving textile industry was first to go as it competed with its own and was far superior in quality. The British would import raw materials from India, manufacture them in factories in Northern England and then re-import them into India and elsewhere, where they would be sold. This was all carried out by the Honourable East India Company with many investors belonging to the British Parliament. They were supported by an army of 260,000 by 1800. The British transferred enormous wealth from the sub-continent, £18,000,000 each year from 1765-1815, and this helped drive their industrial revolution. While, the British were not as maniacal as the Spanish, they nevertheless showed a complete disregard for Indian lives, and between twelve-thirty million Indians died of starvation during the Raj. While extremely ruinous El Nino weather systems caused problems, many of these Indians did not die of food shortages. In fact, the grain silos were full, and India actually exported 320,000 tonnes of wheat to Europe between 1877-1888, the worst year of the famine. The silos were under armed guard and people perished outside as traders bet on the rising prices of grain. The railways could have delivered the grain to the hungry, but they too, were more interested in the higher prices the global grain market could provide. The Viceroy at the time, Lord Lytton, when asked to intervene in the markets to help those starving, insisted on the *"prohibition of interference in private trade."* He was also asked to suspend taxes on Indians but refused as he was desperate for funds to fight his war in Afghanistan. Those unable to pay had to sell off their last assets just to meet his demands. Unbelievably, even by the Raj's record, Westminster passed the Anti-Charitable Contributions Act in 1877 that imprisoned those who offered financial assistance to Indians in need. As Indians starved to death, the wheat merchants in London were making a killing, quite literally. The British finally agreed to "help" by setting up workcamps where completely malnourished people were forced to carry out back-breaking labour in return for inadequate rations. Many more died in these camps from the mixture of extreme labour and minimum nutrition (380). This is not taught in British schools and is hardly mentioned in the West, perhaps because it highlights the atrocities carried out in the name of profit. Today, we currently grow enough food to feed a population of ten billion and yet we have close to a billion who go hungry every night while neon-lit supermarkets and convenience stores are stocked full of unnecessarily wasteful and nutritionless, plastic drenched garbage. Those who can afford to pay the highest price receive the goodies. Ignoring the huge loss of life, by the time the British left in 1946, India's share of manufacturing exports had fallen from 27% to just 2% and Britain had climbed up the back of India to become one of the richest countries while India had become one of the poorest (711). This theft of India had enriched the British elites still further and the gap between those desperate to amass power and fortune and those longing simply to be self-sufficient grew wider and wider.

In 1791, as nations around the world were entering the servitude of others, the people of Haiti were fighting to finally expel their French colonial masters. Unhappy with this, the French decided to cut the country off from the rest of the world. In 1825, fourteen French warships were sent to the island to offer Haiti their "independence" if they agreed to pay 150 million Francs ($21 billion today). The money was intended to cover the lost income the plantation owners had endured. Until 1947, whenever the Haitians declined to service this odious "debt", French gunboats were sent in. The country continues to be one of the poorest in the world today. Haiti was one of the first countries to find itself at the mercy of unsavoury debt and while Haitians were faced with both structural and physical violence, today, the specter of debt is usually enough to keep a nation of people in servitude, not to foreign states, but to the corporations who require their raw materials (380).

South America and India were not alone in being ensnared in the capitalist net. In the century after the Napoleonic War ended in 1815, a long period of peace in Europe was experienced. This became known as The Hundred Year Peace or Pax Britannica. Unfortunately, it also coincided with mass violence outside Europe's borders where European nation states simply outsourced the suffering. The African continent, Middle East and Australia were overrun by the end of the eighteenth century and during the nineteenth century, China finally succumbed to the tyranny of the British Empire. The vast riches on offer were enough to entice investors to fund this violent expansion, but religion played its part in justifying the whole shameful affair. Christian missionaries occasionally condemned the killing of indigenous people but more often than not they acted as partners in crime. On the Pacific Islands of Tahiti, The London Missionary Society created its own police force that banned the singing of local songs, prohibited dancing, and cut down the local breadfruit trees to end their self-sufficiency. All across the globe, formerly self-sufficient people who were living sustainably were forced to join the Great Leviathan and our march towards the cliff face we are on now speeded up. When James Cook visited Tahiti in 1769, he commented, *"It would have been far better for these poor people never to have known us."* I think it would be fair to say that the rest of planet Earth felt similarly about their encounters with the pompous, despotic, and insatiable European invaders. Cecil Rhodes, who to this day, has a statue erected at Oxford University, was characteristically hubristic when he stated:

"I contend that we are the first race in the world, and that the more of the world we inhabit the better it is for the human race. If there be a God, I think

that what he would like me to do is paint as much of the map of Africa in British red as possible."

This racism was an important part of the European expansion as it offered a defence for the vile atrocities committed by the colonialists. Mr. Rhodes had been granted a license to extract the natural resources of southern Africa by her royal highness, Queen Victoria; not of southern Africa, but of England. His British South Africa Company didn't bother to add an "Honourable" prefix to the official name and went to work stealing gold, copper, and diamonds from under the African soil. This didn't happen without resistance and to counter this, in 1885, the British invented the machine gun which they put to good use. This was just the latest, and most deadly weapon, to roll off the armament's conveyor belt, bankrolled by previous land grabs and other people's resources. The copper that was looted made its way to Europe's capital cities where it enabled them to enjoy the use of electricity for the first time. Of course, the media paid little attention to the tens of thousands who had been machine gunned to death to provide this light in the darkness.

Speaking to the French Chamber of Deputies in 1884, the two-time French Prime Minister, Jules Ferry, spoke openly about his feelings:

"Gentlemen, we must speak more loudly and more honestly! We must say openly that indeed the higher races have a right over the lower races. . . . I repeat, that the superior races have a right because they have a duty. They have the duty to civilize the inferior races. . . . In the history of earlier centuries these duties, gentlemen, have often been misunderstood, and certainly when the Spanish soldiers and explorers introduced slavery into Central America, they did not fulfill their duty as men of a higher race. . . . But in our time, I maintain that European nations acquit themselves with generosity, with grandeur, and with the sincerity of this superior civilizing duty."

354 · SIMON WHALLEY

On 15th November 1884, fifteen of Europe's nation states got together to decide the fate of all of Africa. In what was known as the Congress of Berlin, they decided to continue with their exploitation of Africa's resources and people (712). As we saw in an earlier chapter, this exploitation continues to this day, only now under the guise of free market economics. Not only was a great crime committed but the skills and knowledge of many indigenous people around the world were lost to the rampaging Europeans who in their arrogant drive to "educate" others, disregarded their own ignorance. Today, as the last forests are felled and the last survivors of a forgotten human age make their exits stage left, we are once again ignoring the knowledge of our ancestors in exchange for synthetic well-being, driven by the thirst for power and greed of a few.

One reason the Europeans gave for their decision to divide Africa was that it would avert a war between the colonizing powers. Sadly, even after the annexation of Africa, in 1914, the royal families of Europe decided war was nonetheless now necessary. After the assassination of the Archduke Franz Ferdinand of Austria-Hungary, the three cousins and offspring of Queen Victoria, the leaders of Russia, Britain and Germany declared war on each other and decided that they would sacrifice millions of lives in order to fuel their family feud. Germany's Kaiser Wilhelm joined with Austria-Hungary on one side and Britain's George V and Russia's Nicholas II formed an alliance with France, Belgium, and Serbia on the other. By the war's end in 1918, twenty million people were dead and twenty-one million were wounded. A month before the war's end, German sailors in Kiel began refusing to fight against the British navy. The powers were already discussing an armistice and the sailors were exhausted and hungry and saw no reason to risk their lives for a lost cause. They formed councils and within a week the movement had spread throughout Germany. They demanded an end to militarism and private capital and wished for the workers to govern themselves. As you can imagine, this was not popular, and the German empire tried in vain to find soldiers willing to shoot their fellow citizens. A few months later, Friedrich Ebert, the first social-democratic leader of the German Reich turned to the only people willing to do the job: far-right radicals. The Freikorps, as they were known were sent into Berlin where they drove into the working-class areas and destroyed the Bavarian Soviet Republic, killing thousands of unarmed people and torturing many more. Many of these men wore swastikas on their arms. The seeds had been sewn (380).

It was not only the seeds of fascism that were sewn in WWI. The war also had an unseen and potentially disastrous consequence for the future of life on Earth, the end of a drive towards solar energy. Utilizing the heat of the sun had long been practiced by humans with the ancient Greeks designing their homes to capture the sun's rays during the winter months as far back as the 5th century BC. By the 1760s, a rectangular glass box was used to heat water by Swiss scientist Horace de Saussure. Throughout the 19th century designs flourished and between 1880 and 1914, more than fifty articles on solar energy were published in the popular science magazine, Scientific American, alone. In 1890s California, where coal was in short supply, but sunshine was plentiful, water tanks were designed to use the heat of the Sun and by the start of the 20th century, more than 1,600 solar heaters had been sold. The future for solar looked bright. By 1908, a former engineer at Carnegie Steel Company created the world's first thermal collector which collected hot water throughout the day and could provide warm water from an insulated tank until the next morning. While this progress was on a relatively small scale, another American had rather more grandiose ideas. Frank Shuman began experimenting with solar motors in 1906 and by 1907 he had built a prototype in his garden. He planned on creating a 10,000-horsepower steam engine that covered sixty acres of land. In September 1911, he wrote in Scientific American:

"The future development of solar power has no limit. Where great natural water powers exist, sun power cannot compete; but sun-power generators will, in the near future, displace all other forms of mechanical power over at least 10 per cent of the earth's land surface; and in the far distant future, natural

fuels having been exhausted it will remain as the only means of existence of the human race."

Shuman was positive that within decades, solar power would be the main source of power across the globe. Together with British engineer A.S.E. Ackermann, he created a new business, Sun Power Company in 1911. They built an experimental plant in Meadi, Egypt, from locally sourced materials and the British colonialists were overjoyed at having a low-cost power source for the irrigation of Egypt's cotton crop near the Nile River. Shuman was soon offered a 121 sq. km plantation in British Sudan where he could continue his progress. Germany followed suit by offering him $200,000 to design a similar solar system in East Africa. Shuman was now dreaming of a 52,000 sq. km solar plant producing 270 million horsepower in the Saharan Desert. This would have been equal to all the fossil fuels burned around the world in 1909. Unfortunately, for Shuman and all the species of Earth, the warring families of Europe had made the calamitous decision to start the planet's first world war and in 1914, Shuman's solar plant in Meadi was dismantled for parts and scrap metal. Shuman himself died in 1918 and after the war, with a drop in fossil fuel prices, the momentum for solar energy was lost. Without the self-inflicted calamity brought on by Europe's arguing royal elites, our economies could have been powered largely by the Sun for the best part of a hundred years and the climate crisis would not exist. Regrettably, a decade later, the cost of fossil fuels dropped even further as the impacts of the 1929 Great Depression were felt (713).

Theft of the Global South

The European colonization of Africa and the forced removal of human resources may have ended but the continent continues to prop up our neoliberal economic system through providing cheap access to natural resources, mostly to the detriment of its ecosystems and human population.

It is not just people in the global north, however, that benefit from this theft of the global south. There are also winners in the developing nations too, although they are essentially batting for Team Global North. In what is called the resource curse, it has been found that the nations with the most resources are often home to the poorest people. The consultancy firm McKinsey estimate that 69% of people living in extreme poverty are living in countries where oil, gas, and minerals are a key part of the economy. Take for example Nigeria and Angola which are the first and second largest oil and gas producers in Africa. Here, the number of people living on less than $1.25 a day is 68% and 43% respectively. How can this be?

The author Tom Burgis explains in his book, The Looting Machine: Warlords, Tycoons, Smugglers, and the Systemic Theft of Africa's Wealth, that the practice of economic rent is partly to blame. Poorer nations usually provide licences to foreign companies to extract their resources and are paid in dollars. As this money is essentially unearned, it provides a bag of riches for the ruling class who have no need to tax the people to fund their operations. This means they do not need the approval of the people anymore and can act in their own interests. He goes on to explain that education spending then declines but military spending increases in order to protect the elite who often rule for many years as there is no incentive to leave office. Of course, foreign banking interests are all too happy to assist these kleptocrats in transferring this unearned money to safe foreign tax havens and the people remain poor. Africa is now home to 165,000 extremely wealthy Africans who hold a combined $860 billion (714). Our economies in the global north are literally fuelled by resources from the south. Take France for example which is run on 75% nuclear fuel that relies largely on Niger for uranium. In France for every woman that dies in childbirth, one hundred will die in Niger. Another example is the smartphone industry. In Finland and South Korea, the homes of Nokia and Samsung, the average life expectancy is eighty, but in the Democratic Republic of Congo, where the minerals used to make the phones are extracted, people rarely make it past fifty. To make up for a short work life, children start working from the age of seven. Sixty percent of the planet's cobalt is mined here, largely unregulated, with kids working in tunnels liable to collapse and breathing cobalt dust deep into their lungs as protective equipment is deemed unnecessary. The waste from these mines seeps into the river systems and contaminates the drinking water (715). In the rich north, we could, of course, pay higher costs at the register to provide adequate wages and protective equipment, but then we would have less money to waste on stuff we don't need and the mirage of our lives improving would disappear, and with it, the dream we have been fed since we were born. That if we just keep spending to keep the giant machine ticking over, everything will slowly but surely get better and better for everyone. Just don't stop buying things whatever you do.

In Africa, there are at least twenty countries that depend on re-source extraction for 25% or more of their exports. In total, 66% of Africa's exports are resource based whereas the figure in Europe is 11% and 15% in North America. In Latin America the number is 42%. In Nigeria, where 68% of the people live on less than $1.25 a day, oil, and gas exports account for a whopping 97% of total exports and in Angola where 43% live in extreme poverty, 98% of export revenue comes from oil and gas with the remainder coming from diamonds (716). Africa is not poor because of itself but the systematic looting of its resources that has been going on since the nineteenth century. It is not the west propping up Africa but the other way around. So much so, that according to research published in the journal New Political Economy, the north drains more than $2 trillion a year from the south. In total, it is estimated that around $62 trillion was syphoned north between 1960 and 2020. If this money had been invested in the south, based on average growth rates in the south, it would be worth $152 trillion today (717).

This theft has an extremely pernicious impact in that African states today do not have sufficient finances to decarbonize their economies or mitigate the severe effects of the climate crisis and they often have to rely on loans and foreign aid in times of need, when they should be some of the richest nations on Earth.

Child Labor often hides beneath the surface. Be on the look-out for it. You are the one sailing in the cruise-ship.
by David Blackwell. Licensed under CC BY-ND 2.0.

Tax Havens

S urrounded by abject poverty, with billions of dollars of unearned wealth, it is essential that African kleptocrats have somewhere safe to hide their unjustified bonanza. Fortunately, just such places have existed since the 1920s and are known as tax havens. As income inequality grew exponentially in the roaring twenties, the rich started looking for places to hoard their wealth safely away from the people who made them rich. Starting in Switzerland and Liechtenstein, tax havens spread to Jersey, Bermuda, the Bahamas, Panama, the Cayman Islands, and the U.S. state of Delaware, to name but a few.

Under the pretence of keeping their money safe from government expropriation, it was estimated in 2014 that $500 billion (30%) of African wealth, itself expropriated from the people, is now sitting in various safe havens around the world (718). This dirty money is joined by notes belonging to the soccer superstar Lionel Messi, the King of Saudi Arabia, President of the United Arab Emirates, former Emir of Qatar, Former leaders of Ecuador, Australia, Pakistan, Jordan, Moldova, Ukraine, Georgia, Iceland, and Qatar. Illustrating his France first priorities was Jean-Marie Le Pen, former leader of the far-right party Front National. The eldest daughter of former Philippine President; and one of the world's richest thieves, Ferdinand Marcos, also takes advantage of the secrecy jurisdictions (719). The list is fairly exhaustive and is a who's who of rich politicians, socialites, businesspeople, royalty and athletes. It also includes her Royal Majesty, Queen Elizabeth the Second, who has $13 million invested in the Cayman Islands and Bermuda (720). When the Panama Papers: 11.5 million secret documents, were released to media organisations in 2016, then British Prime Minister David Cameron promised to *"sweep away"* tax secrecy. It is of little surprise that nothing much has changed as his own father was found to be managing a multi-million-pound investment fund in the Bahamas that helped many rich citizens avoid paying any tax. When asked whether the Cameron family still had money invested in the fund, the Prime Minister's spokeswoman said *"that is a private matter"* (721). Interestingly, the U.S. Federal Trade Commission disagrees as they state on their website *"free and open markets are the foundation of a vibrant economy"*. Clearly, there is nothing open about tax havens. It's estimated that governments lose between $500-$600 billion a year in lost tax revenue with around $200 billion being lost in low-income countries. American Fortune 500 companies have stashed approximately $2.6 trillion offshore, and they have been joined by individuals whose tax haven holdings amount to $8.7 trillion. The total amount of money being hidden from the tax authorities in these tax havens is estimated at $36 trillion (722).

If you were wondering about government duplicity in the existence of tax havens, perhaps this paragraph will help clear things up. It emerged in 2013 that some of the largest accountancy firms in the world, including Deloitte, Ernst & Young, KPMG and Pricewaterhouse-Coopers, were helping the U.K. Treasury to draft financial legislation and then helping their rich clients to evade their tax responsibilities to the people of the United Kingdom. They were even found to be advising their clients to adopt tax schemes that had only a 25% chance of courts judging to be legal. There is a revolving door between the treasury and the big four accountancy firms, and these firms profit to the tune of £2 billion ($2.4 billion) a year, all at the expense of the British public who lose much needed billions in funding for schools, hospitals and infrastructure (723) (724).

While many argue that these tax havens are necessary to protect individual wealth from over-zealous governments, this is not a view shared by 300 economists who wrote the following in 2016:

"The existence of tax havens does not add to overall global wealth or well-

being; they serve no useful economic purpose. Whilst these jurisdictions undoubtedly benefit some rich individuals and multinational corporations, this benefit is at the expense of others, and they therefore serve to increase inequality (718)."

Thirty-six trillion dollars could fund the annual transition to a carbon free economy for the next ten years while the 500-$600 billion lost annually would help nations mitigate against the worst effects of the crises.

Mapped: The World's Biggest Private Tax Havens
https://www.visualcapitalist.com/

Make to Break

On September 4th, 1929, the stock market crashed, and global GDP dropped by an estimated 15%. The crash was brought on by a drought in the U.S. that lasted throughout the 1930s as well as banking failures and over production of consumer goods. That fateful day, now known as Black Tuesday, was the stimulus for a change in consumer product design. While perceived obsolescence - consumers believing their products were old and wishing to replace them themselves - had been instigated by Alfred P. Sloan, the President of General Motors (GM) in the 1920s, a decade later, producers weren't about to wait for consumers to decide to replace their products of their own volition; they were about to take matters into their own hands. They came up with an idea, now known as planned obsolescence. With many of the era lacking a decent standard of living while warehouses were full to the brim of supplies, producing more commodities wasn't very appealing for manufacturers. Consumers, facing extreme financial hardships, were using their products for as long as they possibly could before replacing them. This was not acceptable to the corporations who in their quest to end the depression and get their products into the hands of even more consumers, hit on the grand idea of making things to break. Writing in 1932, Bernard London laid his plan out in a paper published by the University of Wisconsin (725):

"I would have the Government assign a lease of life to shoes and homes and machines, to all products of manufacture, mining, and agriculture, when they are first created, and they would then be sold and used with the term of

their existence definitely known by the consumer. After the allotted time had
expired, these things would be legally "dead" and would be controlled by the
duly appointed governmental agency and destroyed if there is widespread
unemployment. New products would constantly be pouring forth from the
factories and marketplaces, to take the place of the obsolete, and the wheels of
industry would be kept going and employment regularized and assured for the
masses."

This form of leasing products and homes was not adopted by the
U.S. government, but it was eventualized by corporations making light
bulbs and stockings and more recently, printers and cell phones. The
details were not, however, released to consumers. Today, when we buy
products, we know two things; that they will only last a short time and
that this isn't a problem because we will want the new version when it
comes out anyway. This is the combination of both planned and perceived obsolescence, and it hasn't happened by chance. Manufacturers
employ psychologists whose job it is to make us desire their newest
offering, in what Edward Bernays described as Propaganda. While we
are dreaming of the newer version of our latest product, our current
product is approaching the end of its intentionally very short life.
This combination is great for corporations' bottom line and whichever
country is the new location for Dickensian novels that highlight the
unimaginable hardships of the working class. What though, happens to
all the waste? Today, it is either incinerated or sent to landfill, often in
poorer countries where they have no facilities capable of disposing of
the trash they receive. This idea may have made sense in 1932 where Roy
Sheldon and Egmont Arens wrote:

"We still have tree-covered slopes to deforest and subterranean lakes of oil
to tap with our gushers."

It does not, however, make any sense today, other than to make
huge profits. In the 1930s, they may have had lakes of oil and lush
forests, but what they didn't have then was the threat of climate and

biodiversity breakdown and the accumulation of an ocean's worth of plastic. Not content with making cars that intentionally broke, GM also set about dismantling the competition. In the 1930s, U.S. cities were connected by 17,000 miles of streetcars, off-street railways, and trackless trolleys, powered by a centrally generated electrical grid. This affordable and relatively clean transport system stood in the way of GM's project to get Americans travelling by privately owned automobile. They conspired with tire maker Firestone and Standard Oil to purchase the public transport companies in forty-five U.S. cities using pseudonyms (380). The transport systems in these cities, including New York and Los Angeles, were then gradually shut down with rails being covered in concrete and trams replaced with buses which were made by GM, had Firestone tires, and were powered by Standard Oil. The buses were not as popular as the trolleys and many turned to the cars being produced by GM (726). They marketed these cars as freedom and independence. The American Dream, in essence, was actually built on lies and conspiracy. By the 1950s, the length of car ownership had dropped from five years to two and even this was too long for the Head of Design at General Motors whose target was just one year (727). Today, on average, Americans replace their cell phone every 22.7 months, fast fashion items are made to be worn just ten times, and on average only 1% of consumer items are in use after six months (728).

The Rejection of John Maynard Keynes's International Clearing Union

As we have seen, the global south is already suffering the worst consequences of the multiple crises we have caused. And we have also seen that many nations are trapped in debt repayments to their northern creditors. Due to the nature of these debts, many nations are unable to adapt or mitigate the worst of the impacts of our rapidly warming world. For instance, according to Jubilee Debt Campaign, thirty-four nations in the global south, on average spend five times more on debt repayments than they do on climate adaptation. This system of never-ending indebtedness didn't need to be this way.

After the destruction of WWII, the need for new rules in the international monetary system led to a conference in Breton Woods, Massachusetts, in July 1944. Delegates from forty-four nations were present, but, somewhat unsurprisingly, it was the delegates from the U.S. and U.K. who stole the limelight with contrasting visions of the future. In the British corner was the Eton educated John Maynard Keynes, and in the American corner was Harry Dexter White, the son of Jewish Lithuanian immigrants.

Benn Steil explains in his book: The battle of Bretton Woods, that Keynes built the case for an imaginative bank called the International Clearing Union (ICU). The bank issued its own currency: the bancor, and provided nation states with an overdraft equal to half their average trade over a five year period. Where the ICU and our current system were aligned is that nations that built up a trade deficit were required to pay interest on their deficit as well as reducing the value of their currency to prevent capital flight. Where the systems detached was that nations with surplus trade were equally obliged to balance their books. This was achieved by placing a 10% interest rate on their surplus, and they were required to increase the value of their currency to incentivise capital export. Countries with surpluses would thus be motivated to buy deficit countries' goods. At the end of the fiscal year, any surpluses were confiscated. In this way, nations were encouraged to clear the debts of those nations with deficits. The idea gained popularity with the following from the Daily Herald:

"Here at last is something which breaks away from the doctrines of the past . . . an entirely new approach to the problem of international monetary arrangements. It will hardly commend itself to the Bank of England. For it departs from the rigid orthodoxy of that institution. The plan puts gold in its right place . . . [it] puts decisive control over vital external operations in the hands of the Government . . . [it] aims at setting up an international authority which is responsible to Governments instead of private banking interests. It provides the control through which alone we can avoid the disastrous recurrence of trade slumps and booms."

While this was the British proposal made at the talks, The American Treasury: fronted by White, refused point blank to entertain the idea. They instead proposed the International Stabilisation Fund (ISF) which burdened the deficit nations alone. The ISF later became the International Monetary Fund (IMF) and the massive indebtedness of the global south had been set in motion. Within decades, the U.S., the largest creditor nation had become the largest debtor nation, but they had been given veto power over any IMF decisions, and the IMF ensures the American economy does not falter no matter how indebted it becomes by requiring that all foreign exchange reserves are kept in U.S. dollars.

Greed is Good

While corporations in the U.S., and later, around the globe settled on planned and perceived obsolescence as the answer to stimulate the post 1929 crash economy, Franklin D. Roosevelt, took a different approach. Between 1933-1939 he introduced the New Deal which helped alleviate the suffering of the poorest and also led to the golden era of U.S. prosperity. There were those, though, who despised his interference in the markets and considered this a sin, akin to the emerging threat of Nazism. Interfering in the lifespan of products, however, was not seen in such poor light. In 1938, these men met clandestinely in Paris to discuss an alternative path. The idea was to be known as neoliberalism, and its founders were the Austrians Ludwig von Mises and Friedrich Hayek. They believed that any form of government interference should be avoided, unless of course it *consists of giving the people not what they want but what they will learn to want,*" as was described in Walter Lippman's book, the Good Society, written in 1937, which acted as inspiration for the early Mont Pelerin Society and has since become conventional wisdom in neoliberal circles (729). The outbreak of World War II put the dampeners on this excitable group of middle-aged white men, but they would be back when the fighting was over.

In 1944, Friedrich Hayek published his book: the Road to Serfdom, which explained that government interference would lead to complete authoritarianism and would crush individualism. The book struck a chord with many who were concerned by the specter of communism, but it was most popular, unsurprisingly, with the rich, who were angry at their relatively high taxes and strong regulations that protected vulnerable workers. The Mont Pelerin Society attracted the finances of these men and got to work building a network of think-tanks, including the Cato Institute, the Heritage Foundation, the Institute of Economic Affairs, the Centre for Policy Studies, and the Adam Smith Institute. They also funded new departments at the Universities of Chicago and Virginia, who together with the think-tanks, developed and promoted the theory of neoliberalism to a wider audience. Unfortunately for the movement, the adoption of Keynesian economic theory to redevelop societies after the horrors of World War II prevailed and the 1940s ended with the introduction of free health care in many countries and stronger government planning. The Mont Pelerin Society, and neoliberalist disciples, would bide their time producing materials and influencing the wealthy, waiting for a better moment when they could reappear with the answer to the world's problems. In 1960, Hayek wrote another book, the Constitution of Liberty, in which he argued that the rich and powerful were the most heroic of all people and that limiting their greed was an attack on the liberty of human existence. He painted a picture of the extremely wealthy as philosophers of their time, pushing the boundaries, not of thought, but of expression to spend. He even took his nonsensical thought to new levels by arguing that those who inherited wealth should be free to influence thought and opinion, taste, and beliefs. If this is true, then surely, the aim of all neoliberals would be to offer economic freedom to all members of society, so they too, could spend their time influencing thought and opinion, taste, and beliefs? Sadly, this opportunity to influence should only be the preferred past time of those more inclined to steal their wealth from others. Hayek was equally opposed to the conservation of resources, trade unions, progressive taxes and in general, the welfare of citizens. It sounds like what he was actually about was the totalitarian power, not of the state, but of the rich.

As Europe and the U.S. enjoyed post war economic success, neo-liberalism struggled to gain a foothold, despite massive funding from the wealthy. That continued up until the 1970s when Keynesian policies hit a snag: stagflation. Until now, governments had reacted to low

370 - SIMON WHALLEY

unemployment by increasing the amount of money in circulation and reducing it when inflation rose. Now, both unemployment and inflation were rising simultaneously, in part due to the oil crisis brought on by the Arab-Israeli war in 1973. Governments were now looking for an alternative to the heavy investment period inspired by John Maynard Keynes. This was the moment the Mont Pelerin Society and its anti-tax, anti-regulation and pro-privatization proponents had been waiting so patiently for. The most famous advocate for neoliberalism was Chicago University's Milton Friedman, and he remarked that *"when the time came that you had to change ... there was an alternative there to be picked up."* It was indeed picked up, by both American Democratic President Jimmy Carter and the Labour leader in the U.K., Jim Callaghan. The complete package was not adopted in full at first, but it was about to change the face of life on Earth, starting in the freedom loving Pinochet's Chile and spreading north through South America to the U.S. and across the Atlantic until it would become the all-pervasive global driver of greed and gluttony that it is today.

The language extolled by Milton Friedman and his fellow Chicago Boys, as those from the Chicago School of Economics were known, is like a torrent of synonyms for freedom. Every sentence constructed is replete with liberty this or emancipation that. What they don't do is give the full meaning for this freedom. They leave it out intentionally because when they give the full picture, it isn't quite as enticing to potential conformants. When they use words like freedom, liberty, and emancipation, they are talking about the freedom, liberty, and emancipation for those with money to do whatever they like. If this is at the expense of the freedom, liberty, and emancipation of others, then this is not just acceptable but enviable. Another term that better suits Friedman's system is monetary fascism, but the term doesn't sound as appealing as "free market". Whereas traditional fascism promotes national industry in order to strengthen the state, monetary fascism uses the power of the state to put the interests of money and the financial class above all else, including the nation. As we have seen, these monetary fascists were waiting in the wings with an economic package they promised would provide individual freedom instead of social welfare for all. What they needed was an opportunity to test out their hypothesis to the full on a society at large. They were brimming with ideas but had nowhere to experiment until they hit on the idea of a coup d'état. They couldn't do this in the rich countries of the northern hemisphere where tanks and troops rolling through the streets were never extremely popular; so instead they decided to overthrow the democratically elected leader of Chile, Salvador Allende. If overthrowing a democratically elected government of a foreign nation doesn't strike you as the basis for freedom, don't worry because Friedrich Hayek made his feelings well know when he stated:

> *"My personal preference leans toward a liberal dictatorship rather than toward a democratic government devoid of liberalism (730)."*

In 1973, the U.S. backed just such a right-wing authoritarian military dictatorship to overthrow Allende with an orgy of violence displayed throughout the country. No longer was structural violence to suffice, now physical violence was deemed necessary. They installed a puppet leader called General Augusto Pinochet to replace Allende who had done the unthinkable and tried to improve education, nationalize major industries, and improve the lives of the working class. Monotheistic religion was not necessary for this coup as the new religious dogma was to be "free-market" capitalism. Of course, the trade unions had to go and with them the freedom of workers to bargain effectively for a living wage. The freedom of the rich to suppress wages was now deemed more important. Regulations also had to make way as they hindered corporations' freedom to pollute and endanger workers; and this freedom was judged more important than the freedom of people to have clean air and water, and a safe place to work. Taxes also had to vanish as they suppressed the freedom of the rich to amass vast fortunes and this freedom was considered more important than the freedom to distribute wealth in a way that would galvanise society most effectively. In general, the freedom of a few was believed to be more important than the freedom of the masses to democratically elect a leader of their choice. This freedom of democracy, which the United States presented to the world, was to be suppressed by the very bastion of freedom itself, the United States of America. In the years prior to the coup, the CIA and other U.S. "trainers" were on the ground in Chile, gearing Chilean troops up for a war against communism, which effectively was a war against the people of Chile (731).

On 11th September 1973, as the CIA backed tanks and troops rolled through the streets of Santiago, President Allende took his own life. It seems he knew what was coming. After grasping control. Mr. Pinochet turned Santiago's stadiums into torture chambers where 2,279 people "disappeared" and 31,947 were tortured. The torture provided free of charge included waterboarding, electrocution, suffocation, having heads shoved into buckets of urine and excrement and being hanged by the feet or hands and beaten (732). A further 1,312 "troublemakers" were sent into exile and 200,000 escaped for political reasons (733).

After opening the Chilean market to foreign imports, getting rid of price controls and cutting government spending by 10%, the Chicago Boys assured Pinochet that the free market would balance itself but that is not what happened. Inflation rose to be the highest in the world at 375% which was twice as high at any time under Allende. Milton Friedman visited Chile many times and during his visits he would congratulate Pinochet on a job, thus far, well done, but what he wanted was further deregulation and less government spending. In what he termed "shock treatment", he urged Pinochet to embrace the free market wholeheartedly and reduce government spending by a further 25% in the next six months and fire thousands of government employees who he assured Pinochet would be rehired by the soon to be booming private sector. In reply, Pinochet assured Friedman that *"the plan is being fully applied at the present time."* Within the first year of this "shock treatment", the economy shrank by 15% and unemployment rose from 3% under Allende to a staggering 20%. After removing price controls, a family now had to spend 74% of its income on buying bread whereas under Allende, bread, milk, and bus fares accounted for just

17% of income. By 1980, government spending had been cut to half what it was under Allende and almost 500 state owned companies and banks had been sold off for a pittance. By 1982, unemployment had increased to 30% and Chile's debt had risen to $14 billion, and its assets were now safely in the hands of foreign owned corporations. Fortunately, for the Chilean government, the state-owned copper company, Codelco, had not been sold off and this single company was now providing 85% of Chile's export revenue (731).

Eventually, the economy did stabilize, but not for more than a decade after the "shock treatment" and after finding its feet, the situation was far better for the few than it was for the many. By 1988, 45% of the population was living under the poverty line while the richest 10% had seen their income rise by 83%. A few decades later and inequality in Chile was the eighth highest on Earth (731). By 2019, the top 1% earned 33% of the country's wealth. Chile now has the highest levels of inequality in the Organisation for Economic Cooperation and Development (OECD), and this led to the mass demonstrations across Chile in 2019 (734), and ultimately to the appointment of socialist leader Gabriel Boric in late 2021, with a remit to rewrite the constitution to focus on the climate crisis. Naomi Klein described the situation in Chile splendidly in her book, The Shock Doctrine:

> "In Chile, if you were outside the wealth bubble, the miracle looked like the Great Depression, but inside its airtight cocoon the profits flowed so free and fast that the easy wealth made possible by shock therapy-style "reforms" have been the crack cocaine of financial markets ever since. And that is why the financial world did not respond to the obvious contradictions of the Chile experiment by reassessing the basic assumptions of laissez-faire. Instead, it reacted with the junkie's logic: Where is the next fix?"

It is abundantly clear that neoliberalism is not about the greatest growth for all, but the greatest growth for a few. Growth was higher prior to the 1980s than it has been since the introduction of free market ideology, yet the wealth of the richest has grown exponentially to the

374 - SIMON WHALLEY

day in 2020 when Jeff Bezos pocketed $13 billion in a 24-hour period, as the global economy tanked. Naomi Klein's question of where the next fix is, was answered in junkie fashion by Margaret Thatcher and Ronald Regan who thought the model of Chile would be perfect for the United Kingdom and the United States respectively. First though, the Chicago Boys had some unfinished business in South America.

After the unbridled success of the Chicago Boys in transferring the wealth of the Chilean people into the hands of a few Chilean capitalists and the international corporations who were circling like vultures, other South American countries followed suit with similar results. Brazil, Argentina, and Uruguay all opened up their economies and sold off their assets at the behest of the neoliberal disciples who had now moved into key government positions within those countries. Within a year, wages in Argentina had plunged 40% and whereas people living under the poverty line was 9% and unemployment just 4.2% before the coup, neighbourhoods now had no running water and diseases were prevalent. In Uruguay, real wages dropped by 28% and the streets of Montevideo filled with people searching for food. Argentina was soon operating its own torture camps scattered throughout the country where it is estimated 30,000 people were "disappeared". Likewise, Uruguayan Chicago Boys took to torture, with people avoiding the barracks on Montevideo's boardwalk to avoid the sounds of the screams coming from within. Brazil was not immune to the use of torture to extract information and terrify the population into obedience with the U.S. police officer, Dan Mitrione, training police officers in Belo Horizonte to hone their skills on homeless people. The former defence minister of Allende, Orlando Letelier, himself tortured for a year by Pinochet before being released, wrote in 1976 that Milton Friedman shared responsibility for the torture and killings perpetuated by Pinochet as there was no other way to enforce the kind of changes he advocated. A month after publishing the letter, Mr. Letelier's car was blown up in Washington D.C. and he was killed along with a U.S. colleague. The assassins, sent by Pinochet, had arrived in the country on false passports and with the knowledge of the CIA. The torture in South

America was accompanied by the burning of leftist books, the closure of newspapers and magazines and the banning of strikes and political meetings (731). In their quest for global ideological hegemony, the neoliberal order now had a problem on its hands. How could they export their model to the rich north in a slightly less nefarious way than the torture camps and "shock-treatment" that blighted Latin America. This was to be the moment for Thatcher and Regan to put their distinctive mark on neoliberalism.

Pinochet and Kissinger
Ministerio de Relaciones Exteriores de Chile. is licensed under CC BY 2.0.

Thatcher and Reagan: A Conservative Romance

The nineteen eighties brought together an unlikely romance between the daughter of an English greengrocer and a Hollywood
actor. This romance may have lacked the intimacy usually associated
between a man and woman with such a shared dedication and confidence in their beliefs; but their ability to work in tandem would put
many more traditional partnerships to shame. There are two reasons
the U.S. and U.K. are singled out here. Firstly, they are the two G20
countries that have followed the Mont Pelerin Plan most diligently,
and secondly, the U.S. is the single superpower on our planet and what
happens there is usually transposed elsewhere, much like it was by the
Greeks, Romans, Dutch, Spanish and British empires. This is ultimately
what happened with neoliberalism, and it was the two countries with
the *"special relationship"* that led the way.

Mrs. Thatcher was first to get her feet under the desk of power by
winning the Conservative leadership contest in 1975. She was a woman
on a mission and wasted little time in making her ideology known.
At a meeting where one of her male colleagues was explaining what
he saw as the merits of Conservativism, she allegedly reached into her
bag, grabbed a copy of Hayek's book The Constitution of Liberty and
slammed it on the table before saying, *"This is what we believe."* The next
decade was to be as tumultuous a decade as any in British political
history.

What it was that she believed was that trade unions had to go, taxes had to be slashed, regulations protecting the environment and workers' rights had to go, and that any and all national assets had to be sold off. In his book, Out of the Wreckage, George Monbiot explains that she attempted to do all this in her three terms in office but that those in her own party stood in the way of her completely dismantling the welfare state and making everyone pay for education and health care. She shouldn't have worried because her slogan, *"There is no alternative,"* has now been widely adopted by both the red and blue wings of politics throughout the world.

Today, for many, it is more easily acceptable for people to imagine the end of humanity and life as we know it on this planet than it is to imagine an end to the neoliberal system which governs our lives.

To admit that this system of stripping individuals of their rights and giving them to corporations instead, is fundamentally at odds with who we are as a species is blasphemy. Removing workers' rights to unionize, forcing people to pay for the education that is so badly needed by the corporations who benefit, and destroying the planet in an insane quest for eternal growth is not only at odds with our own human nature but our survival on this planet. Mrs. Thatcher, or Maggie as she was commonly known, was a tough cookie, but she could not complete her attempt at changing society forever, by herself. She needed assistance and this came from the right-wing think tanks, mass media, universities and other foundations, flush with billions from business interests betting on the kind of returns enjoyed by the mass benefit scroungers in Latin America in the previous decade. They disseminated a propaganda story of epic proportions that completely flipped reality on its head.

The collective bargaining through unions that had provided workers the right to a two-day weekend, an eight-hour workday in safer conditions, and a pension when they retire was now an evil monstrosity that threatened their survival. The National Health Service (NHS) that had been treating anyone and everyone who got sick since WWII was now threatening the health of the very people it cured. People were now being told that if they went alone, without the assistance of their fellow citizens, that the free market would enrich everyone, and the government would be superfluous. Superfluous that is, with one caveat, that the security state would now need to be greatly enlarged to provide the police and surveillance necessary to keep control of those who resisted such radical attacks on the structure of society.

After one of his many trips to Chile to check on the progress being made by the genocidal torturer General Pinochet, Friedrich Hayek wrote a letter to his friend Margaret Thatcher urging her to adopt the Chilean model for the unknowing U.K. population. To her credit, on this occasion, the lady was not for changing. Even though, she described what was going on in Chile as, *"a striking example of economic reform from which we can learn many lessons"*, she was not convinced that "shock-therapy" and death camps were the best foot forward in Britain's case. She was also immensely unpopular at the outset of the eighties and worried that too much, too soon, would damage her in the polls. Her first three years in office had been poor with little of her project completed and her own party in open mutiny. Unemployment and inflation had doubled on her watch and her approval rating was the lowest of any British Prime Minister, at just 25% (731). She was saved, ironically, by the neoliberal regime who had seized control of Argentina from the popular Isabel Perón in 1976. In 1982, as Thatcher slashed the top rate on tax, she reduced government spending to balance the books and the foreign office was forced to agree to hand back Hong Kong to China and negotiate the sale and lease back of the Falkland Islands, referred to as the Malvinas in Spanish. The islands were regarded by Britain as a nuisance as they were costly to protect while the Argentines also had little motivation to fight for them other than the national embarrassment of having the British in their waters. Argentinian write Jorge Luis Borges likened the war to two bald men fighting over a comb (731). With this in mind, all British navy vessels were ordered to return to home territories to save money, not to benefit society, but so the rich could pay less in taxes and the lease and sale of the Falklands went to the Commons for a vote. The idea was roundly rejected, and the Argentinian junta saw their opportunity to strike. Pressure on the islands began to build and on 31st March, Thatcher was urged by the foreign office to strengthen British naval presence in the area. She refused, but as it became clear that Argentina was going to invade, she became concerned that her job was on the line and her stance changed from Chamberlain indecisiveness to Churchillian conviction, and like Churchill, her war-time leadership saved her reputation and allowed her more time to complete her ideological mission of forcing British citizens to compete against one another for survival.

After the Argentinian invasion began on April 2nd, 1982, British warships were already on their way, much to the chagrin of her friend in the White House, Ronnie, who supported their regime change. By the time the war was over in June 1982, the U.K. had increased its debt by £3 billion and 255 British service people and 649 Argentinians had lost their lives (735). The war might have ended for Britain, Thatcher declared, *"but it is not yet a nation at peace."* She had a new lease of life with an upwelling of jingoistic national support raising her approval rating to 59% (731). The Iron Lady had unfinished business.

As right-wing authoritarian states around the world began to fall, Milton Friedman was becoming anxious that given the opportunity, poor people would vote for those promising to redistribute wealth from the top to the bottom. He understood that people would not vote for a party if it was not in their self-interest, and this is why he championed autocracy rather than democracy. One after the other, U.S. backed right-wing authoritarian states in Iran, Ecuador, Peru, Bolivia, and Nicaragua fell to democratically elected leftist governments. The pressure was on the Chicago Boys to prove that people would actually vote against their self-interest, democratically.

In the United States, Friedrich Hayek, who had been thwarted in his attempts to persuade Richard Nixon to follow his Chilean experiment on U.S. soil, finally had his man in the White House. On January 21st, 1980, Ronald Wilson Reagan began his work on behalf of the Chicago Boys to reverse the policies of the New Deal. Though Friedman was a friend of Nixon and a regular visitor to the Oval Office, Nixon was worried about his re-election chances should he agree to the radical changes he advised.

Regardless, Nixon did put many of Friedman's advocates into government positions, including George Shultz and Donald Rumsfeld. Nixon's refusal to acquiesce to Friedman's policies had angered and hurt the latter who had earlier opined, *"Few presidents have come closer to expressing a philosophy compatible with my own (731)."* His anger and frustration were about to dissipate with his friends and followers now in power on both sides of the Atlantic. The Mont Pelerin Society, and neoliberalism, were finally about to get their opportunity in the northern hemisphere.

The Irish cowboy actor turned California politician had moved from the west coast to the east and had his feet securely under the Oval Office desk at 1600 Pennsylvania Avenue. As he looked affectionately at the work being done by his neoliberal compatriot in London, he too was keen to implement Chicago Boy policies as soon as possible. Like Thatcher, he believed that government was too big, and he set about reducing spending and slashing taxes in an attempt to close the budget deficit. In his inauguration speech, he was quite succinct with his diagnosis when he said

> *"Government is not the solution to our problem, government is the problem."*

What happened in the two countries with the special relationship, was soon exported elsewhere. With the collapse of the Soviet Union in 1991, those sitting astride the levers of power in formerly nationalized Russian industries were instant billionaires. Russian Oligarchy was born. Throughout Eastern Europe, communist states turned to neoliberalism overnight and just as quickly, the controlling class under former dictatorships were now once again in control, only this time, the masses were at the mercy, not of a dictator, but the markets, which promised complete freedom. Unfortunately, these markets were actually run by real human beings, and they tended to idolize Milton Friedman who in turn idolized the markets. Friedman now believed that *"Only a crisis—*

actual or perceived—produces real change. When that crisis occurs, the actions that are taken depend on the ideas that are lying around." The crisis in Europe was the end of communism and the neoliberal hawks were once again waiting in the wings, as they were in February 1989 when Chinese students demonstrated in Tiananmen Square. When Nelson Mandela walked free in 1991, another shock wave was being released. The African National Congress (ANC) was attempting, not just to get into power, but to enshrine its Freedom Charter into policy. The charter had been compiled in true democratic fashion, with 50,000 volunteers dispatched into the townships and countryside to collect the demands of the people. The charter enshrines:

"The right to work, to decent housing, to freedom of thought, and, most radically, to a share in the wealth of the richest country in Africa, containing, among other treasures, the largest goldfield in the world. "The national wealth of our country, the heritage of South Africans, shall be restored to the people; the mineral wealth beneath the soil, the Banks and monopoly industry shall be transferred to the ownership of the people as a whole; all other industry and trade shall be controlled to assist the well-being of the people."

Of course, this was not popular with the Chicago Boys, the IMF, World Bank, and especially those of the ruling National Party. The latter used all its slyness, developed over decades of brutal apartheid, to hand control over trade policy and central banking to the former, who duly ignored the peoples' demands and plunged the country to the bottom of the inequality league table (731).

One of the main problems Nelson Mandela and his ruling *ANC* party were to find was that they were now responsible for servicing the apartheid era debt, roughly $4.5 billion a year. That the oppressed were now being held responsible for paying off the debt that their oppressors had built up while oppressing them, is bad enough, but when you compare it to the paltry $85 million that was awarded to the 19,000 victims and the families of apartheid killings and torture, it looks even more odious (731).

The physical violence provided by colonialism has been replaced by structural violence in the form of debt provided at massive interest rates that can never be paid off. Around the world, thirty-four of the poorest nations are now stuck paying back ever-increasing interest on loans and are either in a debt crisis or tipping into one. The rich northern countries went from stealing other nation's resources to buying them very cheaply as these borrowing nations have no choice but to sell their raw materials to pay off their debts. They are stuck in an ever-increasing cycle of debt repayments that can never end. In return for the debt, countries are forced to reduce social spending and reduce food subsidies. This impoverishes the people even more than they already were and relegates dealing with the climate and biodiversity crisis to the bottom of a very long list of important obligations. Poorer countries are now spending five times more on servicing their debt than they are on the climate crisis, and by 2025 this is expected to rise to seven times (736) Sudan is one of the countries facing such problems. It owes former colonizer, the United Kingdom, £861 million with around 80% (£684 million) made up of interest payments. In a constant attempt to service these debts without defaulting, as it did in 1984, it has to reduce public spending, impose austerity on the people and cut subsidies. These policies have led to violent protests. The charitable old United States is offering a bridging loan to Sudan so that it can apply for even more debt from the IMF which will continue the cycle further. The U.K. has offered £40 million in aid that the country would have no need for, were it not for the interest payments (737). It is also worth noting that many of the problems Sudan currently faces stem from British intervention from 1899-1956, such as the divide-and-rule policy which resulted in *"distrust, fear, and conflict between the various Sudanese peoples"* (738).

It should be fairly clear at this point that the free markets we've been told to love aren't guided by an invisible hand as much as they are guided by the hands of the extremely wealthy. As the planet is trashed so they can get richer, the rest of us lose our connections with one another, suffer from anxiety and depression, and the most vulnerable among us end up in private prisons as punishment for not being able to cope. Neoliberalism has been a complete disaster for every species on this planet, including ours. The most toxic and ruthless humans on our planet have also been rewarded for committing, what surely, in the coming decades will be seen as crimes against humanity, and far worse.

"Wende Museum of the Cold War (4853)"
by Ron of the Desert is licensed under CC BY-ND 2.0.

Bringers of Death

When we discuss the worst atrocities committed by humans, we usually include Hitler, Stalin, Mao, Pol Pot, and the rather less cited Churchill. We do so because of the enormous amount of suffering these men caused to others and then we multiply their suffering by the sheer number of victims. This illustrious group usually come out top of the Genocide Premier League (GPL). It is difficult to identify who is in top spot of the GPL due to the inaccuracy of available figures, but whoever it is now, they are about to be pushed back into second place. When it comes to the looming climate and biodiversity crisis, it has been estimated that by 2030 alone, 100 million human beings may perish, 90% of them in the global south (739). It gets worse; Dr James Lovelock, Fellow of the Royal Society (FRS), and one of the planet's most eminent climate scientists, argues that the multiple crisis hitting our species may leave only 500 million alive by the end of the century. He is supported by Professor Kevin Anderson, director of the Tyndall Centre for Climate Change. Other climate scientists, like Johan Rockström from the Potsdam Institute for Climate Impact Research concur that Earth might only be able to support half a billion people should we hit 4°C of warming (740). For anyone harbouring thoughts of the temperature being halted at 1.5°C, the former director of the United Nation's Intergovernmental Panel on Climate Change, Robert Watson, argues that 3°C is the absolute minimum we are heading for (741). Of course, what we do, or don't do, in the next few years will affect how much the temperature increases, how quickly, how many people die, and how many species go extinct. The concern is that for now, it is still largely business as usual and greenhouse gas levels are still rising, not falling. Some will claim that these deaths cannot be attributed to any one person, country, or corporation. In the following pages, an argument will be made that there are people and corporations, and indeed countries, that must be held accountable, if we are not to make a mockery of even the limited justice system we currently have. We will begin as Reagan begins to put the Chicago Boys' policies into practice.

As Ronnie settled into his presidency, a research paper was published in the journal Science. The paper, which came out on 28th August 1981, stated that the planet had already warmed by 0.4 °C and that the cause was increased carbon dioxide in the atmosphere. The scientists from the National Aeronautics and Space Administration (NASA) believed that in the 21st Century, we would start to see drought-prone regions emerge in North America and central Asia as well as the erosion of the West Antarctic ice sheet, and with it, rising sea levels. They further predicted that the Arctic Sea ice would melt, and the Northwest Passage would open up. Reagan, of course, took this on board and started reducing the U.S.'s emissions. If only the last sentence had been true, the topic of this book might have been something completely different. Alas, what actually happened was that the Department of Energy removed their funding from the scientists, led by James Hanson; and throughout the 1980s, the government continued to alter his findings to suit their narrative. It seems the fossil fuel industry were not very happy with James Hanson's message. Finally, towards the end of Reagan's eight-year stint as the most powerful man on Earth, Hanson testified to congress, and this time when his findings were once again censored, he asked Senator Al Gore to question him about the changes the Office of Management and Budget had made to his testimony. The year was 1988 and this was the moment that global warming really entered public consciousness in a meaningful way (742). If the fossil fuel corporations had acted on his testimony, the world today could have been a very different place. Sadly, they had no intention of changing their business models or leaving their assets in the ground. In fact, they had known all about the greenhouse effect and its link to their product for a very long time.

Way back in the 1850s, the British physicist John Tyndall had shown that CO_2 could increase temperatures as infrared radiation from the sun was absorbed by CO_2 molecules and this extra energy caused the molecules to speed up which in turn increased the temperature. Decades later in 1896, the Swedish scientist Svante Arrhenius estimated how much the temperature would rise, given a doubling of CO_2 from burning fossil fuels. For this, the ancestor of Swedish climate activist, Greta Thunberg, was awarded the Nobel Prize for Chemistry in 1903. These warnings, like many others, were ignored and coal continues to power our economies in 2022.

Back in 1966, the industry itself was certainly aware of the dire consequences its product was having once emitted into the atmosphere. Writing in the trade publication, Mining Congress Journal, the then president of Bituminous Coal Research Inc., stated quite clearly:

"There is evidence that the amount of carbon dioxide in the Earth's atmosphere is increasing rapidly as a result of the combustion of fossil fuels. If the future rate of increase continues as it is at the present, it has been predicted that, because the CO_2 envelope reduces radiation, the temperature of the Earth's atmosphere will increase and that vast changes in the climates of the Earth will result. Such changes in temperature will cause melting of the polar icecaps, which, in turn would result in the inundation of many coastal cities including New York and London".

The coal industry was in a position to stop billions of deaths, but it chose to ignore the warning and the article only came to light in 2019, when an engineer at the University of Tennessee found it in a box of old journals marked for disposal. The largest coal company on the planet, Peabody Energy, also knew that its product was polluting the air. In the same journal, they said that pollution standards to protect health had a place and that the situation was urgent (743). Peabody went on to fund more than two dozen groups that raised doubts about the legitimacy of climate change data before declaring bankruptcy in 2016 (744).

In 1965, the President of the United States had been made aware of the problem. The President's Science Advisory Committee had requested Roger Revelle at the Scripps Institution of Oceanography to report to them on the impact of carbon dioxide caused warming. This was still four years before Neil Armstrong landed on the moon so the impacts weren't fully understood yet, but what Revelle was sure of was that sea levels would rise and:

"By the year 2000 there will be about 25% more CO_2 in our atmosphere than at present [and] this will modify the heat balance of the atmosphere to such an extent that marked changes in climate ... could occur."

While Revelle's prediction was fifteen years early, it wouldn't have mattered as the U.S.'s ongoing war for capitalism in Vietnam was the focus of then President Lyndon Baines Johnson, but he did mention it in congress later that year, saying:

"This generation has altered the composition of the atmosphere on a global scale through ... a steady increase in carbon dioxide from the burning of fossil fuels."

After Johnson was replaced by Nixon, the later introduced the Environmental Protection Agency, but the CO_2 issue was not at the forefront of Nixon's agenda. It was, however, studied by an elite group of scientists later in the decade. The advisory group, known as the Jasons, were asked by the Department of Energy, to look at the research being done by the department around CO_2. They agreed to look and focussed their attention on CO_2 and climate. In 1977, they reported back that a doubling of CO_2 from 270 PPM would raise global temperatures by 2.4°C (4.3°F). More troublingly, their climate models predicted an increase at the poles of between 10-12°C (18-22°F) should the amount of carbon dioxide double (745). The oil giant Exxon was also carrying out its own independent research in 1977 and their findings synced nicely with the Jason's data. Their senior scientist, James Black, said:

"In the first place, there is general scientific agreement that the most likely manner in which mankind is influencing the global climate is through carbon dioxide release from the burning of fossil fuels."

A year later, Mr. Black informed Exxon's management committee that a doubling of CO_2 would increase temperatures by 2 to 3°C (3.6-5.4°F) and that we had a time frame of five to ten years before difficult decisions around energy strategies would become critically important (746). This looks convincingly like Exxon knew what it needed to do in 1978, but they, like the coal industry, decided to mislead the public in order to continue making dividend payments to their investors.

In the public domain, the Jasons, while revered widely, were largely made up of physicists, so the government asked a prominent climate scientist at the Massachusetts Institute of Technology (MIT) to look into the Jason's report. Jule Charney, the most admired meteorologist in the U.S., put together an expert team, including James Hansen, and their findings echoed those of the Jasons. There would be warming if CO_2 levels continued to rise but the sensitivity of the Earth's systems were unknown, meaning just how much warming and how fast was difficult to gauge. They stated that with a doubling of CO_2, the most likely temperature rise would be 3°C (5.4°F) and any CO_2 increase beyond that would increase the temperature further. Politicians, trans-fixed on their next election cycle were hardly excited by this prediction and when told of the fact that CO_2 might double in the next fifty years, they would tell the scientists to come back in forty-nine years. The problem with this was that one of the MIT findings was that the oceans had been absorbing most of the heat and that by the time we noticed large atmospheric changes, it would likely be too late to do anything about it. Coming back in forty-nine years wasn't an option. The White House had already asked the National Academy of Sciences to look into the situation before the MIT results were published. This time, they turned, oddly, to an economist mostly known for his work on Game Theory, not climate science. Thomas Schelling agreed with the previous findings that temperatures would rise but that how much and how fast were not known. In April 1980, the team wrote a simple letter, and in it, the strategist argued Earth could experience completely new climatic conditions by 2050 or it could look much the same. His advice to the White House was to fund more research but not do anything now as we would be able to mitigate and adapt at a later state. This was the scenario parodied in the movie Don't Look Up, when Meryl Streep's U.S. President character tells the scientists to" sit tight and assess" as a meteor careens towards Earth. Schelling argued that the price of fossil fuels would rise in the future and consumption would decrease as a result. In essence, he was suggesting that the free markets would solve the problem for us. The group's findings were that:

"The near-term emphasis should be on research, with as low a political profile as possible," and that *"We believe that we can learn faster than the problem can develop (745)."*

This was not the position of the oil industry itself. At a 1980 meeting of the American Petroleum Institute (API), representatives from nearly every major U.S. and multinational oil and gas company, including Exxon, Mobil, Amoco, Phillips, Texaco, Shell, Sunoco, Sohio as well as Standard Oil of California and Gulf Oil, concluded that by 2005, there would likely be 1°C (1.8°F) of warming and by 2038 that would rise to 2.5°C (4.5°F) and by 2067, the temperature would rise by 5°C (9°F) and cause *"globally catastrophic effects."* Their findings were not made public (747).

In the early 1980s, both Democrats and Republicans realised that action was needed immediately before the temperature changes were physically noticeable. Speaking to industry executives in 1981, President Reagan's acting chairman of the president's Council for Environmental Quality, said:

"There can be no more important or conservative concern than the protection of the globe itself (741)."

Action to prevent global warming in the early eighties had bi-partisan support but as Reagan got comfortable in the White House, this support waned and while scientists and environmentalists lobbied hard behind the scenes, the United States of America decided that the future of life on Earth was worth a roll of the dice.

Another of the world's largest oil companies, Shell, had been com-missioning research into the greenhouse effect since 1981. In a study in 1986, they found that the doubling of CO_2 in the atmosphere may happen far earlier than others were predicting. Shell's scientists were in private conceding that the doubling may happen by 2030 and that temperatures would rise accordingly between 1.5 and 3.5°C (2.7-6.3°F). In the report, published in May 1988, they even went as far as analysing their own contribution to this doubling and stated that their oil, gas, and coal products were responsible for 4% of the carbon emissions in 1984. Shell's experts warned that:

"If this warming occurs then it could create significant changes in sea level, ocean currents, precipitation patterns, regional temperature, and weather. These changes could be larger than any that have occurred over the last 12,000 years. Such relatively fast and dramatic changes would impact on the human environment, future living standards and food supplies, and could have major social, economic, and political consequences (748)."

Again, Shell, not only refused to act on their own findings, but they, along with the coal industry, oil industry, and Exxon, went on to fund misleading research in order to maximise their profits.

In the same summer of 1988, the United States was experiencing extreme heat and drought like conditions in the Midwest that were ruining crops. Politicians were beginning to wonder whether CO_2 emissions were already affecting the climate. This was to be James Hanson's time to shine. He testified that:

"The global warming is now large enough that we can ascribe with a high degree of confidence a cause and effect relationship to the greenhouse effect."

He stated that there had been a half degree Celsius warming since 1980, compared to the 1950-1980 average, and that the chance of this warming being naturally caused was just 1% (745). The U.S. media reported the hearings widely and the timing could not have been better with a change of leadership in the White House on the immediate horizon. In a 1988 campaign speech, George H.W. Bush spoke of the urgency to enact far reaching climate policies. The wannabe successor to Ronald Reagan had a nice play on words to excite his supporters in the primaries:

"Those who think we are powerless to do anything about the greenhouse effect forget about the 'White House effect. As president, I intend to do something about it (749)."

In 1989, the Intergovernmental Panel on Climate Change (IPCC) had its very first meeting and President Bush sent his secretary of state, James Baker along. Later in the year research initiatives were proposed in Congress and it looked like, the U.S. was finally about to take the climate threat seriously. Of course, we all know that that was not what happened. The United States did not take the crisis seriously, and neither did any other government. Since James Hansen's seminal testimony, we have emitted more carbon dioxide into the atmosphere than all prior human history. The fossil fuel industry helped to make sure of that.

This wasn't the first time that an industry lied to its customers about the dangers their products posed. In 1964, the U.S. Surgeon General, Luther Terry, concluded what many already knew, that smoking cigarettes caused lung cancer. The industry had a few choices. They could have made their product safer; they could have changed their business model, or they could have lied to their customers in order to get them to continue to purchase their murderous product. Like any good capitalists, they decided to lie, and in the process, continue killing their customers. They hired a group of public relations companies who created a huge misinformation campaign that raised doubts about the surgeon general's findings. In the early 1990s, the fossil fuel industry found itself in a similar position. It was now widely believed that burning oil, gas and coal was causing the Earth to warm and the consequences would be far greater than the threat of tobacco smoke. The industry followed the strategy of the tobacco companies and hired the very same strategic public relations companies. They then spent hundreds of millions of dollars funding right-wing think tanks that hired scientists with PhDs and attacked the facts around climate change. As with tobacco, the scientists didn't need to prove anything. All they had to do was plant a seed of doubt in the mind of the average citizen who was too busy to pay close attention and the industry could continue pumping oil out of the ground and profiting from the slow methodical mass murder their product was knowingly causing.

In 1990, Dr Brian Flannery, on Exxon's payroll, but representing the International Petroleum Industries' Environmental Conservation Association, attempted to derail the IPCC's first report that stated global carbon emissions must be reduced by 60-80%. He argued that too much scientific uncertainty existed, but his attempt failed, and the report moved forward (750). Remember, this is the same company whose chief scientist had warned them they had only until 1988 at the latest to address the critical situation. In 1991, Western Fuels Association began disseminating information to journalists and university libraries. Their materials repositioned global warming as theory; not fact, and espoused the positive benefits of increased carbon dioxide. In 1992, Exxon formerly joined the American Legislative Exchange Council (ALEC) who lobbied at the state and federal level to weaken any action. Three years later, a Mobil climate expert warned the Global Climate Coalition (GCC) that the human burning of fossil fuels and the link to warming was well established and cannot be denied. The members of the GCC included BP, Chevron, Exxon, Mobil, and Shell. Maybe Exxon's CEO, Lee Raymond, was drilling for new oil in the Arctic throughout 1995 because in 1996, he told the Economic Club of Detroit that, *"Currently, the scientific evidence is inconclusive as to whether human activities are having a significant effect on the global climate (750)."* Between 2003 and 2007, Drexel University emeritus professor, Bob Brulle, found that ExxonMobil paid $7.2 million to ninety-one institutions, including the Cato Institute and George C Marshall Institute. API donated just under $4 million to the same institutions between 2008-2010 (752). The one thing these institutions had in common was that they either down-played the risks of climate change or denied them outright. Mr. Brulle further found that between 2003-2010, organisations that promoted climate misinformation received $900 million in corporate funding (751). Just in case, you are sceptical of these links, maybe a Cato Institute insider will help assure you of the nefarious work these companies and think tanks carried out. Jerry Taylor worked for the think-tank for twenty-three years, appearing on TV and radio and promising the public that the science was not certain. He comments:

"For 25 years, climate sceptics like me made it a core matter of ideological identity that if you believe in climate change, then you are by definition a socialist. That is what climate sceptics have done."

Is it therefore surprising that only 22% of Republicans believe that climate change is caused by humans (752). The founders of neoliberalism, the Mont Pellerin Society (MPS) have strong links to these right wing think-tanks that include the Hoover Institute, the Heritage Foundation, the American Enterprise Institute, and the Heartland Institute. Koch Industries, which is run by MPS member, Charles Koch, have donated $100 million to fifty-four groups connected to MPS members. The MPS groups have been responsible for the spreading of climate change denial for decades and in 2013 the non-profit organisation Overland wrote:

"Neoliberalism is a coherent political movement embodied in the institutional history of the global network of think tanks: the American Enterprise Institute, the Cato Institute, the Institute of Economic Affairs, the Institute of Public Affairs (the key Australian node of the network) and their dedicated

spin-off counter-science think tanks. All can be traced back to the Mont
Pelerin Society, the central think tank of the neoliberal counter-revolution,
founded in 1947 by Friedrich Hayek and Milton Friedman (753)."

It appears that those who continue to support neoliberalism are supporting the certain deaths of hundreds of millions of human beings and perhaps billions.

It isn't enough to sway public opinion, politicians too, must be "swayed". U.S. political spending by the energy industry has risen ten-fold since 1990 and accounts for $140 million a year. It cannot shock anyone that taxpayers still subsidise the industry by $20 billion annually. It is impossible to exaggerate the nonsensical situation we are in. We are currently paying hundreds of billions of dollars a year to the very same companies who have intentionally lied to us in order to keep us buying their products. These products are largely responsible for the sixth extinction and kill millions of people every year. Imagine we were still subsidising tobacco, so it was cheap to smoke even when we knew of the damage it was causing. Would anyone stand for that? The fossil fuel companies are now rubbing their hands with glee as the Arctic ice retreats summer after summer. They can see the day, not far off, when they can begin drilling for oil in the northernmost region of planet Earth, where there will no longer be any ice because of the impact of these companies' products.

We can see why" think-tanks" and political parties still support the industry, but why the media? Between 1988 and 2002, 53% of stories on climate change gave equal time to the scientific consensus as they did to the climate deniers and a further 35% gave space to deniers in a less than equal way (745). The science was long ago settled. When that happens, no time is usually afforded those who disbelieve scientific certainty. Flat Earthers are not invited on TV to debate whether it was actually possible for Monique the hen to sail around the world. It is understandable that many of these free market loving media owners join in with misinforming the public as those who believe in the neoliberal ideology are more likely to deny that climate change is happening. If believing the science means we need more regulation, not less, it makes sense for these people to simply disbelieve the reality. This has been termed motivated-disbelief. The fossil fuel industry continues to take responsibility for the damage its products are causing in public, but it simultaneously funds think-tanks that sow doubt and confusion. Since the Paris Agreement was signed in 2015, the largest five publicly traded oil and gas companies have invested more than $1 billion of share-holder funds on misleading climate related branding and lobbying. ExxonMobil, Royal Dutch Shell, Chevron, BP, and Total are all limited liability companies, and their employees will expect to walk away from the mess they have knowingly created by simply filing for bankruptcy when the time is right, like Peabody (747). But we should not forget the bosses of these companies knew the consequences and then wilfully misled the public to keep the profits increasing. If they are not guilty of crimes against humanity, then who is?

Exxon Knew Johnny Silvercloud is licensed under CC BY-SA 2.0.

Mass Surveillance

"What the government wants is something they never had before. They want total awareness. The question is, is that something we should be allowing?"

These are the words of former Central Intelligence Agency (CIA) and National Security Agency (NSA) contractor, Edward Snowden. Snowden became a household name when he relinquished a six-figure salary and comfortable Hawaiian life to inform the world that the U.S. government was spying on global citizens around the world. On the evening of 4th June 2013, very few people on planet Earth went to bed knowing their online behaviour was being tracked by security agencies, but on June 5th, the world was waking up to what had been happening since October 2001.

Edward Snowden had decided the world needed to know that their governments were recording their online lives and on 20th May 2013, he flew to Hong Kong with four laptops and a trove of highly classified encrypted files. He had arranged to meet the Guardian journalist Glenn Greenwald and filmmaker Laura Poitras to entrust them with delivering the news to the world. With the help of the Guardian's Ewen MacAskill, they duly did so with headline after headline outlining the immense surveillance system that had been built with our taxes, but without our permission. On 6th June, The Guardian ran the story about Prism which claimed the NSA had a backdoor to data held by Google, Facebook, Apple, and other tech giants. No need to request access, just help yourself from the data buffet. A day later, it emerged that the U.K.'s version of the NSA, GCHQ, also had access to this data through the Prism program. As the world's collective jaw dropped, Edward Snowden made the decision to go public, even though he worried the U.S. government would use this to claim he was acting out of narcissism. On 12th June, his video was played around the world and the young twenty-nine-year-old self-claimed geek from Maryland who had long enjoyed literally being in the shadows was now centre stage in one of the most incredible news stories of all time. A day later, he fled the Mira Hotel in Hong Kong and entered into a short period of hiding while the NSA frantically tried to understand what he knew. It turns out, he knew a lot. He knew that whether a mail you sent to a friend, something you purchased online, a website you visited, a friend you called or a question you asked Google, a record was being kept safely away on giant servers in billion-dollar facilities in the desert. He also knew this information would be available night or day to security operatives and police around the world. He knew that the English-speaking countries of the U.S., Britain, Canada, Australia, and New Zealand were cooperating together in what is known as the Five Eyes, and he knew that other countries around the world were complicit in this data collection of their very own citizens, in exchange for information being shared with them by the NSA.

On 14th June, Snowden was formally charged with espionage and as a 30th birthday present, on the 21st June, the U.S. government requested his extradition from Hong Kong. Rather than spend the rest of his life in prison, he wiped the memories of his four laptops and destroyed the cryptographic key which meant he could no longer access the materials even if he was being tortured. On 23rd June, Snowden left Hong Kong on his way to Ecuador via Moscow, Havana, and Caracas. He had been denied political asylum by the majority of western democracies who were being threatened by the United States and instead was relying on the generosity of anti-imperialist Ecuadorean leader, Rafael Correa. Due to the very real threat of the plane carrying him being forced to land in U.S. allied countries and him being duly extradited to the U.S., he had to choose very carefully which countries he could travel through on his way to safety. Thus, in the absence of a direct flight from Hong Kong to Ecuador; Russia, Cuba and Venezuela were his necessary route. Unfortunately, Snowden was never to reach the safety of Ecuador. While transiting the Sheremetyevo airport in Moscow, the United States revoked his passport, meaning he could not travel any further. He had been purposely trapped in Russia by his own government who had absolutely no idea what information Snowden was carrying. In order to spite him, they had risked handing their arch nemesis the entire secrets of their all-encompassing surveillance operation. Fortunately for them, Edward Snowden was a patriot who had volunteered for active service in Iraq after the 2001 terrorist hijackings and was in no mood to switch allegiance to an even more repressive police state. Snowden was to spend the next forty days in the airport before being handed temporary asylum in Russia, where he lives to this day. In case any readers consider it far-fetched that the U.S. would have grounded his plane should he have travelled through U.S. friendly nations, consider the fact that on 1st July, the plane carrying the President of Bolivia, Evo Morales, was forced to land in Vienna, Austria, because it was suspected of carrying Snowden. The plane was searched thoroughly before being able to take-off and deliver the democratically elected leader of Bolivia back to his capital city, La Paz (754).

After leaving Snowden no choice but to apply for asylum in Russia, of course the U.S. government did what they do best in orchestrating a smear campaign against him, insinuating that he had travelled to Russia to sell state secrets. They didn't offer any evidence because they didn't need to. With the help of the media, they can just put it out there over and over again until eventually it sticks. All the mainstream media has to do is post the question "Snowden: hero or traitor?" and the government's job is done. From CNN to ABC, NBC, the New Yorker, the New York Times, and the Washington Post, they all ran stories implicating that Snowden was in effect, a traitor. Forget the fact that the government, a supposedly democratic government at that, had been spying on its own citizens for over a decade, without their awareness, let alone their consent. That's of little importance. Forget the message, just focus on the messenger. Surely, the job of the press is to report the truth and if in doing so, this makes you a traitor, what kind of country are you living in? But, if Edward Snowden is a traitor, then aren't the newspapers who made a profit from publishing the information too? And, if ever the role of the BBC as an arbiter of truth and justice needs to be questioned, the case of Snowden is it. The head of BBC News at the time was former Rupert Murdoch man, James Harding. The former Times editor considers BBC News to be the best news organisation in the world with curious, inquisitive journalism at the forefront of his plans for the organisation fighting for the public interest. Fast forward, past the PR, when asked if it would have been in the public interest to be told that their own government was spying on them with their much-needed taxes, the head of BBC News answered "no". He was being asked whether he would have run the stories if the BBC had been approached first, rather than the Guardian. He stated that running the stories would have been "campaigning" journalism (755). This was the biggest story of a generation and included high ranking lying to congress, not to mention the outright tyranny of spying on your own citizens and the head of BBC News was saying it would have been "campaigning" journalism to report on it. In 2020, an appeals court ruled that the mass surveillance of U.S. citizens' telephone records was unlawful, but if it was left to the BBC, we wouldn't even know it was happening.

The media and government like to pretend there is nothing to worry about. That if you haven't done anything wrong, you don't need to worry about your communications and web history being recorded. Snowden has this to say in response:

"Arguing that you don't care about the right to privacy because you have nothing to hide is no different than saying you don't care about free speech because you have nothing to say."

If the government is so blasé about privacy, perhaps they won't mind their tax-paid meetings with the fossil fuel industry or Rupert

Murdoch being made public? Surely, they have nothing to hide. This mass surveillance is only necessary to keep us safe they like to pretend. Safe from who? Environment ministers? Some of the documents provided by Snowden show that the NSA was actively monitoring communications between key countries before and during the Copenhagen climate summit in 2009 (756). Here, the U.S. acts like the arch villain Goldfinger from the James Bond movie. In the movie, Goldfinger has Miss. Masterson help him win a card game by using binoculars to look at his opponent's hand from the safety of her hotel room balcony. Of course, James Bond has been made all but redundant due to this system of all-encompassing surveillance.

Who needs physical spies, when wherever we go in public, we are watched by cameras and the smart phones in our pockets record our every move. They log who we meet and where and who we speak with and for how long. In the safety and privacy of our own homes, every website we visit, how long we spend there, who we connect with and what we think about certain topics, everything is recorded and stored by our own governments. More insidiously, our governments would have access to our health condition, sexual preferences and even fantasies by looking at our internet records. If ever they needed to discredit someone, they felt was a threat, they would be able to rely on all of our internet records at a mouse-click.

George Orwell envisaged a world of surveillance, but even he couldn't have foreseen the system of complete control that has been built up around us. In 1984, a TV screen on the wall might or might not have been watching you. In 2022, we have TVs on the wall, computers on the desk and Amazon speakers in the kitchen and they are all capable of listening to or watching our every action. In effect, we are giving up our freedom to enjoy apparent security. Perhaps we should listen to one of the founding fathers of the United States, Benjamin Franklin, who had this to say on the topic:

"Those who would give up essential liberty, to purchase a little temporary safety, deserve neither liberty nor safety."

Capitalism Versus
Democracy

Not content with the assets already stripped from the public sector, the corporate machine wanted even more. There was just one thing standing in its way, election funding rules. In the global model for democracy, the United States, it costs more than $1 billion to become president, up to $42 million to become a senator and over $1.7 million to become a member of the house (757) (758). This money must come from somewhere, and that somewhere is usually corporations that want something in return. While the wealthy had long enjoyed outsized reach in U.S. politics, it was a disastrous supreme court decision in 2010 that allowed corporations unlimited financial access to politicians who would ultimately be asked to oversee their actions. The supreme court had decided that corporations were U.S. citizens and that money equated with speech. Arguing that their first amendment rights to free speech were being violated by a limit on their contributions, corporations went from spending less than $200 million on elections in 2008 to $1 billion in 2012 and over $1.6 billion in 2020. With this kind of money gushing through the corridors of power, how can anyone expect for the democratic will of the people to transpire? The decision also impacted the fledgling, but bipartisan, climate movement. Until the fateful court decision, Democratic Senator Sheldon Whitehouse explains that:

"There was a constant steady heartbeat of Republican climate change

activity in the Senate, then, after the ruling, no piece of carbon dioxide
regulation legislation has managed to get a single Republican co-sponsor in
the Senate."

The reason Mr. Whitehouse gives for this change of heart is the power of the fossil fuel industry weaponry aimed at any dissenting Republicans (759). This same industry that receives billions in taxpayer subsidies then uses this money to purchase politicians who continue to give them the subsidies, in what is a cynical cycle of death. As the planet tips through points of no return, our politicians continue to fund the perpetrators of the biggest crime against humanity and ecocide imaginable. The machine that began rumbling thousands of years ago in ancient Greece is not done yet. There are still some forests left to burn, there is still ancient sunlight deep underground, there is still wildlife left to be killed, and there are many more people to be exploited. The machine will not rest until it has devoured everything in its path, and then it will devour itself. And as Thatcher matter-of-factly said, "there is no alternative". This has been conveniently drip-fed into our consciousness over the past forty years. An entire generation has grown up believing that we have tried every kind of system and that what we have now, is the best we can do. As the clock continues its countdown, the time for change is upon us and if you have gotten this far, you must need some positivity. In the final chapter, solutions will be discussed and alternatives to Thatcher's mantra will be described. The game is not yet over, but it will take a mammoth effort from all concerned citizens of planet Earth if we are to avoid the extermination of life as we know it.

Dear Indy,

You are part of a generation too young to be able to do anything about the extinction event unfolding, and even worse, unless the adults reading this take action, you will be part of the generation that suffers the worst consequences of your ancestors' actions.

If the adults that read this before you, took heed and then action, you may be reading these pages with some hope. If they took neither heed, nor action, these words will be heavy for you to read, and will likely fill you with dread.

Our situation, while seemingly insurmountable, wasn't that difficult to solve. Our ancestors have risen to many challenges in our shared history. I'll give you a few examples. Until the early 20th century, women could not vote. They had no say in their present or future. Fortunately for all your female friends, women in the U.K. decided not to accept this. In 1903, they formed a group and demanded voting rights. The media called them suffragettes and they embraced the term. They used civil disobedience, chained themselves to fences, smashed windows and even blew things up. They weren't successful over night, but they never gave up, and in 1918, women over thirty were allowed to vote. The age was lowered to twenty-one in 1928.

When slavery ended in 1865, African Americans weren't yet treated equally under the law. Right up until the 1960s, African Americans continued to fight for equal rights. They were discriminated against, beaten up, and even killed because they happened to have darker complexion. Fortunately for all your American friends of colour, their ancestors refused to accept this injustice. Many people of colour, including children, were beaten, or lost their lives in order that African Americans today can vote, and even become President.

Your father and his generation had their own challenge to face. By the time you read this, you will know what kind of people we were. Were we like our ancestors who truly loved their children and stood up for your future, or were we too lazy, entitled and scared to fight for your right to a clean and safe planet? It is far from over and like Nelson Mandela, another man who never gave up, once said:

"It always seems impossible, until it's done."

FOUR

How to Get Out of
This Mess

Let's start the chapter by discussing hope. The word itself describes what we want or expect to happen. Well, we can hope all that we want, and nothing will ever change. How many people hope their football team wins at 4:45pm on a Saturday evening, but then end up supporting the losing team. Hope is the worst thing for us to be doing right now. Hope only exists if you yourself are acting. As the wise man Gandhi once said, *"be the change you wish to see in the world."* Hope is a kind of faith and survives by expecting others to do something. Hoping that politicians do the right thing will not affect change. Hoping that God saves us will not affect change. Hoping that technology saves us will not affect change. The only thing that will affect change is a global awakening, or as it has been described, a revolution of consciousness. And that begins with each and every one of us. If enough of us decide that we are unwilling to be the generation that causes the first known extermination event in history, then we will affect change. The first step we need to take is to accept that as convenient as our lives have become, they are not sustainable in any way, shape, or form. Once we have accepted this rather inconvenient truth, then we can begin to move forward down a safer path. We need to talk about the problems we face, to normalise them, rather than ignore them in preference of comfort. In our world of paranoia, it is easier for people to believe their friends and colleagues than it is scientists and politicians, so get out there and initiate a conversation about the seriousness of our situation.

In her wonderful book, Saving Us: A Climate Scientist's Case for Hope and Healing in a Divided World, climate scientist Katharine Hayhoe discusses why talking about these problems is likely the most impactful thing we can do. In the U.S., only 35% of people talk about the climate crisis just once in a while. She argues that we talk about the things we care about, and if we aren't talking about these problems, then it looks like we don't care. The best marketing strategies are simple messaging that is repeated over and over by trusted messengers. So, we need to keep talking about it, even if we sound like a stuck record. Possibly after the seventh or eighth discussion, people will be paying attention. It is hard, as I have found out myself, but keep at it. Humans are much more likely to respond to personal stories over pages of data, so find a common interest and discuss how it will be affected. If you're a surfer, discuss that most beaches will disappear or that the ocean is dying. If you snowboard, talk about the disappearance of snow. If you are a wine aficionado, talk about grape harvests declining. If you are a parent, discuss your kids' likely future in a conflict ridden world.

To support the grassroots activism approach, recent events from the sporting world provided a wholly unexpected blueprint for change. When twelve owners of the richest football teams in Europe decided that they weren't rich enough and that they wished to launch a European super league that would forever enshrine them as the richest, and prohibit any other teams from challenging their superiority, football fans across Europe, and beyond, rose-up in one unorchestrated cacophony of dissent to say, "no chance". Within 48 hours of the super league announcement, all six of the English clubs had withdrawn from the planned competition and the project lay in ruins. A week later, fans furious at their owners' secret plans continued to protest, both inside and outside stadiums, to make their discontent known. The success of supporters over the wealthy club owners and banks willing to finance the project provides an excellent example of what can be achieved when enough people put their minds to something. The power has always been with the people, but we have long been played off against each other in a classic case of divide and rule. Here, football fans forgot their tribal rivalries and spoke as football fans united. Thatcher said there was no such thing as society. Football fans would argue there is, and we need more of it, not less.

Another example of how to improve our world is provided by cryptocurrencies. Love them or loathe them, they succeed by harnessing the combined computer power of millions of interconnected computers around the world. While the raw processing power of millions of computers is used to power crypto, our species is being driven by a select few who are not always the most deserving. Take the U.K. as an example, one in five members of parliament attended Oxford or Cambridge university (Oxbridge) (760). While this isn't necessarily a problem, 1,310 students at Oxbridge made their way there via just eight schools, six of which were privately run and charge up to £58,500 per year. A further 1,220 students came from 2,900 schools. In total, 7% of U.K. students study privately, but 42% of Oxbridge pupils benefitted from a private education (761). A staggering 52% of senior judges went from private school to Oxbridge and 39% of cabinet ministers also had a private education. The media is made up predominantly by those able to afford the best education with 43% of editors and broadcasters and 44% of newspaper columnists being privately educated and a third going on to Oxbridge (762). This is a cause for concern because those in positions of power have little in common with the electorate they represent and the best educations at the best universities end up going, not to the worthiest, but to those with the richest parents. This was highlighted with the college's admission scandal in the U.S. where wealthy children who had gained the best private education, had personal tutors, every advantage possible, and yet they still had to cheat their way into universities. In a capitalist world intent, apparently, on competition, this makes absolutely no sense. In a world where we want everyone to have equal opportunities in life, it makes even less. This inequality in education, unsurprisingly, leads to inequality and division in society and studies show that higher government spending in public education consistently lowers income inequality (763). This income inequality is also a driver of crime. While Thatcher and Reagan's solution to solving crime was to lock people up, a slightly better idea is to reduce poverty and inequality. This is according to chief constable of the U.K.'s Merseyside police, Andy Cook. When asked how he would spend £5 billion to cut crime, he answered that £1 billion would go into law enforcement and the rest would be spent on tackling poverty (764).

If we wish to tackle both income inequality and harness the entire brainpower of the population, then public education spending needs to increase, and private schools should be abandoned. Lovers of competition can't complain with this as they believe strongly, above all else, that fair competition is vital in the markets.

While we are discussing education, schools and universities must have a giant role to play in the coming decade and beyond. While it is certain that this current crop of school children does not have time to solve the problems we have created, they can play an important role, but for them to be able to do so, education around the problems we face, needs huge improvement. By the time school children today graduate from university, we will have reached 2030 when emissions need to have at least halved. By the time they get into positions of power, we will have likely heated the planet by 2°C (3.6°F) and there will be no turning back. School children, however, already are showing adults how to do things. Greta Thunberg's Fridays for Future movement has propelled the climate crisis to the forefront of people's minds. While it is obvious that, alone, schoolchildren striking from school will not create the change necessary, they can, and are, raising the issue at home with their parents, and it is their parents, and grandparents, who will be more effective. A worldwide strike orchestrated to force through demands would certainly hit those in power where it hurts. For this to happen though, climate change education must drastically improve as few educators seem to understand the precariousness of our predicament. Indeed, seven out of ten U.K. teachers feel they have inadequate training to teach about the crisis. A promising 92% of teachers are concerned but 41% said it is rarely, or never, mentioned in their schools (765). There are growing calls to include climate change across curricula, including food science, religious education, maths, English and art. For this, educators need to be provided with professional development and training. Additionally, academics are calling on schools and universities to declare a climate emergency and to reduce their emissions as fast as possible (766). The climate crisis is a moral issue and schools must do everything possible to provide the skills necessary for a changing climate while doing all they can to ensure their students have a future where they can hone those skills. The ability to pass tests will have limited benefit in a world with scarce food and water.

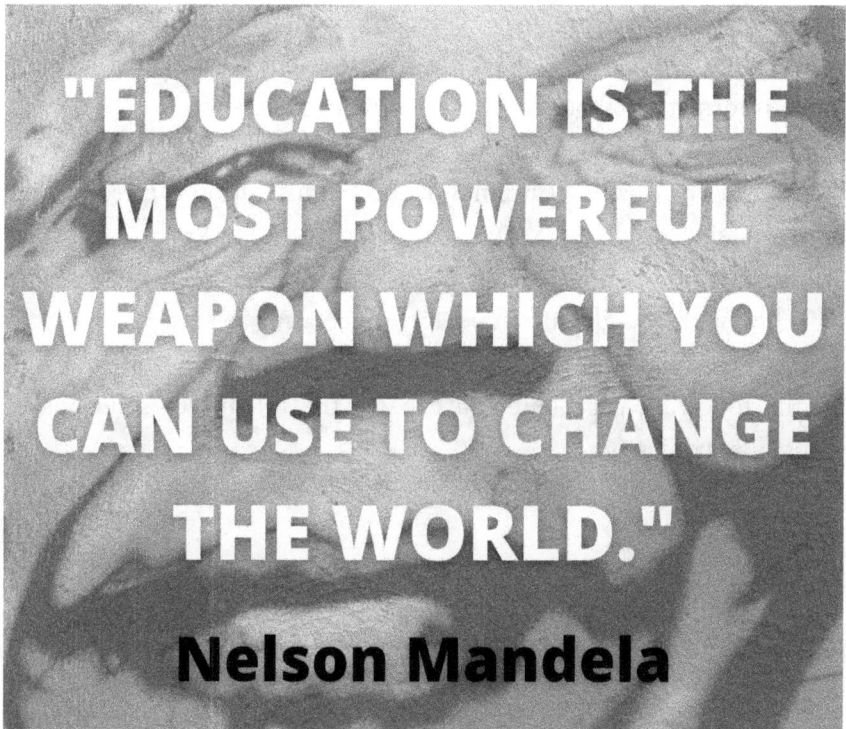

"EDUCATION IS THE MOST POWERFUL WEAPON WHICH YOU CAN USE TO CHANGE THE WORLD."

Nelson Mandela

Nelson Mandela painted portrait P1040799
by thierry ehrmann Attribution 2.0 Generic (CC BY 2.0)

Project Drawdown Top 20

The first place to visit when you want to understand potential solutions to the climate crisis is Project Drawdown. It is a non-profit founded in 2014 by Paul Hawken. In 2017, a team of global experts from their specialised fields published the world's first book of solutions to not simply slow down climate change, but to reverse it. The solutions are constantly being updated as the situation changes, and new data emerges. Project Drawdown's website is the place to go for authoritative information about solutions to the problems we face. The book lists eighty of the most effective changes we can make, and it lists them in order of how many gigatonnes of CO_2 each solution can reduce our total emissions under two separate scenarios. Scenario 1 is roughly in line with temperature increases of 2°C (3.6°F) by 2100, and scenario 2 is in line with the global temperature increasing by just 1.5°C (2.7°F) by 2100. The top twenty solutions by gigatonnes of reduction are listed on the following page, but you can see the full list of solutions at https://drawdown.org/

SOLUTION	SCENARIO 1	SCENARIO 2
Reduced Food Waste	88.5	102.2
Plant-Rich Diets	78.33	103.11
Family Planning and Education	68.9	68.9
Refrigerant Management	57.15	57.15
Tropical Forest Restoration	54.45	85.14
Onshore Wind Turbines	46.95	143.56
Alternative Refrigerants	42.73	48.75
Utility-Scale Solar Photovoltaics	40.83	111.59
Clean Cooking	31.38	76.34
Distributed Solar Photovoltaics	26.65	64.86
Silvopasture	26.58	42.31
Peatland Protection and Rewetting	25.4	40.27
Tree Plantations (on Degraded Land)	22.04	35.09
Methane Leak Management	19.81	4.5
Temperate Forest Restoration	19.42	27.85
Concentrated Solar Power	18	21.51
Perennial Staple Crops	16.34	32.87
Insulation	15.38	18.54
Regenerative Annual Cropping	15.12	23.21
Tree Intercropping	15.03	24.4

Geoengineering

A s the crisis gets worse, the calls to tech our way out of the mess will grow ever more fervent. Already, this is the preferred solution of the billionaire class, who see no reason to halt the megamachine that is ultimately causing the extermination event we are witnessing. We still have time, if we act immediately, to avoid complete collapse, but if we act slowly, the need for drastic action will arise and this will be in the form of giant geoengineering projects. While they will be sold as a panacea for our problems, they will do more harm than good and once started, there will be no turning back. One such project is the giant space umbrella proposed by James Early in 1989. The idea is that the asteroid that killed the dinosaurs blocked out 90% of the Sun's rays, resulting in a massive temperature drop. Rather than wait for another asteroid, we could mimic the affect ourselves, the logic goes. The project entails, somehow, getting a 2,000 km-wide glass shield into position where it is balanced between the Earth's gravity and the Sun's. That position is around a 1.6 million km away. The shield would be so enormously heavy that it would need to be made on the moon. As a work around to the weight problem, an astronomer called Roger Angel has suggested producing sixteen trillion flying space robots on Earth, each weighing one gram. These space robots would deflect sunlight by forming a cylindrical cloud 96,560 km wide. They would need to be regularly "nudged" to avoid them crashing into each other. If this doesn't sound outlandish enough, it has even been proposed that we simply move the Earth further away from the Sun. This would be done by causing an explosion equivalent to five thousand million million (yes two millions) hydrogen bombs (767).

As comical as these "solutions" seem in 2022, by 2042, we may see these as necessary. By far, the most widely touted of these projects, and the most likely to be attempted is solar radiation management (SRM). The aim of this project, funded partly by Bill Gates, is to mimic volcanic explosions. After the Philippine's Mount Pinatubo erupted in 1991, around fifteen million tonnes of sulphur dioxide was injected into the stratosphere, between 10km and 50km in altitude. This sulphur dioxide then mixed with water and created a layer of sulphuric acid droplets which scatter and absorb incoming sunlight. Stratospheric winds then carried this layer around the globe. The effect was that over the next fifteen months the global temperature dropped by 0.6°C (1.1°F) (768). Of course, as this was a one of event and CO_2 continued to be pumped into the atmosphere, once the aerosols naturally dispersed, the temperature rose again. The man-made version of the Pinatubo effect is to send high altitude airplanes into the stratosphere 4,000 times a year to inject sulphur dioxide. As a bonus, the sunsets would apparently be incredible, but on the flip side, scientists claim SRM would cause a "calamitous drought" in the Sahel region of Africa, projected to be home to 196 million people by 2050. We have experienced the global impact of another large volcanic eruption. After the Indonesian volcano, Mount Tambora blew its top in 1815, the following year, Europe experienced widespread harvest failure. Three of the four large dry spells in the Sahel region in the 20th century followed large volcanic eruptions in Alaska and Mexico the previous years. These droughts created ten million refugees and killed 250,000 people (769). It isn't just Africa that might suffer the effects of SRM. It has been suggested that to negate the impact on the Sahel, sulphur should be added in the tropics, but scientists warn that this will then cause a decrease in monsoon rains in South Asia, home to two nuclear powers and over a billion people already suffering the effects of drought. According to the Intergovernmental Panel on Climate Change, SRM will also cause damage to the ozone layer which has been in a long recovery since the 1980s, due to human activities (770).

The concern for many scientists is that as the temperature continues to rise, populations will demand that their leaders take drastic action, and twenty or thirty years down the line, this may be a last resort. For this reason, research needs to be carried out now, so countries do not act blindly. Such research was carried out by the Swiss National Science Foundation in 2020, and they found that while the temperature would drop, the interference would impact precipitation, flood, and drought patters around the world. While it may become necessary, if we fail to act now, climate scientists are clear that it should be avoided. Emeritus Professor at Imperial College London, Joanna Haigh, said of these research findings:

> *"The results of this study indicate that solar geoengineering can in no sense be viewed as a sensible rescue plan due to the potential to severely impact on temperature, precipitation, floods and drought patterns across the globe (771)."*

In 2022, a group of climate scientists wrote an article titled: Solar geoengineering: The case for an international non-use agreement. In the paper, the concerned scientists explain that *"solar geoengineering at planetary scale is not governable in a globally inclusive and just manner within the current international political system,"* and they call on governments and the U.N. to create an International Non-Use Agreement on Solar Geoengineering.

Other research has looked at the long-term effects of SRM and the findings do not look good. It has been estimated that stratocumulus clouds would gradually thin and break up, in turn causing global warming of 5°C (772). Further problems arise because SRM is relatively cheap at $2 billion a year, so cheap that most countries could afford to carry out their own projects independent from each other (770). This would create a wild west free-for-all where countries attempt to whack-a-mole and could descend into complete climate chaos, causing mass starvation and triggering global wars. Fortunately, there are less dangerous solutions at hand.

Energy

In the world of solutions, there are few that gain complete consensus, but one such anomaly is renewable sources of energy; wind, solar and tidal. We can all agree that we need to stop burning fossil fuels for energy. In many countries renewable sources of energy are now cheaper than fossil fuels and they tend to employ more people as an added bonus. The drawback with simply switching to renewables and continuing to consume ever more energy is that they still require the mining of resources to produce solar panels, wind turbine blades and batteries, they still produce toxic waste through their production and this waste can end up polluting rivers and killing wildlife. It is estimated that the extraction of critical and rare earth minerals would need to increase sixfold by 2040 if we are to reach net zero by 2050. Electric cars require six times more mineral inputs that combustion engine automobiles and onshore wind farms require nine times more mineral supplies than gas-fired power stations. Lithium alone would need to increase by forty times to power the green transition and keep warming within 2°C (3.6°F) (773). As was discussed earlier, dams for hydroelectricity also have a detrimental impact on river ecosystems. Huge amounts of water are also used in the production process and large amounts of concrete are used for production while coal is still burned to produce solar cells (774). In short, even renewable energy causes damage to the Earth, just not as much as fossil fuels. An obvious answer to this problem is for us to move to renewable sources and then simply use less energy. On an individual level, this has been drummed into us by parents and teachers for decades, but on a commercial level it would mean a drop in consumption. This is where we can feed two birds with one scone. As we saw in the previous section, planned, and perceived obsolescence have been used by marketing companies for almost 100 years to encourage us to keep buying more and more products and increase their profits. Government bans on planned obsolescence would reduce consumption drastically. Imagine buying a computer that is modular so that when a piece breaks, you can buy a replacement part instead of buying an entire PC. This would not be difficult to legislate. The initial product price would be higher, and companies could make money from selling replacement parts. This would also lead to less pollution and less resource use.

Another positive lesson comes out of Germany where, since 2013, the power grid in many municipalities has been returned to public ownership after being sold to Vattenfall and Eon in the 1990s. The logic behind this move is that publicly owned utilities are more likely to take the climate crisis seriously and would be able to accept smaller profit margins with more available for investment in green technology (775). German citizens have gone even further by forming energy cooperatives where local citizens invest in wind farms and solar parks. By 2016, farmers and private individuals accounted for 42% of installed renewable power generation (776). In late November 2019, the small town of Bordersholm removed itself from the grid completely. Although the disconnection was a trial and only lasted an hour, it provided proof that renewably generated systems could power a town of 8,000 people independently from the national grid (777).

Public Ownership of the Energy Beneath our Feet

As we saw in the Bringers of Death section, the fossil fuel industry is potentially complicit in the deaths of hundreds of millions of people, ecosystem collapse and the extinction of entire species. It is beyond doubt that they lied for decades to keep us buying their products even as CO_2 continued to concentrate in the atmosphere and drive super storms, record wildfires and intensified ice melt at both poles. The question, surely then needs to be asked, why are we not only continuing to buy their products, but also using our hard-earned taxes to make them artificially cheap so it is harder for renewable forms of energy to compete? As we will see in a later section, global fossil fuel subsidies amount to around $5 trillion a year, and just twenty-five of the largest fossil fuel corporations reaped profits of $205 billion in 2021. This is at a time when gas prices have been surging around the world. Rather than reinvesting these profits to help lower prices for struggling consumers, the companies have been doing all they can to divert the profits to their investors, many of whom are not in need of any financial help. This is the insane situation we find ourselves in. We are diverting our money, which needs to be spent wisely on a rapid transition to renewable forms of energy, into the coffers of the very same corporations who have profited, and continue to profit, from climate breakdown. The only people smiling are the industry leaders. Chevron said 2021 was one of its *"most successful years ever,"* and Shell CEO Ben Van Beurden said 2021 was a *"momentous year (779)."* I don't think he was referring to the record CO_2 emissions that rose by 6% on the previous year (780). The first four months of 2022 saw 100 fossil fuel companies reap $100 billion in profits and BP's CFO said *"Certainly, it's possible that we're getting more cash than we know what to do with (781)."*

The fact that these gentlemen can boast of success as their products cause climate catastrophe says a lot about the world we live in. Because of the inaction of the last decade, we now face no choice but to race towards a renewable transition at a colossal pace, yet these companies spend just a fraction of their resources on renewable energy. According to the international Energy Agency (IEA), clean energy investments by oil and gas companies accounted for just 1% of total capital expenditure in 2020. Compare this to the public spending on research and development (R&D) of low-carbon technology which amounts to 80% and we can see a startling difference emerge between what governments and corporations are prepared to do (782).

With the Russian invasion of Ukraine being funded by our planetary addiction to Russian gas, and the second Iraq war largely fought over oil, it is even clearer that we need to leave fossil fuels where they belong, deep underground. Sadly, this is not the plan of our leaders who are simply planning on increasing domestic exploration. While it is clearly in the interest of every country to become energy self-sufficient, further investment in fossil fuel extraction will lock us into dirty energy for decades. This is at odds with the International Energy Agency (IEA) whose executive director, Fatih Birol, said in 2021:

"If governments are serious about the climate crisis, there can be no new investments in oil, gas and coal, from now – from this year."

Mr. Birol went on to say that a there is a growing chasm between what governments say and do (783). The secretary general of the U.N., António Guterres, added that *"Investing in new fossil fuels infrastructure is moral and economic madness (781)."* Research published in Nature states that 60% of oil and gas reserves and 90% of available coal must be left in the ground if we are to have just a 50% chance of remaining within the

1.5°C (2.7°F) carbon budget (784). You would think that our elected representatives would heed this advice. Sadly, elected by the people, they might be, but represent the people, they do not. As we have seen, they are in the pockets of the very industry they need to regulate, and the U.S. is planning on increasing liquefied natural gas (LNG) exports to the E.U., from fifteen billion cubic meters to fifty billion cubic meters by 2030 (785). Remember, this is the year we need to have almost halved emissions, and this is a watered-down conservative estimate. It total, an investigation by the Guardian found that fossil fuel companies were planning on spending $103 million a day until 2030 on what are being called 'Carbon Bombs'. The 195 projects planned will each contribute 1GtC and cost over $3 trillion. Should these projects be completed, combined they will add around 195GtC to the atmosphere when our current budget for staying within 1.5C is between 170GtC (67% chance) and 380GtC (50% chance).

While the fossil fuel industry on both sides of the Atlantic used the COVID-19 pandemic to increase their taxpayer subsidies, what could have been done is that these ecosystem destroying corporations could have been, once again owned by the people, so they could become part of the solution and not the largest cause of the problem. In the U.S., it is estimated that nationalizing the three largest fossil fuel companies would cost around $420 billion, just 10% of the amount spent on bailing out the corporate sector in 2020 due to the pandemic (786). We have seen that we need to leave the vast majority of known fossil fuel reserves in the ground to have just a 50% chance of remaining within 1.5°C (2.7°F), but fossil fuel corporations have no interest in doing so because this will reduce profits and shareholder returns. Many will say it isn't feasible, but we only have to rewind our memories to 2008 to see that General Motors, Chrysler and the Royal Bank of Scotland were returned to public ownership. If our governments are at all sincere in their pledge to prevent disastrous global warming, it is clear that business as usual will not suffice. To achieve the rapid decarbonisation of the economy we so badly need, we must leave fossil fuels in the ground, and this isn't in the financial interests of the fossil fuel industry who are setting their sights on the Arctic where the summer sea ice will soon be gone, and they can begin drilling in the pristine waters at the top of our world.

In his article, A History of Nationalization in the United States: 1917-2009, Thomas Hanna highlights the plausibility of renationalising the fossil fuel under our feet:

> "The United States actually has a long and rich tradition of nationalizing private enterprise, especially during times of economic and social crisis. Importantly, this approach has often been deployed when private companies are hindering national efforts to address a crisis (either through obstruction, incompetence, or incapacity). This history of nationalization, along with other robust government economic interventions, suggests that far from being a non-starter, a public takeover of the fossil fuel industry should be considered an eminently plausible and viable policy option for dealing with the forth-coming climate crisis (787)."

No one who has read this far will doubt we are in a crisis, and few will honestly contend that the fossil fuel industry has not obstructed attempts at a clean energy transition. Most will also agree that these companies should not be profiting from the destruction of our climate and the suffering that will come from the continued use of their products and services. Renationalisation will also remove the corruption between fossil fuel interests and politicians which has thus far prevented any meaningful change. To succeed in leaving fossil fuels in the ground, they need to belong to the people.

Transport

When it comes to transport, many believe that the future involves electric cars zipping silently around, and private electric planes connecting skycrapers. While we certainly need to stop driving gasoline powered automobiles, is it really a good idea to swap out 1.4 billion gas guzzlers for 1.4 billion lithium gulpers? That's a lot of lithium, a lot of cobalt, a lot of rubber, a lot of tarmac, a lot of steel and a great deal of energy, powered by renewables like solar panels, produced by burning coal. Something that would benefit society and the environment would be the reintroduction of free or subsidized electric public transport systems, ideally powered by citizen owned cooperatives. We would be able to turn the tarmac into public spaces with trees that suck up carbon and provide a home to wildlife, have less cars on the road, more room for kids to play and drastically reduce the number of traffic related deaths from the current 1.3 million, which is the biggest killer among those aged 5-29 years of age (778). Faced with the rising cost of energy in April 2022, Germany did just this; they reduced monthly public transportation costs from as much as €107 to just €9.

When it comes to short-haul aviation, small electric planes are making headway and should be able to replace kerosene fueled planes in the near future. When it comes to long-haul trips, currently aviation is not coming close to finding a solution; and instead is reliant on offering passengers the option to offset their carbon in exchange for flying. An alternative solution would be transnational investment in state-of-the-art magnetic levitation (Maglev) train networks that travel at 460kph (286mph). Connecting the Eurasian and African continents by high-speed railways would reduce the need for many long-haul flights and passenger times would not be greatly affected. A train journey from Shanghai to Paris would take around twenty hours whereas a flight takes twelve hours and twenty minutes. As airports are usually an hour outside of cities, the difference can be reduced to just six hours. Similarly, a flight from London to Cape Town takes around eleven hours and forty minutes. This journey could be done in twenty-one hours by high-speed rail. Imagine the experience of seeing twenty countries on your journey, as opposed to fluffy white clouds. Those living on the American continent would also benefit from being connected by high-speed rail. Flying from New York to Los Angeles by plane currently takes six hours; by rail that journey would only take eight and a half hours. When you account for transport to and from airports, the journey difference is negligible. Flying from Los Angeles to Buenos Aires takes almost fifteen hours and by train; this journey would take less than twenty-two hours.

Unfortunately, we would still need to connect the other continental land masses with the Americas and Australasia. Sea transport, which is a major cause of fossil fuel emissions (2.2%), and also impacts three other planetary boundaries: ocean acidification, biodiversity and chemical pollution, could provide a solution here: wind energy. Sail ships are making a comeback, and the sails today can be made of photovoltaic cells which combine the power of the Sun and wind to push through the waves. Japanese company Eco Marine Power is working on such a solution: the Aquarius MRE, while Swedish company Wallenius Marine is developing the world's largest wind-powered cargo ship: the Oceanbird, which will be able to travel at ten knots and traverse the Atlantic in just ten days.

As with most industries, the solutions are there; we just need government policies to get us over the line. That's where you come in, and we'll get to that in a bit.

Less Children

Another concern in the coming decades is that the poorest and most vulnerable in society will be scapegoated by those who have overseen the collapse we are witnessing. Those fleeing from uninhabitable lands because of lack of water, extreme heat, rising waters and starvation will need somewhere safe to live. Our leaders and those in the media will no doubt use language associated with vermin and pests to describe them as they struggle to survive. It is imperative to remember that those benefiting the most from the destruction are those we need to hold to account; not the most vulnerable in society.

Many like to excuse themselves from action by blaming the rising human population for the crises we face. It is clear, the planet will benefit from policies aimed at limiting the global population. This will only happen though, if health care in developing nations improves, and grinding poverty declines. People need a safety net to help them when they are too old to work; otherwise, they will continue to rely on their children in old age. This is not going to happen without debt cancellation as poorer countries are currently crippled by debt. The only way to limit population growth on our planet is to share resources and wealth more evenly.

It is also just as clear from the map below that the richest of us are responsible for the majority of emissions. This will be discussed in later, but one solution to the coming migrations is for those of us in the rich countries who have benefitted the most to voluntarily have less children. If we all decide to have two children or less, then our population will fall naturally, as is already happening in Europe and Japan. A side effect of this falling population is that it becomes difficult to find workers to pay for the retirees, as Japan is now finding out. This doesn't need to be a problem though. As the populations in rich countries fall, desperate climate refugees could take their place in the workforce and help to pay the pensions of the older generations. Again, I can already hear the groans of many who don't wish to allow foreigners into their country. The simple fact, though, is that under a capitalist system, this is inevitable. The aim of capitalism is to increase wealth and raise living standards, and this involves education. Once women can become educated, they have less children. Educated women have children later in life, they tend to marry equally educated men who also wish to have fewer children and they are more inclined to stay together to raise their offspring which leads to less likelihood of early pregnancies (788). This is what is causing the population decline in many countries. If we simply close our borders, who is going to pay our pensions if there are no young workers to pay taxes? Offering climate refugees a safe haven is not only the morally correct thing to do, but it also happens to be a practical solution for one of the problems that is arising.

Per capita CO2 emissions, 2020

Carbon dioxide (CO_2) emissions from the burning of fossil fuels for energy and cement production. Land use change is not included.

Our World in Data

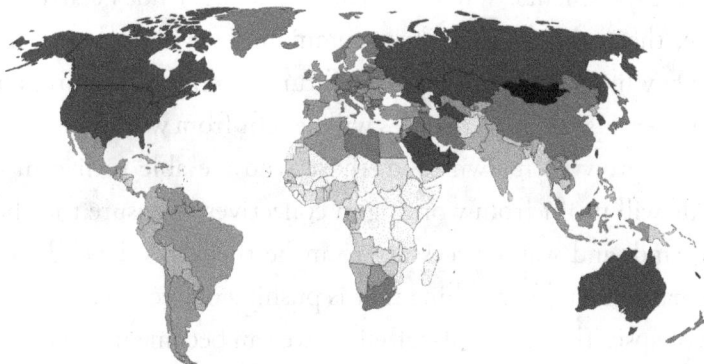

No data   0 t   0.1 t   0.2 t   0.5 t   1 t   2 t   5 t   10 t   20 t

Source: Our World in Data based on the Global Carbon Project   OurWorldInData.org/co2-and-other-greenhouse-gas-emissions/ · CC BY

Sustainable Communities

Just like the Anabaptist movement, and many others throughout history, sustainable agrarian movements are springing up around the world. From Denmark, Iceland, Scotland and Italy in Europe to Egypt and Tanzania in Africa, sustainable ecovillages are attempting to remove themselves from the megamachine as much as possible. While this is extremely difficult to do completely at present, it is possible for small groups of people, with less than 1,000 citizens to design their own communities that focus on regenerative lifestyles. There are more than 10,000 such communities around the world where cooperation, self-sufficiency, renewable energy, and ecological buildings are the aims of the collective. While they are often distant from each other, they are now connected through the Global Ecovillage Network (GEN) which is active on five continents. While for many, this might not be attractive or feasible, the reality is that these communities will be best placed to survive any likely tension caused by future scarcity of resources like food and water. Just as those in cities will benefit from working in street collectives to grow their own food, those who are able to live in the countryside will profit from working in collectives to ensure they have the skills, land, and water necessary to make them less dependent on the extremely vulnerable machine that is pushing our ecosystems to the brink of collapse. The more self-sufficient we can become in the coming decades, the better we will be able to deal with any crises that emerge, and as we have found throughout the COVID-19 pandemic, crises can emerge when we least expect, and their likelihood will increase as our planet warms and biodiversity is lost.

Half-Earth

In the prehuman age, extinction rates were around one-to-ten species per million each year but after our rapid population boom and even faster expansion across the globe, the extinction rate is now around 1,000 per million and accelerating. In a last gasp attempt to prevent further biodiversity loss and the complete collapse of ecosystems around our planet, the U.N. Biodiversity Conference is urging countries to conserve 30% of their land by 2030. They have branded this project as 30x30, and it is hoped it will also include 30% of national territorial waters. According to PEW Charitable Trust, more than 100 nations have publicly committed to the target:

Albania, Angola, Antigua, Armenia, Australia, Austria, Bangladesh, Barbados, Belgium, Belize, Benin, Bhutan, Botswana, Cabo Verde, Cambodia, Canada, Chad, Chile, Colombia, Comoros, Cook Islands, Costa Rica, Côte d'Ivoire, Croatia, Cyprus, Czech Republic, Democratic Republic of Congo, Denmark, Dominica, Dominican Republic, Ecuador, Ethiopia, Federated States of Micronesia, Fiji, Finland, France, Gabon, Germany, Ghana, Greece, Grenada, Guatemala, Guinea, Guinea Bissau, Guyana, Honduras, India, Indonesia, Ireland, Israel, Italy, Jamaica, Japan, Jordan, Kenya, Kiribati, Liberia, Luxembourg, Marshall Islands, Mauritania, Mexico, Monaco, Mongolia, Montenegro, Morocco, Mozambique, Namibia, Netherlands, Nicaragua, Niger, Nigeria, Niue, Norway, Pakistan, Palau, Panama, Papua New Guinea, Peru, Portugal, Republic of Congo, Republic of Maldives, Romania, Rwanda, Saint Kitts and Nevis, Samoa, São Tomé and Príncipe, Senegal, Seychelles, Slovakia, Slovenia, Solomon Islands, South Korea, Spain, Sweden, Switzerland, Togo,

Tonga, Trinidad and Tobago, Tuvalu, Uganda, United Arab Emirates, United Kingdom, Vanuatu.

With only 17% of land and just 8% of ocean now benefitting from some form of protection, the aim is a good start (790). It shouldn't be seen as the finished article though, we can and should protect much more of nature if we are to protect ourselves. Edward O. Wilson states in his book Half-Earth: Our Planet's Fight for Life that if we removed 90% of natural habitat then you could expect to find 50% of species survive but when you remove 10% of that final habitat, you would see most, or all surviving residents disappear. Humans have indeed wiped out 90% of habitat in many areas of our wonderous planet and we are balancing over the cliff edge, staring down into the abyss while desperately clutching at whatever foliage is left growing out of the soil. Wilson estimates that the final section of habitat could be destroyed by a group of lumbermen in as little as a month. He advocates that we set aside half of Earth to conserve most species who have survived evolution's path thus far and that failure to do so would mean we fail in our own attempt at surviving the crises we have manufactured. He explains that half is the number necessary to provide a full representation of all the existing ecosystems and they would provide adequate habitat for all surviving species to continue their evolution. On the face of it, with a rising population and rising demands on the land, this seems impossible but in the following section, we will look at how it can be achieved (789).

Agriculture

As we saw in section two, the way we farm and eat is having a disastrous impact on our planet and biodiversity, as well as our health. Around half of our ice-free land is currently used for agriculture and as we need to set aside 50% for nature, the best place to start is here.

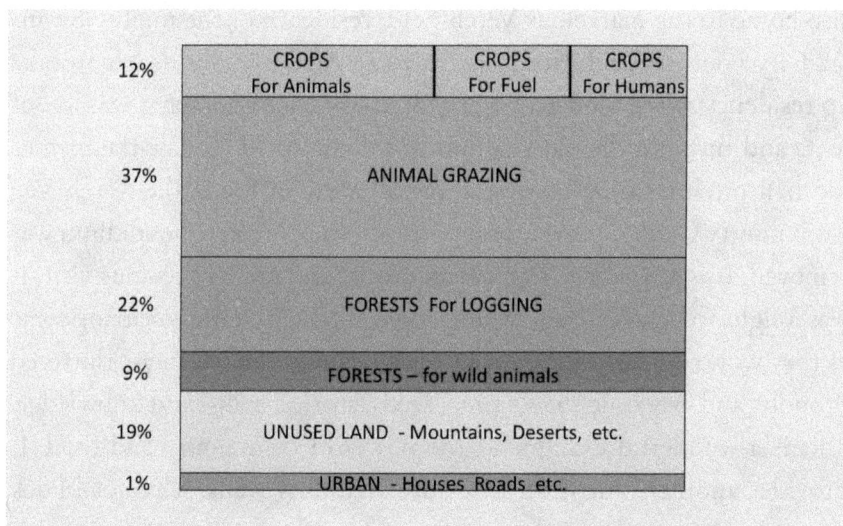

12%	CROPS For Animals / CROPS For Fuel / CROPS For Humans
37%	ANIMAL GRAZING
22%	FORESTS For LOGGING
9%	FORESTS – for wild animals
19%	UNUSED LAND - Mountains, Deserts, etc.
1%	URBAN - Houses Roads etc.

How the ice-free land area of the planet is distributed for different uses.
2019 IPCC Special Report. Figure courtesy: Rebecca Allen, Climate Healers.

Additionally, it is impossible to refute that our current system of factory farming animals is hugely immoral. Fortunately, there are solutions at hand that will be more ethical, healthier, less impactful and have an additional benefit of bringing us back together. To illustrate this point, we will visit Swansea in South Wales. Here, the average

garden is around 203 m² (791) The standard garden in Swansea today
will consist of a lot of chopped trees in the form of decking, some
stones or gravel, grass, a BBQ, perhaps a trampoline, and in the corner,
the latest trend, a hot tub. Generally speaking, the garden has become
an outside room rather than a garden. This relatively small space can
easily be reverted back to a productive garden that would be enough
to feed ten people during the harvest season (792). This would not only
reduce expenditure, provide a space for bees and other pollinators, like
birds, but it would also get families outside working together. It would
increase national food self-sufficiency and when the next pandemic
hits, we would have some food out the back. It's hard to find such a
win-win situation for all. It gets better though. The reason Swansea is
used as an example is, in addition to being the place I grew up, it was
also home to the marvelous Vetch field, residence of the mighty Swansea City Football Club from 1912 to 2005. After the Swans left to take
up residency at the modern Liberty Stadium, fans took what was left of
seats and turf and the old stadium was demolished. Today, the former
football pitch is being used to grow all sorts of food. The Vetch Veg
Community Garden project began in 2011, shortly after the stadium was
removed. It is described as an urban utopia and on my previous visit, it
was a sight to behold. There were people of all ethnicities chatting away
as they worked. Flags of the various nations represented there, fluttered
proudly, and everyone shares time, seeds, stories, skills, and knowledge.
This is a wonderful example of the power of community and food. It
provides another blueprint. Instead of building walls, why not knock
the walls down and form a street garden where everyone gets their
hands dirty. This would help to alleviate the loneliness that has crept
into our lives the past forty years. Thatcher was wrong. There is such a
thing as society, and we need to get it back.

Now, any committed carnivores out there may not be licking their
lips at the prospect of a garden full of veg. Perhaps they would prefer a
communal slaughterhouse where we can all sit and chat as the screams
of animals permeate the chilly air. There is, however, a growing understanding of the damage animal agriculture is causing to the planet and

plant-based diets are listed as the number two solution to reversing climate change by Project Drawdown *(793)*. The obvious reason for this is that there are a lot of farmed animals (ten times the human population), and farmed animals use a lot of land and consume a lot of food and water, both of which are about to become very scarce due to a warming planet. In the case of cows and sheep, they also produce a lot of methane.

For these reasons, it makes sense to eat the food we feed to animals ourselves and cut out the middleman. Fortunately, for humans, and unfortunately for farmed animals, we won't have to get used to eating grass as this is not the staple food of factory farmed animals; who by and large eat the same food we do, soybeans, fish meal, corn, and palm oil kernels. According to research in the journal Science, moving to plant-based diets would reduce farmland by 75%, an area the size of the U.S., China, the E.U. and Australia combined. All this land could then be reforested and left so nature can do her thing *(600)*.

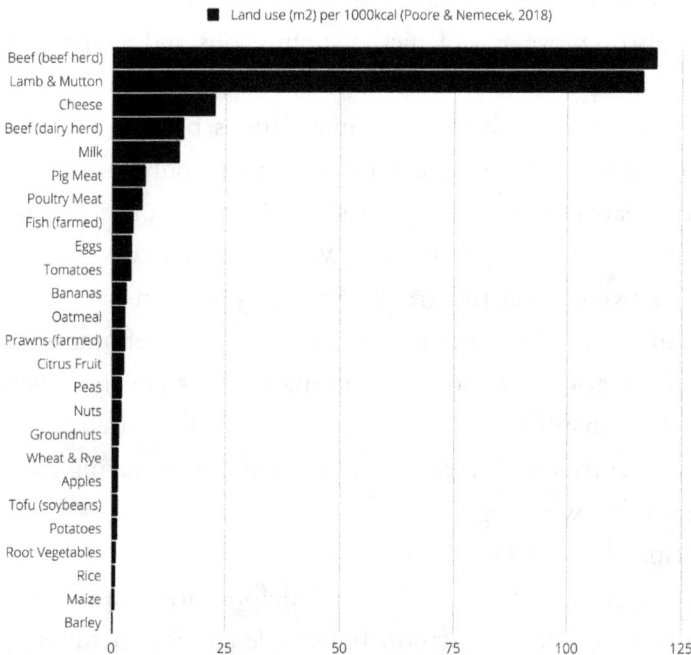

Poore, J., & Nemecek, T. (2018). Science, 360(6392), 987-992.

The previous chart shows that growing peas and soybeans uses almost 100 times less land than for cows and sheep, and five times less land than chickens and pigs (794). In one compassionate hug, we would have met our half-Earth aspirations, job done.

The added benefit for humanity is that all these newly planted trees would not only provide habitat for wildlife but suck tonnes of CO_2 out of the atmosphere and store it safely. James Hanson estimates that, in addition to halting emissions, we need to remove 12.5% of CO_2 from the atmosphere (795). Research published in Nature in 2020 suggests that if we reverted just 15% of land currently used for animal agriculture to forests, we could sequester $299GtCO_2$ by 2050. This is approximately 30% of the CO_2 increase since the industrial revolution (796). A further paper published in the Journal of Ecological Society in 2021 stated that changing to plant-based diets would be equivalent to sequestering over 2000 $GtCO_2$ through regenerating soils and vegetation, returning atmospheric greenhouse gas levels to the 'safe zone' of 350 ppm and at the same time healing the planet through restoring biodiversity (797). It would also remove 37% of methane emissions and as methane is a massively more potent greenhouse gas than CO_2, this would have a huge impact in a relatively short timeframe. This is because whereas CO_2 lingers in the atmosphere for centuries, methane only lasts for around ten-twelve years before converting to CO_2. Prior to converting to CO_2, it has a global warming potential (GWP) of seventy-two times that of CO_2 over a twenty-year timeframe. As it only stays in the atmosphere for a decade, it would be better to use a ten-year timeframe and when we do this, it has a GWP of 130. That means it is 130 times better at trapping heat than CO_2 and when you account for this, according to the research author Dr Sailesh Rao, we would be removing around 17 $GtCO_2e$ just by switching diets (797).

Additionally, with 60% of extinctions being caused by eating meat, the single biggest thing we can do to halt deforestation and biodiversity loss is also change our diets (798). The simple fact is that we use 43% of Earth's non-ice-free land to produce animal products and they provide just 37% of protein and 18% of calories while we use a mere 3% of land to produce crops that provide us with 82% of calories and 63% of our protein (600). It is an incredibly inefficient system that has led us to the edge of complete ecological collapse.

Afforestation will be relatively cheap, whereas the alternative option to removing CO_2 from the atmosphere has been estimated at $100 billion per billion tonnes (1Gt), and we are putting more than forty times that amount into the atmosphere every year (799). Using machines to do the job will not help us restore forests or protect biodiversity. In fact, they will require massive amounts of water, energy, and resources to produce and operate. They will also take decades to scale up and their efficiency is not yet known. Of course, the latter is the preferred option of corporations who see no profit to be made from forests, unless they are being cut down. As we saw, our oceans are also in a perilous state, largely caused by the industrial fishing industry which replaces trillions of fish with huge amounts of plastic fishing gear, kills massive numbers of sharks, turtles, dolphins, whales, and seabirds as bycatch and removes the prey of large oceanic mammals. Switching diets would allow our oceans to bounce back from the brink and provide us with the natural services we have taken for granted; like oxygen to breathe and a place we can realign with our ancestry. It isn't just non-human biodiversity that needs protecting, human biodiversity is also at stake as the forests are lost. Providing indigenous people with tenure over the forests would allow traditional practices to continue and they could protect the forest that is protecting them. Project Drawdown predicts this would sequester between 8.69-12.93 gigatonnes by 2050 (793).

As if this isn't enough, plant-based diets also help to save lives, lots of them. According to research from Oxford University, a switch to plant-based diets could save as many as eight million lives by 2050. It would also save around $1.5 trillion in healthcare and climate related damages (800). When analysing the health and environmental impacts of food, the research found that those foods with the least environmental impact were also the healthiest. Marco Springman from Oxford University says, "We now know pretty well that predominantly plant-based diets are much healthier and more sustainable than meat-heavy diets (801)."

Foods that are good for the environment are also good for our health

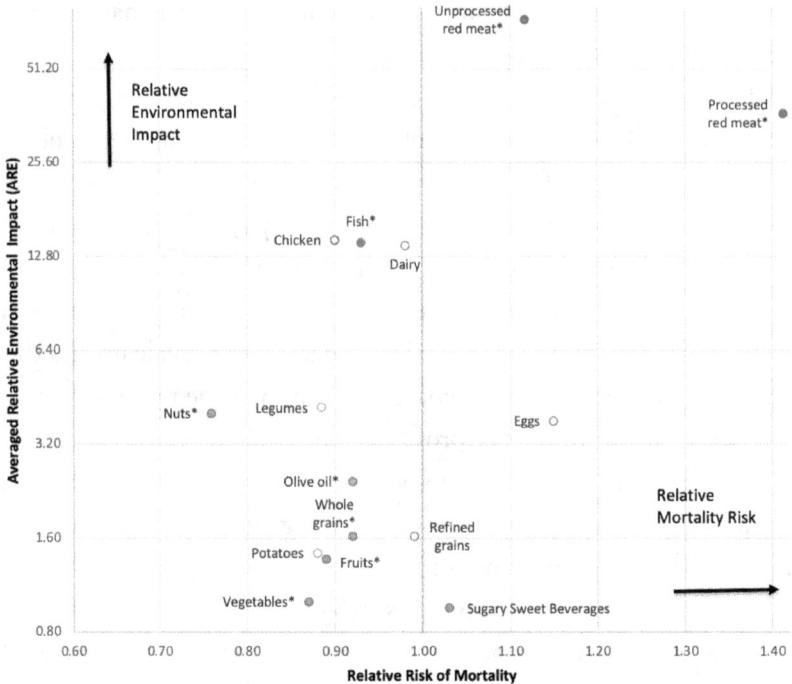

Clark et al, PNAS, 2019. Note: Foods linked to a statistically significant change in mortality risk are denoted by solid circles. Those not linked are denoted by open circles

While plant-based diets provide us with the biggest positive environmental impact, veganism will help us atone for the mass violence we have inflicted on non-human animals. As we saw in sections two and three, the pain and suffering we have caused other species is gargantuan in size. While many justify their actions by saying that other animals also kill animals for food, the truth is that we do so for enjoyment, not necessity. Others like to argue that plants feel pain too. Well, if this is the case then we need to minimize plant suffering as well as animal suffering. As we have seen, plant-based diets use far less plants than meat-based diets so for anyone concerned with plant sentience; plant-based diets are the answer to this conundrum as well. No one can reasonably argue that human beings living in developed societies have any need to eat animals, drink their milk, wear their skins, or steal their eggs. If this was the case, then Hindu people would all be dead, as would the Seventh Day Adventists and a whole range of individuals, from sporting greats like Carl Lewis, Lewis Hamilton, and Venus Williams to famed historian Yuval Noah Harari. Hollywood would also be deprived of such prominent actors as Brad Pitt, Woody Harrelson, Joaquin Phoenix, Natalie Portman, and Peter Dinklage. The music world wouldn't b e the same without Mob y Morrisey, Prince, Paul McCartney, Stevie Wonder, John Joseph, Russell Simmons, or Akala. No, there is no argument that we need to kill and enslave animals for our survival. Quite the opposite is true, we need to stop breeding them into existence in order to preserve our own.

Moreover, if we start to teach our children that all animals deserve respect, regardless of how they look, this is likely to produce more compassionate and caring citizens. An argument for the continued exploitation of non-human animals is an argument against peace on Earth as we cannot have peace when we are slaughtering tens of billions of innocent non-human animals and forcing sentient beings to endure torture at the hands of scientists. When humans hope for peace, what they really mean is peace for humans. They rarely stop to consider why the most violent and destructive species on Earth deserves peace while they sub ject all other species to unimaginab ly violent ab use. As the Russian writer, Leo Tolstoy, once said:

"As long as there are slaughterhouses there will always be battlefields."

Another area where food impacts the environment and society is food waste. With half of all food in the U.S. never b eing eaten, 20% in Europe and glob ally a third, it is easy to see why, according to Project Drawdown, reducing food waste is the biggest solutions to reversing climate change (793). This waste occurs at the demand side and in supply. At home we can ensure that we eat all the food we buy and at the production side, it must be mandated that edible food cannot b e thrown away. With close to a b illion humans undernourished, throwing good food in the b in is an immoral act. France b ecame the first c ountry t o e nact such a l aw in 2016 and now s upermarkets must donate any unsold food to charities and food banks (802). We can solve the problem of food being left in fields to rot by supermarkets ending their insistence on all fruit and veg being a uniform size and colour. Nature doesn't work that way. While we are reducing food waste, we must all insist on bringing about an end to harmful pesticides and herbicides which are killing the soil, insects and ultimately us. Some will contend that this isn't possible, that we need these chemicals to maximise harvest. The truth is the only thing being maximised is Monsanto's bottom line. We already grow enough food for ten billion humans, but we feed seventy billion animals in addition to the human animals around the globe. Shifting to organic plant-based diets would free up so much space for afforestation while causing no declines in crop yields. Research published in Nature Sustainability in May 2022 analysed data from thirty experiments in Europe and Africa and found that crop yields were not reduced when Ecological Intensification (EI) methods were used instead of using harmful chemicals. The methods, which include increasing crop diversity and adding fertility crops and compost, were found to positively increase yields of staple crops; especially when using low amounts of Nitrogen fertilizer. Our soils would be strengthened, biodiversity would rebound, and our health would improve as a result. We can no longer ignore that what we eat has an outsized impact on our planet, our health, and all non-human animals. The choices we make three times each and every day from now will decide the kind of world we want to live in, or whether we want to live at all.

Forest Cemeteries

A dditionally, rather than chopping down trees to use as coffins when we are cremated, which emits our carbon. Why not start planting trees over our buried bodies in huge memorial forest parks?

The wood we use for caskets alone is responsible for the cutting down of four million acres of forest while we pour an incredible six billion tonnes of concrete into cemeteries and in the U.S. alone, enough fossil fuel is used for cremation to drive a car halfway to the Sun (803). The human body has also accumulated large amounts of mercury which is released when burned and settles in the ground (804).

Companies are popping up around the world to offer forest burials where a tree is planted over the body and the family receive the GPS coordinates for future visits. In Australia, these burials are being offered for $1,100 AU to fund the conservation of koala habitats which have been decimated by 80% (805). In California, where burials are heavily regulated, companies are offering to mix cremated remains with soil and water and plant a tree over the remains for between $970 and $30,000 depending on the type of tree (806).

These forests are intended to reintroduce the natural carbon cycle where our bodies go back to the soil and help rejuvenate life. They would also provide a future deterrent for our descendants if they ever reverted back to our destructive ways. Rather than visit concrete cemeteries, we could go cycling and walking through forests to visit the tree where our family members are buried. This practice of forest bathing has been found to have positive physiological effects in research conducted in Japan where Japanese workers are some of the most stressed and fatigued on our planet. Spending time in nature reduces blood pressure, improves the immune system, and alleviates depression (807). We are also more likely to want to protect nature if we spend time there. I for one wish my body to go back to the soil and provide sustenance for life.

A typical extended family cemetery in
Takehara, Japan

Individual and Collective Action

To deflect attention away from personal change, some people like to say that whatever individual choices we make will be useless without government action. While this is certainly accurate, the opposite is just as true. Whatever our governments decide will be useless without individual support. An excellent example of this two-way street comes from France where Emmanuel Macron tried to introduce a green tax on fuel in 2018. While on paper, this seems like an obvious way to reduce fuel consumption, unfortunately, French people don't exist on paper. Many have no choice but to drive long distances and felt they could not afford to pay a higher price at the pump. A movement quickly sprang up and the gilets jaunes, as they were known, took to the streets, more than a quarter of a million of them. The French government soon ended their plan.

In the consumerist society we live in, how and where we spend our money is sometimes the most effective way to affect change. This was certainly the case in South Africa during apartheid. When Western consumers decided to boycott South African products in the 1980s, apartheid soon collapsed. The movement had existed since the 1950s, but it was successful only when enough people took action. The idea of boycotting was not novel in the 1980s, it had been tried and tested by Mahatma Gandhi earlier in the century. To undermine cruel British colonialism, Gandhi started a boycott of British products, the most pronounced at the time was British clothing. Gandhi's movement reached hundreds of millions who began wearing Indian made cotton instead of the cheaper mass-produced British products. It helped rejuvenate the rural economy which had been intentionally shattered by British rule. If we follow these lessons, then individual changes are necessary to affect systemic change, although it must be stressed that individual changes will not be enough, government action must be forced. If we know what needs to be done though, why wait until the government enacts laws when we can start right away, as we put pressure on the levers of power.

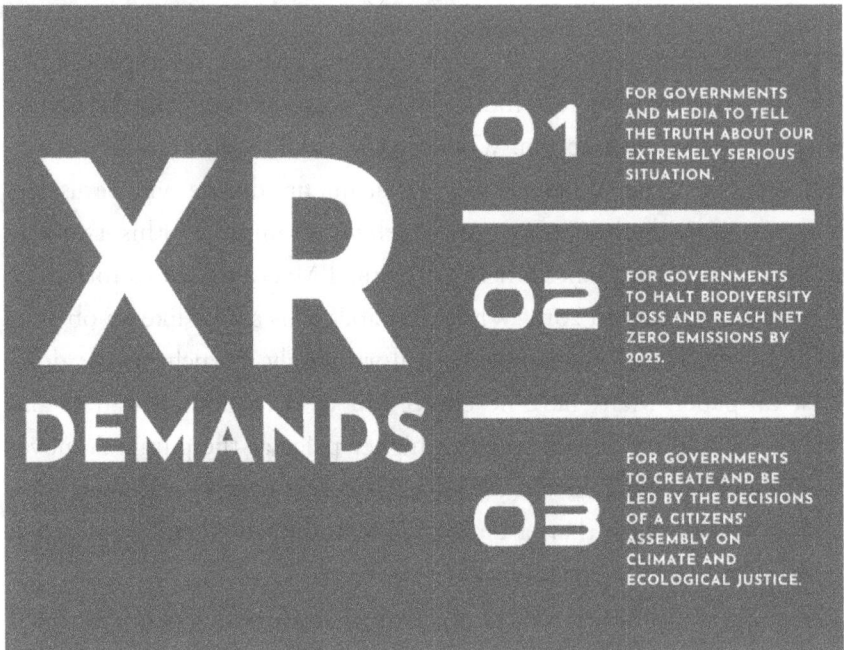

XR DEMANDS

01 FOR GOVERNMENTS AND MEDIA TO TELL THE TRUTH ABOUT OUR EXTREMELY SERIOUS SITUATION.

02 FOR GOVERNMENTS TO HALT BIODIVERSITY LOSS AND REACH NET ZERO EMISSIONS BY 2025.

03 FOR GOVERNMENTS TO CREATE AND BE LED BY THE DECISIONS OF A CITIZENS' ASSEMBLY ON CLIMATE AND ECOLOGICAL JUSTICE.

This is where community plays its part. The EU's slogan is "Strength in Unity", and this must be our slogan too. We need to join forces with environmental groups and work collaboratively. This has already begun and there are millions of people waiting for you in the streets. Fridays for Future is calling out the hypocritical politicians; while Extinction Rebellion (XR) has been taking a more direct route on streets all around the world. This peaceful movement is based around research into societal change over the past 100 years. It was found that non-violence was twice as likely to bring about change than violent revolution. Non-violent campaigns had a 53% success rate while violent conflict only worked 26% of the time.

More interestingly was that when 3.5% of the population got behind a cause; they never failed. This means that the U.K. needs just two million people to join XR and change will be very likely. In the U.S., the figure is 11.5 million, in Japan 4.2 million, Brazil 7.4 million and Egypt 3.5 million. As the situation worsens, XR numbers are guaranteed to keep growing and that critical mass of 3.5% will be reached sooner or later. What should we demand if we are successful? XR's three demands are listed above.

While the U.K. government declared a climate emergency in the aftermath of XR activities in central London, and even set up a Citizens' Assembly, it hasn't done anything to reduce emissions, other than its disastrous response to COVID-19. If enough of the population decide to play their part, then this might change quite abruptly. Some in the media appear to belatedly understand that the ends justify the means. In April 2022, Harry Cockburn wrote in the Independent that *"The fastest way to get Extinction Rebellion to stop is to listen to them,"* and urged readers to redirect their frustration from the protestors' blockades to the government whose plan to expand fossil fuel extraction is not in line with their own net-zero goals (808). Writing in iNews, James Dyke had this to say about XR:

> *"These protestors may not be the heroes we all want, but they are the heroes we all need. We all benefit from a safe and stable climate. Listen to their demands and you find them sensible and supported by the science (809)."*

Carmody Grey wrote in the Tablet that *"The media and political establishment which condemns them is living in a fantasy land"* and added *"we should be glad that at least some people in our society have the courage to risk their lives so that the rest of us will notice that we are about to walk off a cliff (810)."* Even U.N. Secretary-General António Guterres sided with protestors by stating *"Climate activists are sometimes depicted as dangerous radicals. But, the truly dangerous radicals are the countries that are increasing the production of fossil fuels (811)."*

While XR's 2025 target (set in 2019) would certainly be difficult to reach, what the mainstream media fail to mention is that there is nothing stopping us from reaching net zero emissions by 2030, other than a lack of political will. The technologies already exist (812). If it's possible to get to net zero by 2030, then why are we taking a massive risk and aiming for 2050? With every .1°C of temperature rise, people are going to die, and ecosystems will be lost. The ongoing pandemic has increased the need for a green new-deal to help get people back to work and we could use the opportunity to retrofit boilers, put solar panels on all rooves, and install wind turbines and tidal lagoons. How quickly governments act really does depend on how quickly we get to 3.5%. They won't act with the required urgency unless forced to. This is where collective action comes in and it begins with you, the individual.

"OUR PROBLEM IS CIVIL OBEDIENCE. OUR PROBLEM IS THE NUMBERS OF PEOPLE ALL OVER THE WORLD WHO HAVE OBEYED THE DICTATES OF THE LEADERS OF THEIR GOVERNMENT AND HAVE GONE TO WAR, AND MILLIONS HAVE BEEN KILLED BECAUSE OF THIS OBEDIENCE. AND OUR PROBLEM IS THAT SCENE IN ALL QUIET ON THE WESTERN FRONT WHERE THE SCHOOLBOYS MARCH OFF DUTIFULLY IN A LINE TO WAR. OUR PROBLEM IS THAT PEOPLE ARE OBEDIENT ALL OVER THE WORLD, IN THE FACE OF POVERTY AND STARVATION AND STUPIDITY, AND WAR AND CRUELTY. OUR PROBLEM IS THAT PEOPLE ARE OBEDIENT WHILE THE JAILS ARE FULL OF PETTY THIEVES, AND ALL THE WHILE THE GRAND THIEVES ARE RUNNING THE COUNTRY. THAT'S OUR PROBLEM."

HOWARD ZINN

Howard Zinn graffito on Valencia, the Mission
Jeremy Weate is licensed under CC BY 2.0.

Fund the Election Cycle

Donations (bribes) to political parties must end if we are to free ourselves from the disastrous corporate decision making that has gotten us into our predicament. A staggering $1.6 billion was spent on the 2020 U.S. Presidential election. This money was given to candidates from corporations who will want something in return. Now, corporations will argue that individuals can and do donate to political parties and candidates and therefore, corporations should be able to as well. Notwithstanding that a corporation is not actually a real-life thinking and breathing human being and therefore should have no such rights in a functioning society, let's humour the corporation and say okay, let's just ban outright any political bribes; whoever they are from. No more money may pass between business and politics or individuals and politics, or trade unions and politics. That way, corporations cannot complain they are being unfairly treated.

How would this work? Well, during each election cycle, any political party that garners a specific percentage of the vote will be allocated a specific budget and gain identical airtime in the media as the other parties who garnered the required share of the vote. It's extremely simple and something that we should all be pressing for as a matter of upmost immediacy.

In the U.S., it could be that any party that receives 500,000 votes shares in a budget provided by taxpayers. For those who argue taxpayers shouldn't pay, the obvious answer is they pay a whole lot more in the current system. Whether through open corruption, legal tax evasion or externalised costs pushed onto taxpayers by corporations doing all they can to minimize their own costs,;we all lose out substantially by allowing this free trade between the corporate and political class. We must be under no illusions; this season of open political bribes is fuelling inequality, climate catastrophe, ecosystem collapse, and ultimately, the breakdown of society.

Citizens United Money Globe
DonkeyHotey is licensed under CC BY 2.0.

Global Strike

While the younger generation, led by the inspirational Greta Thunberg, have been busy building a global movement, many of their parents have been busy positioning their heads in the sand. And while everything these young people are doing is amazing, we need to be honest and admit that they alone are not going to change government policy. It is extremely easy for our elected officials to applaud them and nod in agreement when they chastise the supposed adults in the room for their complete dereliction of duty and gross irresponsibility. They simply move on as if nothing has happened and the world spins round another seven times before the next strike.

What would be much harder for our elected officials to ignore would be the parents of these children joining them on the streets each and every Friday until governments start to treat our situation like the emergency it so very clearly is. These strikes could and should be for as long as is necessary for us to direct government policies in the right direction, so our children have a safe future: free from starvation and conflict.

Again, many will see the word 'strike' and feel uneasy. Again, also understandable as governments have portrayed strike action, and indeed unionisation in general as the enemy of the people. The reality is the exact reverse. Without strike action and the unionization of workers, our lives would be so much harder. If you enjoy only having to work an eight-hour work day, thank strike action. If you enjoy only having to work five days, instead of six, thank strike action. If you enjoy a safe workplace, thank strike action. The list goes on, and it is easy to see why they have been so effective at affecting change. Our current system is built around the old adage of 'time is money', and if people aren't sat at their desks or tools in hand, then someone somewhere is losing money. This is the weak underbelly of our current system, and it would be extremely simple to exploit.

An organized global strike for climate with tens of millions of adults and children joining hands around the planet to say 'enough is enough, no longer will we allow our futures to be destroyed for the profit of a few.' A mass coming together of parents with love in their hearts, and placards in their hands would put enormous pressure on governments to act. Business leaders would be in politicians' ears demanding that they get people back to work. The government would face two choices: armed brutality or acquiesce to the strikers' demands. If they choose the first option, the police or armed forces would be faced with becoming violent towards their neighbours, families and friends. If they chose this route, many more people would join the peaceful parents on the streets. It is more likely the second option would be chosen. Of course, the government would likely say one thing and do another. When this happens, further strikes would be called until policies are enacted and action taken.

If any parent is reading this, one thing you can take for granted is that our governments are not going to take the required action unless we get involved. We all teach our kids to be responsible for their actions. It is high time we started to lead by example. Strike action might be our last and best hope at affecting the changes our children so very desperately need.

Millionaire and Billionaire
Wealth Tax

O ur politicians, flush from millionaire/billionaire campaign "con-tributions," like to pretend that the cost of dealing with the climate and biodiversity crises is prohibitively expensive and pretend that we should wait until the cost of solutions come down before we act, but is this really true?

In their 2019 Emissions Gap Report, the U.N. states that transition-ing to a zero-carbon economy by 2050 will cost between 1.6-$3.8 trillion annually (813). While this is an enormous amount of money, to put it into context, $16.7 trillion was spent globally by May 2021 on COVID-19 measures (814). It is also worth noting that should global temperatures warm by 2°C, then the cost to the global economy is estimated at be-tween 10-$20 trillion a year, a fivefold increase (815). The cost of action will rise each year governments drag their feet. It will never be cheaper than $1.6-$3.8 trillion.

Now we have seen what it will cost, the big question should be where this funding comes from? One source of funding could come from mili-tary budgets which total approximately $2 trillion a year and contribute around 6% of global CO_2 emissions (816). As military budgets are sold to citizens as a defense strategy, their use to defend us from runaway climate chaos is clearly warranted. A further source of funding, and one that is equally justified, is a global wealth tax on millionaires and billionaires. Why is this justified? Well, research is clear that emissions are closely correlated with wealth, and as we live in a world where we are taught to be responsible for our actions, it seems fair that the more you have contributed to the problem, the more you should pay to solve it. Research by Oxfam and the Stockholm Environment Institute found that the richest 10% (over $38,000) were responsible for 49% of emissions, the richest 5% contributed 37% of emissions while the wealthiest 1% (over $109,000) created around 15%. The poorest 50% of humanity contributed almost nothing (817).

% Share of 2015 CO2 Emissions

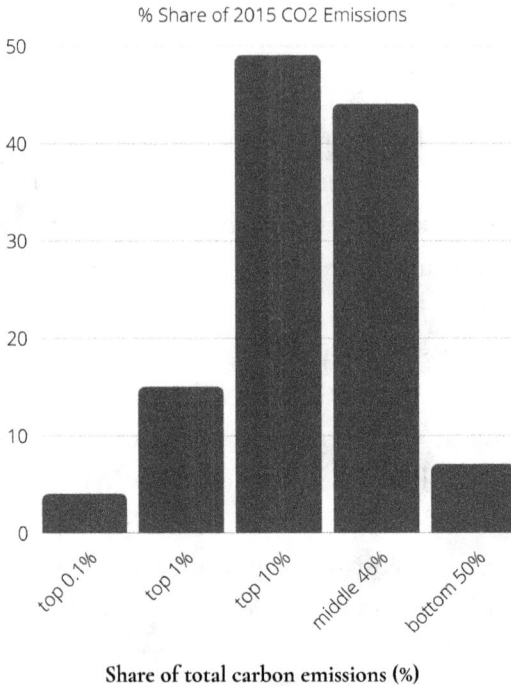

Share of total carbon emissions (%)
OXFAM International and Stockholm Environment Institute

This should be enough evidence to refute the often-cited words that "population is the problem." If the top half of earners lived like the bottom 50% then global emissions would be just 14% of their current total and could be balanced through natural carbon sequestration.

Now that we understand the cost of solving the crisis, and we know that the wealthier you are, the more you have contributed to the planetary crisis unfolding, could we now work out how much the wealthy need to pay in order to fund a transition to a zero-carbon economy? A 2022 report from OXFAM, Patriotic Millionaires, the Institute for Policy Studies, and Fight Inequality has done just that. They calculate that there are 3.6 million citizens with more than $5 million in the bank, 183,000 citizens with more $50 million and there are 2,660 billionaires. According to their findings, if we were to tax these global citizens at 2, 3 and 5% respectively we could raise $2.52 trillion. If we added slightly higher tax rates of 2, 5 and 10% then $3.62 trillion could be raised annually (818). A simple wealth tax on those who have contributed the most and can easily afford it would be enough to fund the U.N. estimated zero-carbon transition. It wouldn't need to saddle future generations with any debt. Should governments start to take our predicament seriously and decide that 2050 is far too late to keep warming within 1.5°C, then we could add the money from the global wealth tax to global military budgets and we would have $5.62 trillion to fund a rapid transition.

GLOBAL WEALTH TAX

	POPULATION	TOTAL WEALTH	2,3,& 5% TAX	2, 5, & 10% TAX
$5 MILLION +	3.6 million	$75.3 trillion	2% Wealth Tax	2% Wealth Tax
$50 MILLION +	183,000	$36.4 trillion	3% Wealth Tax	5% Wealth Tax
$1 BILLION +	2,660	$13.8 trillion	5% Wealth Tax	10% Wealth Tax
TOTAL	3,782,660	$125 trillion	$2.52 trillion	$3.62 trillion

Taxing Extreme Wealth An annual tax on the world's multi-millionaires and billionaires: What it would raise and what it could pay for
OXFAM

There is an extra step that once again won't cost the average citizen anything and is extremely befitting; the reallocation of fossil fuel subsidies worth $5 trillion a year (819). Should we use these subsidies to fuel the growth of renewable energy and plant-based protein alternatives, our climate and biodiversity war chest would total around $10.62 trillion, almost thrice the U.N. estimate for economic decarbonization.

The next time you hear someone say it is impossible to solve the climate and biodiversity crises, let them know that the only thing improbable is governments actually acting without massive grassroots pressure. If we can terraform a hostile red planet 330.47 million km away, we can certainly fund the transition to a zero-carbon economy on Earth.

Universal Basic Income (UBI)

There is no "free-will" without economic freedom. This means that if you are an economic slave then you are faced with a barrage of decisions to make and some have more or less adverse effects on your life, but you will be forced to make one, unless you are economically free, or secure. If you ask Jeffrey Preston Bezos to clean the toilets, he will probably tell you to Prime right off, but should he ask one of his staff, the toilet will be spick and span quite quickly. This is because while he never has to worry about his job, money or security, the rest of us constantly do. Therefore, we need to have a system that looks after all our needs. We need to have economic freedom as the basis. One program that aims to provide economic freedom is called Universal Basic Income (UBI). Everyone receives a living wage each month. If people wish to work on top of that, they can, but under this system no one is expected to follow orders that they see as unfit for purpose just because they are told to. People need to be in a position where they can object to actions they feel are immoral or unprincipled. People may argue that creating such a system goes against natural selection, but the benefits far outweigh the negatives. On the negative side, some will say that paying people not to work creates laziness, but recent studies on the principle of UBI in the U.S., Canada, Finland, Kenya, Spain, Germany, Holland, and Iran do not support this belief. In fact, many people used their income to start their own small businesses or to supplement part-time income. UBI also increased full-time employment and improved happiness levels while reducing anxiety (820). Moreover, people who don't have to worry about their security are able to take more time to follow politics and whether we like it or not; politics effects everything we do. We would also be in a better position to care about both our immediate surroundings and the environment at large. At present, with many people working two or three jobs, little time is available to care about politics or the environment and the corporations who fund our democratically elected, but commercially paid for politicians, benefit at all our expense. For employees witnessing sexual or power harassment in the workplace, they can report the person safe in the knowledge that their future is secure, with or without the job. Likewise, whistle-blowers will be encouraged to raise the alarm of any wrongdoing as their future will not be reliant on their position. Providing a living wage will also make it harder for nations to entice soldiers from poorer backgrounds, and this could potentially lead to less risk of war. Upton Sinclair described the situation crisply:

"It is difficult to get a man to understand something when his salary depends on his not understanding it."

Without a system that secures the economic freedom of all, change will be much harder, and the status quo will lead us ever closer to the sixth extinction.

Democratic Budgetary Allocation

We may live in democracies by name and are able to vote between the red side of a coin or the blue side, but when it comes to the way our taxes are used, both sides of the coin refuse to allow us any say on how our money is used. The only people on Earth who can choose how their money is used are the billionaires who legally cheat their way out of paying taxes and instead choose to donate their money to the charity of their choice. Of course, the "donation" will be tax deductible and they can then bask in the glory as the billionaire owned media applaud their philanthropic achievements. The rest of us have no such option. We pay our taxes as ordered and they are used either for our benefit, or against it. How many of us would really choose to spend money on making bombs and missiles instead of funding schools and hospitals? How many of us would choose to subsidize unhealthy foods over healthy foods? How many of us would choose to subsidize fossil fuels over renewable sources of energy? How many of us would choose to fund a surveillance system that monitors our daily activity? How many of us would choose to fund a security system that supports an elite class over the masses?

These were questions that the people of Porto Alegre, Brazil were asking back in 1989. After reforms in Brazil in 1960 led to economic growth, the wealthy areas had access to basic infrastructure, but the poorer areas didn't. They decided that change was necessary to help alleviate poverty and settled on a process called Participatory Budgeting

454 - SIMON WHALLEY

(PB). This is a truly democratic process where citizens decide how to spend part of a public budget. After its introduction, child mortality decreased by 20% and PB has now spread to over 7,000 cities around the world.

Typically, a committee of residents creates the rules and engagement plan before residents brainstorm ideas in meetings, both in person and online. Ideas for projects are discussed before they are developed into proposals. These proposals are then voted on by the community and the winning ideas are funded by the government.

It's a simple process that could be scaled up and down from the national budget to the smallest village. Wherever a budget exists, rather than leaving our elected officials to spend it as they see fit, the citizens would work together to highlight the needs of the citizenry and then decide together which projects will deliver the best outcome.

With the technology we have now, it would be possible for voting to take place safely using our cell phones as well as in person at town halls and post offices. It would empower people to care about their future and to care about politics. It would give people the long absent agency that we have been wishing for. It would bring us together in collaborative ways where ideas would be discussed and debated, and we would finally have a chance to put our own money into projects that we felt connected to.

It's about time that we all, like billionaires, had the opportunity to choose where our taxes are spent.

Degrowth

This idea will likely be contentious, but it is nevertheless one of the most effective solutions that would enable those in the developing world to enjoy a more comfortable lifestyle while those in the richer nations would be able to enjoy more free time and live less stressful lives. It would also reduce resource use.

The simple but unavoidable truth is that we cannot possibly have unlimited growth on a planet that has extremely rigid boundaries that we have either already passed, or are about to do so. We can also see that the growth in our current system is going to the top 1% while life for the rest of us gets more expensive by the day and our wages remain stagnant. This raises the obvious question, "Why do we need to grow?"

It is certainly clear that those of us who don't have health care, don't have fresh food to eat, or clean water to drink, and work in dangerous conditions need to experience a more comfortable existence. For those of us paddling in plastic, drenched in debt to finance Christmas, and glued attentively to multiple devices on our way to and from work, the need for growth is less apparent. What degrowth means in essence is using less resources. This fact should not prove contentious in any way. It should be clear to everyone that we need to use less energy, less resources, less land, less water, not the opposite. To emphasize the difference between what climate scientists say is necessary and what politicians are prepared to do, in the latest full unredacted IPCC 2022 report, the term degrowth can be found twenty-six times, yet in the redacted watered-down summary for policy makers, it has been removed completely due to political interference (821). Our current economic model is thousands of years old and built around the exploitation of people, the never-ending extraction of resources and military expansion. It is a flawed system that is leading us to extinction. The same IPCC report says in Chapter 4 (p.76), with high agreement:

"Climate change is the result of decades of unsustainable energy production, land-use, production and consumption patterns, as well as governance arrangements and political economic institutions that lock in resource-intensive development patterns (821)."

Where it becomes contentious for some is when degrowth in Gross Domestic Product (GDP) is mentioned. The truth is that if less resources are used and less goods produced then GDP will likely fall. When we discuss degrowth, however, no one is contemplating the average American switching their lifestyle for an Algerian. The idea is that in order to allow the sustainable growth of those impoverished, those of us who have too much can accept slightly less materialism in our lives. If we look at the average global GDP per capita, it currently stands at $17,600. This is equal to $48 per day and high income is classed as anything over $50, so even if we adopted the global average, we would all be classed as upper-middle income or high income (822). According to Jason Hickel from the London School of Economics, all key social indicators represented by the SDGs could be met with just $10,000 per capita, so $17,600 should pose no problem. Additionally, Hickel points out that what is good for GDP might not be good for the population. Using taxpayers' money to make bombs and missiles might improve the manufacturer's bottom line, but it won't improve the average citizens' life and may worsen it if revenge is taken for the use of said armaments. This is called the Lauderdale Paradox and includes privatized health-care. Surely, GDP may increase if health care is privatized but as any American will attest, having private healthcare is hardly appealing if you can't afford to pay for it.

We need to understand that what is good for GDP might not necessarily be good for us. Jason Hickel offers an example of the U.S. versus

Portugal. The U.S. has much worse social outcomes than Portugal, yet GDP is 65% higher in the U.S. He argues that this 65% increase in GDP is a waste of resources as it adds no significant improvement in human well-being (823). Surely, we can accept that somethings do not need to grow to ensure the well-being of humanity. Does the use of single use plastics need to grow? Do we need even more devices to avert our glaze from the surrounding reality? Do we need even more plastic trinkets to tie to our bags? Do we need ever bigger TVs, or faster cars?

The opposing argument, as always, will be that we need to grow the economy now in order that we can afford technological fixes in some magical Disneyfied time in the future when the technology will be cheap enough to scale them up. The truth is we have the answers to the crises now. We don't need to wait. Fortunately, by choosing to adopt more sustainable lives, we will be able to work less as we won't need to waste money on pointless crap that we will throw away after six months. This will allow us more time with friends and family, more time to spend on education, to understand politics, and to care about the world we live in. The following section will explain how degrowth might be achieved.

Global Carbon Allowance Trading System (G-CATS)

While UBI would be a great step forward, especially with 30% of jobs predicted to be lost to artificial intelligence (AI) by 2030, it won't, however, provide the far-reaching response we urgently need to deal with the multiple crisis we face (824). The powers that be have led us to believe that there is no alternative, but the simple reality is that there are a whole host of alternative systems, both political and economic, that would be a huge improvement on what we have now. The reason we aren't moving away from the destructive systems we currently have is that they benefit a small percentage of humans, and these people have corrupted our elected officials and paralysed our thinking through the billionaire owned media. As the British will soon find out, when they took back control from the E.U., all they were doing is placing the power back in the hands of a British elite who have no interest in improving the lives of the masses. Taking back control would mean placing the power in the hands of the people. This is where Citizen's Assemblies are vital. The direction we take should not be decided by one person, or even one hundred. It should be decided by all the people. We need to utilize the brain power of everyone and not just those born into privilege who, let's face it, aren't the brightest, and certainly not the most trustworthy or ethical among us. Every one of us needs to speak out, share our ideas and then decide democratically where we want to go.

To deal with rising CO_2 levels and declining biodiversity, it is abundantly clear that we need to introduce a price on carbon. Fossil fuel companies have pushed back hard against this idea and will continue to fight for the lowest possible price even if they are introduced. The 3.5% rule can help overcome this resistance, but problems will be created, as they were in France, if the poorest are impacted most. To deal with this problem, subsidies need to be transferred from environmentally damaging products to the less damaging alternatives. Free public transport could replace cars for most living in large cities, but many will still require private transport. Electric car subsidies need to be increased to make the transition more affordable. The nuts and bolts of all these problems can be tackled by Citizen's Assemblies but the larger problems still exist, which are rampant income inequality, both nationally and globally, rising greenhouse gasses and the loss of nature. There is a solution that could answer all these problems, but would, without doubt, be controversial. This is a Global Carbon Allowance Trading System (G-CATS).

G-CATS cuts right to the heart of the problem of resource use and income inequality. Under our present system, the more money you have, the more resources you can use. In theory, the richest eight people could use half the world's resources between them. There is nothing to prevent them doing so. It seems difficult to argue that this is either fair, or appealing. G-CATS would put a price on resource use and then resources could be traded with others. It allows for the accumulation of wealth, which many refuse to give up, but it prevents the destruction of ecosystems, as a budget is set at the outset and this budget gets smaller each year as we approach dangerous warming levels. To have a 67% chance of staying within 1.5°C of temperature rise, our budget was 230 billion tonnes (gtCO2) of CO_2 in 2020 and to have a 50% chance, we could safely burn 440 billion tonnes (825). We are currently emitting around 46gtCO2 annually, therefore, to have a 67% chance of staying within 1.5°C, at current emissions, our budget will likely expire in 2025 (826). Remember, this is just a 67% chance and each year, our chances drop. Then remember, that our leaders are aiming to reach zero emissions by 2050, twenty-three years too late.

While existing Emissions Trading Schemes target industry, G-CATS encourages citizens and industry to reduce their carbon footprints in tandem. A holistic view is taken that burning fossil fuels for energy is part of the problem but simply ending fossil fuel use will not solve the climate and biodiversity crises. Climate scientists are warning that we need rapid and far-reaching fundamental transformations. This will include changes to behaviour, technology, innovation, governance, and values (827) .G-CATS aims to profoundly change how we view growth, profit, and development by providing citizens with real time data that highlights the cost of products and services to the environment and society.

There are six phases to G-CATS. In Phase one, G20 nations are separated from non-G20 nations on the basis that G20 countries are responsible for over 80% of global emissions. An independent organization allocates each country within its bloc an equal share of CO_2 based on the bloc's previous year's emissions divided by the population of the bloc. Non G20 nations emitted 10,053,410,000 tonnes of CO2e in 2018 and have a combined population of 2,949,717,878. Therefore, each citizen within the block receives an annual budget of 3.4 tCO_2. Within the G20, the allowance is 7.5 tCO_2. It would be possible to flip this around and provide 7.5 tCO_2 to non-G20 citizens while G-20 citizens would receive 3.4 tCO_2. Again, citizens assemblies could decide this.

In phase two, a price of $160 tCO_2 is set and countries are able to buy and sell emissions. The $160 tCO_2 takes into account the natural and social cost of CO_2 in 2020 (828). This phase allows countries with smaller emissions to sell their share to higher emitting countries. The money received for CO_2 is intended to finance the development of a decarbonized economy in low-income nations. At 2018 emission rates, Afghanistan would be able to raise $5.4 billion by selling its surplus emissions. A high emitter like the United States with per capita 2018 emissions of 15.2 tCO_2 would need to purchase $568 billion of CO_2 through the G-CATS in order to continue with business as usual. Phase two encourages high emitting nations to reduce emissions and low emitters to transition to zero emissions technology. Phase three sees national governments distribute CO_2e allowances equally to every adult with a maximum of two children in each household being given 50% of an adult budget. This aims to encourage families to have fewer children which in turn reduces emissions. Adults are now able to trade CO_2e domestically and this is intended to reduce income inequality and encourage more frugal use of resources. As higher income earners create more emissions, if they wish to continue their high emission lifestyles, they will need to purchase carbon from lower income earners who create less emissions. If a country budget is being depleted, they can purchase CO_2e through G-CATS, but the annual cap is set tightly. The table below shows how citizens in different socio-economic groups could be impacted by a $160tCO_2$ carbon price in the U.K.

United Kingdom								
Percentile points for total income for tax year 2018 to 2019 Percentile point Total income after tax	Weekly Household-income-2021	% of population	Average Emissions 2015 (Oxfam)	Carbon Budget	Carbon Deficit	Carbon Tax	Change in Income after G-CATS
Top 1% (Over £116,000)	GBP 2,000 or more	4	58	7.5	-50.5	160	-8080
	GBP 1,800 or more, but less than GBP 2,000	1		7.5			
Top 10% (Over £46,500)	GBP 1,600 or more, but less than GBP 1,800	2		7.5			
	GBP 1,400 or more, but less than GBP 1,600	3		7.5			
	GBP 1,200 or more, but less than GBP 1,400	4	22	7.5	-14.5	160	-2320
	GBP 1,000 or more, but less than GBP 1,200	7		7.5			
	GBP 800 or more, but less than GBP 1,000	10		7.5			
Middle 40% (Over £22,600)	GBP 600 or more, but less than GBP 800	14	10	7.5	-2.5	160	-400
	GBP 400 or more, but less than GBP 600	19		7.5			
Bottom 50% (Less than £22,600)	GBP 200 or more, but less than GBP 400	26	5	7.5	2.5	160	400
	Less than GBP 200	9		7.5			

Introducing a Global Carbon Allowance Trading System (G-CATS) as an Ecological Alternative to Neoliberalism..
Simon Whalley. ScienceOpen Preprints. DOI: 10.14293/S2199-1006.1.SOR-.PPBMMHI.v1

Phase four involves the rollout of a novel labelling scheme in G20 countries and other high consumption nations that allows citizens to see the environmental and social cost of individual products/services based on the following metrics: emissions, social, biodiversity, longevity, and necessity. Each metric would be analysed by an independently funded organisation who would then give the product a score. The independent organisation would have to be given complete access to all supply chains, regardless of confidentiality. Failure to do so would result in no label being given. This would send a strong message to citizens who could then avoid the product.

Global - Carbon Allowance Trading System

G - C A T S

CO2 Allowances 2021-2030

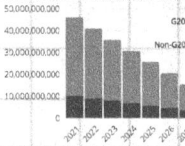

CO2 Deficits 2022

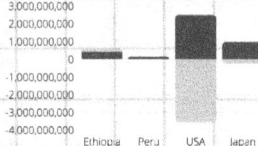

UK Per Capita Emissions by Socio-economic group

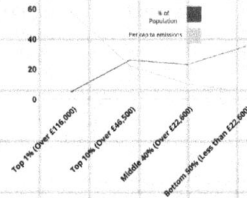

G-20 CO2 Emissions

1. Distribute allowance to each citizen

Non-G20 Nations 3.4tCO2
G20 Nations 7.5tCO2

2. Nations Trade CO2

Nations are able to trade CO2 on a global market for an initial price of $160. This raises money for low emission countries who can use excess funds to transition to zero carbon energy systems.

3. Allowance Distributed Equally

National Governments share CO2 allowance equally between all citizens.

6. CO2 Surplus Allowance Refunded

Citizens/Companies/ Countries with any surplus C02 allowance sell it back for $160 tCO2. Allowances cannot be carried over.

5. G-CATS Application

Citizens use a G-CATS application to balance their allowance. They can buy/sell CO2 if they are running low. They use the application to see the ecological/societal footprint of individual products/services.

4. G20 Labelling Scheme

Products and services are given a G-CATS score based on the metrics on the left. Products/services with low scores will be the most expensive and those with the highest scores will be the cheapest. Subsidies given to products with lowest emission and highest societal benefit. This will further drive emission cuts.

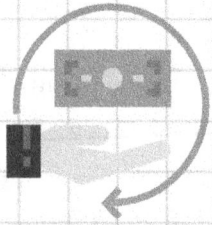

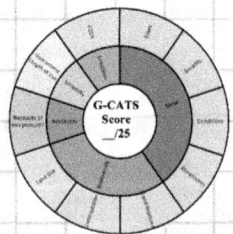

G-CATS Score __/25

www.g-cats.org

This labelling system would encourage manufacturers to improve production to take account of environmental damage and the contribution to society, and it could also be used by governments when deciding who to give subsidies to. Companies that score at least four points in each category could be given a subsidy, and any companies reaching full marks would receive more.

Additionally, subsidies could be based on the previous year's taxes as well as the number of people employed and their conditions. Companies that neglect their tax responsibilities or employees would be impacted by receiving a smaller subsidy.

They would then have to purchase carbon from the market and in doing so, be gently nudged to pay their fair share of taxes. G-CATS would also discourage companies from replacing people with AI to cut costs as they would then have to purchase more carbon. As producing AI is resource and energy intensive, this would encourage less emissions.

During phase five, citizens in G20, and other high emitting countries use a mobile application to find accurate information about products and services they are considering purchasing. The application displays their weekly carbon allowance, and they shop accordingly. This will encourage citizens to purchase the least socially and environmentally damaging products and services while simultaneously encouraging producers and service providers to reduce the damage their products or services cause in order to achieve a high score which will attract consumers.

The final phase would see any citizens selling their surplus CO_2 back to the government for $160. Nations would then be reimbursed for any surplus and emissions calculations would be completed (829).

The COVID-19 pandemic has also taught us a very valuable lesson about what is possible when governments are motivated to act. After ten years of austerity, brought on by neoliberal policies, money was gushing from central coffers like no time before. After a decade of politicians saying that money didn't grow on trees, suddenly, they had found the secret money tree. A lot of it was syphoned off by government cronies,

464 ~ SIMON WHALLEY

but it was also used to keep the majority of heads above water as flights were cancelled, bars were closed, shops shuttered, and where possible, work was done remotely. This all happened overnight, globally. The differences between the pandemic and the climate and biodiversity crises are gargantuan in scale of severity. While COVID-19 is only attacking one species and will be solved quickly by mass vaccinations, the more serious crises will impact every species on Earth and if we fail to act immediately, they will likely cause billions to die, not millions. While some politicians acted more swiftly and efficiently than others, they all, with the possible exception of Brazil's Bolsanaro, acted on scientific data and advice. When it comes to the much more serious threats that face our species, very few, if any, are doing anything other than make insincere speeches about 2050. The difference here is that the response to COVID-19 will likely determine who wins the next election while the lack of response to the climate and biodiversity crisis won't be judged for another decade or more. By that time, this current crop of leaders will be either retired or dead. Anytime we are told that it isn't possible, just remember our response to this pandemic, then consider our response to WWII when industries were turned on their heads, or our response to the war on drugs, or the war on terror. When politicians, and the hands that feed them, see the benefit for themselves, they act, so it is up to us to make them see the benefit of action.

Without doubt, there are many other solutions to the crisis we face. Jacques Fresco's Venus Project inspired the resource sharing in G-CATS and is worth looking at. His system utilizes AI to complete an inventory of the Earth's resources and they are then shared equally among the population. All the jobs are carried out by AI and humanity is freed to follow their own interests. It's a beautiful idea and should be the pinnacle of human accomplishment, however, the short time frame we have raises a doubt as to whether it is a feasible solution now. It will also be derided by those in power as the AI takes over the responsibility of decision making to avert the problems of corruption that have beleaguered human hierarchies throughout our history. The use of AI would also be of concern to a general population raised on movies such as Terminator. George Monbiot's Manifesto for a New World Order suggests taking power down to the smallest possible level with neighbourhoods voting on policy and then a selected member from the neighbourhood representing the neighbourhood at the larger level and reporting back before the group voting again at the larger group. Then a member is selected from that larger group of neighbourhoods, and they discuss at the town or city level and the same is done again until we reach the national level, when once again, a member is selected by the people to represent the nation at the global level. Voters use their mobile phones to vote on the issues and to select representatives. These representatives can be deselected if they fail to accurately represent the people.

G-CATS is promoted here because we live in hyper-polarized times and only have a small window in which to affect massive changes. Those on the political left will argue it isn't enough, and those on the political right will say it is too radical. It is a compromise which should suit those who believe the motivation of capital accumulation is essential, while it also reduces income inequality, encourages employment, the payment of taxes and all the while encouraging the thrifty use of the Earth's resources. Those in power will resist any meaningful change and label any threat to their power as absurd impossibilities, but the truth is the current system they insist has no alternative is built on impossibilities. It is impossible to grow an economy forever. It is impossible to live without nature. It is impossible for wealth alone to make you happy. Everything we do moving forward has to be done together. We need to form strong social bonds so we can learn from each other, rely on each other, and work together for common goals. The result of our four-decade insistence on individuality is that we have become atomised, and alienated from each other. This has enabled corporations and governments to divide us in a culture war while the rift between rich and poor has grown unacceptably large. If we continue on our current race to the bottom, we will create a mass of have nots and a tiny number that enjoy such extreme privilege we can barely imagine.

Conclusion

If you have read this far, you will have seen how our species is causing our planet to become uninhabitable, you will have seen how our behaviour is causing the sixth extinction and you will have seen that as we are fully aware of the situation, calling it an extinction event is slightly disingenuous. We must be honest with ourselves and call it what it is, an extermination of the natural world.

We have seen the damage being done to the planet, we have seen how this is impacting our future, we have seen how our governments have secretly built a surveillance state around us, and we have seen that increased wealth equals increased responsibility for the destruction of the planet. So, why on Earth are we still insisting that we must have more, more, more?

This is not to say that people in low-income countries should be happy with their lot, far from it. As we saw, the wealth of the rich northern nations is in large part stolen from those in the poorer south. While wealth in poorer countries certainly needs to rise to ensure everyone has enough to eat and drink and affordable quality health care, it is in the rich north where we should be questioning the need for infinite growth. If we have enough that we can spend our earnings on crap that we don't need and, in the process, trash the planet, do we really need to be working our collective asses off in order to have more spending power?

We all know that we live on a planet with limited resources. Why do we imagine it is sustainable to continue growing the economy at the expense of the natural world? At some point, a collapse of the ecosystems that sustain life is inevitable. By continuing with this experiment, we would be placing the biggest bet ever taken, and not at the MGM Grand, where the price of losing is walking out the door broke. No, we would be throwing down at the Casino of Life or Death, and the cost of losing would be the loss of all civilizations on the planet, and possibly the sixth extinction which could also include us. Why should we take this enormous risk, when the spoils of winning would only go to the richest in society and the cost of losing would be borne out by the most vulnerable first, and eventually by everyone?

The more we consume, the more we are contributing even further to the extreme wealth inequality that we see in society. Plutocrats aren't using their hordes of cash to help society. They aren't interested in democracy. They aren't interested in ending the climate crisis or reversing the sixth extinction, unless it makes a profit. How do we know this? Just look at what they do with their money. The U.N.'s World Food Program (WFP) estimates they will need $21.5 billion to feed 147 million people in 2022. Everybody's hero: Elon Musk, feels it's more important to spend $45 billion on owning a social media giant. Other billionaires are lobbying governments to reduce their taxes even further so they can throw a tiny amount to charity and use their own media to shower them with praise for doing so. They talk a great game about saving the world but then they encourage us to buy their products which are leading to planetary collapse. If a political party attempts to get these plutocrats to pay their fair share to the people who made them rich in the first place, they are threatened. Here is an exchange between a private equity manager who votes Democrat and a Democratic leader in Congress:

"Screw you! Even if you change the legislation the government won't get a single penny more from me in taxes. I'll put my money into my foundation and spend it on good causes. My money isn't going to be wasted in your deficit sinkhole (830)."

Now, it has to be said that many governments do indeed waste a lot of our money on unnecessary wars, subsidising fossil fuel interests and animal agriculture to name just a few sinkholes. But the billionaire class are the people who are benefitting from this extraction of wealth from the tax paying base to the tax evading captains of industry meaning this private equity manager's anger is possibly misleading. Perhaps, what he is most angry about is money being spent on the poorest in society who he believes are poor because they are stupid and lazy while people like him deserve all their riches as they are clever and hard working. As the salaries of the 1% explode into the stratosphere and everyone else is left desperately trying to hang on to jobs with stagnant salaries, the belief of the 1% is that the middle class need to accept reduced salaries so they can compete with workers in developing nations. This is from the mouth of a Taiwanese born CFO of a U.S. tech company:

"So, if you're going to demand ten times the paycheck, you need to deliver ten times the value. It sounds harsh, but maybe people in the middle class need to decide to take a pay cut (830)."

In her book Plutocrats, Chrystia Freeland shares the true feelings of many more billionaires towards the rest of us. In what is essentially an ode to the richest, she nevertheless highlights their disdain for the people who made them rich. These people may as well live on another planet because they have no idea what life is like for normal people. Fortunately, they seem to want to live on another planet themselves. The President and Chief Investment Strategist of Yardeni Research, Ed Yardeni, stated as much when he said he wished to start a new planet for people like him who dislike government and favour low taxes and fiscal discipline. Of course, he isn't alone. Elon Musk is making plans to colonise Mars and Preston Bezos plans on manufacturing a floating space colony. Both these men benefit from massive taxpayer hand-outs that help to build their infrastructure and train their workers. As the planet becomes uninhabitable, the richest in society who have bene-fited the most from its destruction will be the only ones who can afford the $500,000 tickets. For those who can't afford a ticket, don't worry, you can become a slave on Mars to pay off the loan Musk will generously offer you. The Pay Pal billionaire, Peter Thiel, is content with living in Earth's atmosphere but not on terra firma. Instead, he has joined with other billionaires to propose a sea steading community of wealthy individuals who will be able to live on floating cities where they will pay no taxes and be safe from the scourge of the rest of us who made them rich (831). For those who prefer to remain on Earth, some are already planning their escape by buying apartments in former missile silos in the desert with their own hydroponic agriculture, water production capabilities, indoor shooting range, swimming pool, library, dog park, rock climbing wall, cinema, and most importantly, military grade security and 2.5-meter-thick walls (832). The underground bunkers are equipped with enough resources to last five years after which time the inhabitants expect everyone else to be dead so they can emerge from their bunkers to start over again. Prices start from $1.5 million so it's fair to say, the people who have contributed the least to planetary destruction will not be among the inhabitants.

In essence, what we are doing, by continuing down this path is funding the escape projects of the very same billionaires who are driving the planetary destruction through their outsized egos and greed. With the current trend towards privatization, life on Earth for the rest of us could be one giant Hunger Games. There may be no society left, just individuals fighting for meagre supplies of food and water that are supplied for profit. Readers may feel that this is harsh to the richest in society, who let's face it, have won at the current game, and that is not the real intent. Some of us might act in similar ways should we find ourselves in their position. Our current system is toxic to all those involved. We need to have open and honest discussions about what kind of world we want to live in, and unfortunately, at present, our future is being decided for us by those who are benefitting most from our global crises, and they are following a predictable model that began thousands of years ago along the seas of the Mediterranean.

The COVID-19 pandemic has taught us another important lesson with pharmaceutical companies developing vaccines at breakneck speed in what seems a valiant attempt to save lives. Unfortunately, they only seem interested in saving the lives of those who can afford the vaccines. Clearly, if we wish to end the pandemic and stop new variants emerging, we need to vaccinate every human being, but this is not being done. By June 2022, after eleven billion doses had been administered around the world, only 17.8% of people in low-income nations had received their first dose. In Nigeria, Tanzania, and Sudan, roughly 5% of the population had received a single dose (833). With Pfizer reporting revenue of $81.3 billion in 2021, Moderna $18.5 billion, and AstraZeneca $37.42 billion, couldn't they agree to waive patents for poorer nations to save lives? Of course, this would cut into shareholder dividends, and this goes against hundreds of years of business acumen. It can also be argued that it is in the corporations' interests for the pandemic to persist as long as possible so they can maximise share-holder returns. In our current economic system, what's good for the corporation isn't necessarily good for humanity. This is why, presently, action to reverse climate change and the sixth extinction will only be taken if it is profitable. Planting trees makes no profit for shareholders, so instead money is poured into direct air capture and solar radiation management, which may or may not be scaled up in time, and only deal with one of the two problems. Clearly, systemic change is needed, and our species needs to fundamentally alter the way we behave if we are to continue our evolution.

The solutions provided in this book act simply as evidence to refute the market knows all mantra of "There is no alternative". There are many more just and sustainable alternatives, and it is up to us, both individually and collectively, to make sure that we don't become the generation that is blamed for an extermination event unparalleled in all known history. It is easy to despair and question how change may come, but that question has already been answered many times in our history. Extinction Rebellion has set out a path to change and the most important part of the strategy is citizens' assemblies. This form of direct democracy has been denied us throughout history, whether through communism, socialism, or capitalism. There is a reason for that. Those with power fear the direct will of the people. It is time to utilize the minds of the many, for the many, rather than the few for the few. Our ancestors have consistently risen to the challenge in times of desperation. Now is our time to rise. It is our time to make the world a better place. It is our time to be heroes. Let's get to it!

"NEVER DOUBT THAT A SMALL GROUP OF THOUGHTFUL, COMMITTED CITIZENS CAN CHANGE THE WORLD; INDEED, IT IS THE ONLY THING THAT EVER HAS."

Margaret Mead

"260.365 : mountain mist"
by mac.rj is licensed under CC BY 2.0.

Dear Indy,

I hope the time and effort spent researching and writing this book will contribute something to our fight for survival and look forward to discussing your thoughts and ideas as you grow, and we take on the biggest challenge we have ever faced, together.

It is more than three years since I began writing and many more years of research. So much has changed in this short time, so much so, that I have had to go back and rewrite the first section to include all the new record heat waves, fires and storms that have since occurred. We have also moved prefectures, changed jobs, changed schools, both lost and made friends and taken up new gardening roles. The world has been turned upside down due to COVID-19.

I hope when you are old enough to read and understand these pages, that you will understand what I was doing all this time. I hope you can now see how interconnected we all are, and how we can only truly be safe and happy, if we ensure that all of us are safe and happy. And this doesn't just mean humans. All life on this ball of wonder floating through space are part of the ecosystem, we call home: Earth.

Unfortunately, by the time you are able to read this book, the die may already have been cast and the seeds already sown. When I began writing, I was motivated by a concern for your future, but the more I have learned, the more I am concerned, not just for your future, but my own. The time frame for the looming catastrophes is much shorter than I originally envisaged, and we must work much harder to escape the coming collapse.

We have been given the opportunity to reform our way of living, allowing us more freedom to spend time with family and friends. Nature has set out our challenge and the time has come for us to rise to it. With an abundance of love and rage, I will allow the masterful Dylan Thomas to take us out:

Do not go gentle into that good night,
Old age should burn and rave at close of day;
Rage, rage against the dying of the light.

Acknowledgements

First and foremost, I would like to thank every single animal that I have ever met. Whoever you are, if you've crossed my path, I learned something from you. I would lie to thank Al Gore for first making me aware of part of the problem, and I would like to thank the producers of Cowspiracy for bringing home the reality of animal agriculture, something completely ignored by Mr. Gore.

To Sean Richards and Osama Kassis, thank you for your help with Excel. Much appreciated. A big shout out to Jack Alexander for his generous proofreading advice, even if this project was too big. To Mr. Patterson, thanks for the vignette idea. I have to thank Charles Cabell and Vincent Valdmann for their constant encouragement, and Vincent especially, for his amazing generosity. To the founders of Extinction Rebellion and Greta Thunberg for showing me that I was not alone, although I still feel stuck in the Twilight Zone most days. My appreciation to Sailesh Rao, for his advice, and especially for introducing me to Arizona State University undergraduate Erin Epel who went above and beyond with her fact checking. I cannot thank her enough. To Khin Moet Moet, thank you also for fact checking. You did a great job and you are much appreciated. My appreciation also goes to Luke Sandalls for his cover design in difficult circumstances.

Finally, I need to thank my wife Kaori, and son Indy. At last, it is done. I can't even remember when I wrote the first word, but it has been a long, arduous, and often frustrating journey. You've supported me all the way, and there has been so much missed time that we can never get back. I wish there had been no need to write the book; sadly there was.

From now on, you are getting my full attention. Thank you. I love you both deeply.

References

1. Arsenault, Chris. Only 60 Years Of Farming Left If Soil Degradation Continues. [Online] Reuters, 12 06, 2014. [Cited: 01 10, 2020.] Https://Www.Reuters.Com/Article/Us-Food-Soil-Farming/Only-60-Years-Of-Farming-Left-If-Soil-Degradation-Continues-Iduskcn0jj1r920141205.
2. Factsnet. Homelessness Facts. [Online] Facts.Net, 02 21, 2015. [Cited: 15 01, 2020.] Https://Facts.Net/Issues/Homelessness-Facts.
3. Bronson, Richard 'Skip'. Homeless And Empty Homes — An American Travesty. [Online] Huff Post, 08 24, 2010. [Cited: 01 15, 2020.] Https://Www.Huffpost.Com/Entry/Post_733_B_692546?Guccounter=1.
4. Kentish, Benjamin. 300,000 People In Britain Are Now Homeless, Study Reveals. [Online] Independent, 11 08, 2017. [Cited: 01 15, 2020.] Https://Www.Independent.Co.Uk/News/Uk/Home-News/Homeless-Stats-Uk-Latest-Figures-New-Study-Shelter-Housing-A8042841.Html.
5. The Guardian. Empty Homes In England Up By 11,000 Last Year, Study Shows. [Online] The Guardian, 09 23, 2019. [Cited: 01 15, 2020.] Https://Www.Theguardian.Com/Society/2019/Sep/23/Empty-Homes-In-England-Up-By-11000-Last-Year-Study-Shows.
6. Inequality.Org. Income Inequality In The United States. [Online] Inequality.Org, 20. [Cited: 01 15, 2020.] Https://Inequality.Org/Facts/Income-Inequality/.
7. Oxfam. Just 8 Men Own Same Wealth As Half The World. [Online] Oxfam, 01 16, 2017. [Cited: 01 15, 2020.] Https://Www.Oxfam.Org/En/Press-Releases/Just-8-Men-Own-Same-Wealth-Half-World.
8. Fay, Bill. Demographics Of Debt. Deb.Org. [Online] Deb.Org, 05 06, 2021. [Cited: 07 07, 2021.] Https://Www.Debt.Org/Faqs/Americans-In-Debt/Demographics/.
9. Story Of Stuff Project. Story Of Stuff Project. [Online] Story Of Stuff Project. [Cited: 01 15, 2020.] Https://Www.Storyofstuff.Org/Wp-Content/Uploads/2020/01/Storyofstuff_Annotatedscript.Pdf.
10. Buist, Jacki. How Business Can Help In The Fight Against Plastic Waste. Supply Management. [Online] Supply Management, 2 23, 2018. [Cited: 04 01, 2022.] Https://Www.Cips.Org/Supply-Management/Analysis/2018/March/Business-And-Plastic-Waste/.
11. Conger, Michal. Fact Check: Bloomberg Claims More People Die Of Obesity Than Hunger. [Online] Washington Examiner, 03 12, 2013. [Cited: 01 15, 2020.] Https://Www.Washingtonexaminer.Com/Fact-Check-Bloomberg-Claims-More-People-Die-Of-Obesity-Than-Hunger.
12. Plumer, Brad. How Much Of The World's Cropland Is Actually Used To Grow Food? [Online] Vox, 12 16, 2014. [Cited: 01 15, 2020.] Https://Www.Vox.Com/2014/8/21/6053187/Cropland-Map-Food-Fuel-Animal-Feed.
13. James Dyke, Robert Watson And Wolfgang Knorr. Climate Scientists: Concept Of Net Zero Is A Dangerous Trap. The Conversation. [Online] The Conversation, 04 22, 2021. [Cited: 05 19, 2021.] Https://Theconversation.Com/Climate-Scientists-Concept-Of-Net-Zero-Is-A-Dangerous-Trap-157368.
14. Matthews, Kasia Tokarska And Damon. Guest Post: Refining The Remaining 1.5c 'Carbon Budget'. Carbonbrief. [Online] Carbonbrief, 01 19, 2021. [Cited: 05 19, 2021.] Https://Www.Carbonbrief.Org/Guest-Post-Refining-The-Remaining-1-5c-Carbon-Budget.
15. Iea. Global Energy Review: Co2 Emissions In 2021. Paris : Iea, 2022. Https://Www.Iea.Org/Reports/Global-Energy-Review-Co2-Emissions-In-2021-2.
16. Ipcc. Summary For Policymakers. In: Global Warming Of 1.5°C. An Ipcc Special Report On The Impacts Of Global Warming Of 1.5°C Above Pre-Industrial Levels And Related Global Greenhouse Gas Emission Pathways, In The Context Of Strengthening The Global Response To. Geneva : World Meteorological Organization,, 2018.
17. Hausfather, Zeke. Analysis: When Might The World Exceed 1.5c And 2c Of Global Warming? Carbonbrief. [Online] Carbonbrief, 12 04, 2020. [Cited: 05 12, 2021.] Https://Www.Carbonbrief.Org/Analysis-When-Might-The-World-Exceed-1-5c-And-2c-Of-Global-Warming?Utm_Campaign=Carbon%20brief%20daily%20briefing&Utm_Content=20201207&Utm_Medium=Email&Utm_Source=Revue%20daily.
18. Carrington, Damian. Climate Limit Of 1.5c Close To Being Broken, Scientists Warn . The Guardian. [Online] The Guardian, 05 10, 2022. [Cited: 05 10, 2022.] Https://Www.Theguardian.Com/Environment/2022/May/09/Climate-Limit-Of-1-5-C-Close-To-Being-Broken-Scientists-Warn.
19. Queally, Jon. Common Dreams. Common Dreams. [Online] 09 October 2018. Https://Www.Commondreams.Org/News/2018/10/09/Whats-Not-Latest-Terrifying-Ipcc-Report-Much-Much-Much-More-Terrifying-New-Research.

20. Roser, Hannah Ritchie And Max. Co2 Emissions. *Our World In Data.* [Online] Our World In Data, 2017. [Cited: 04 05, 2022.] Https://Ourworldindata.Org/Co2-Emissions#:~:Text=Since%201751%20the%20world%20has,Needs%20to%20urgently%20reduce%20emissions..

21. University Of Copenhagen. Arctic Permafrost Releases More Co2 Than Once Believed. *University Of Copenhagen.* [Online] University Of Copenhagen, 02 09, 2021. [Cited: 06 15, 2022.] Https://Science.Ku.Dk/English/Press/News/2021/Arctic-Permafrost-Releases-More-Co2-Than-Once-Believed/.

22. Reuters. Scientists Shocked By Arctic Permafrost Thawing 70 Years Sooner Than Predicted. [Online] The Guardian, 06 18, 2019. [Cited: 01 15, 2020.] Https://Www.Theguardian.Com/Environment/2019/Jun/18/Arctic-Permafrost-Canada-Science-Climate-Crisis?Cmp=Share_Androidapp_Facebook&Fbclid=Iwar205wdi320gkioeebtpb_c1bj4bajbb6nmloqmzbivuxwthnujq_Vble7a.

23. Corbett, Jessica. 'This Is Truly Terrifying': Scientists Studying Underwater Permafrost Thaw Find Area Of The Arctic Ocean 'Boiling With Methane Bubbles'. *Commondreams.* [Online] Commondreams, 10 09, 2019. [Cited: 05 12, 2021.] Https://Www.Commondreams.Org/News/2019/10/09/Truly-Terrifying-Scientists-Studying-Underwater-Permafrost-Thaw-Find-Area-Arctic?Fbclid=Iwar3byv75l551x_Okxdtvai9690pigtn8bfgznenszgys_Uwzkheoyjfhng8.

24. Watts, Jonathan. Arctic Methane Deposits 'Starting To Release', Scientists Say. *The Guardian.* [Online] The Guardian, 10 27, 2020. [Cited: 05 12, 2021.] Https://Www.Theguardian.Com/Science/2020/Oct/27/Sleeping-Giant-Arctic-Methane-Deposits-Starting-To-Release-Scientists-Find.

25. —. Melting Permafrost In Arctic Will Have $70tn Climate Impact – Study. [Online] The Guardian, 04 23, 2019. [Cited: 01 15, 2020.] Https://Www.Theguardian.Com/Environment/2019/Apr/23/Melting-Permafrost-In-Arctic-Will-Have-70tn-Climate-Impact-Study.

26. Welch, Craig. The Arctic's Thawing Ground Is Releasing A Shocking Amount Of Dangerous Gases. *National Geographic.* [Online] National Geographic, 02 06, 2020. [Cited: 05 12, 2021.] Https://Www.Nationalgeographic.Com/Science/Article/Arctic-Thawing-Ground-Releasing-Shocking-Amount-Dangerous-Gases#:~:Text=Scientists%20have%20known%20for%20decades,Speed%20up%20global%20climate%20change.&Text=This%20proc ess%2c%20called%20%E2%80%9cabrupt%.

27. Hall, Shannon. 'Freak Weather Event' Sets Antarctic Heat Records. *Live Science.* [Online] 7 04 2015. Https://Www.Livescience.Com/50405-Antarctica-Heat-Records.Html.

28. Bbc News. Antarctic Island Hits Record Temperature Of 20.75c. *Bbc News.* [Online] Bbc News, 02 14, 2020. [Cited: 05 12, 2020.] Https://Www.Bbc.Com/News/World-51500692.

29. Freedman, Andrew. 2018's Global Heat Wave Is So Pervasive It's Surprising Scientists. *Axios.* [Online] 28 07 2018. Https://Www.Axios.Com/Global-Heat-Wave-Stuns-Scientists-As-Records-Fall-4cad71d2-8567-411e-A3f6-0febaa19a847.Html.

30. Samenow, Jason. Another Extreme Heat Wave Strikes The North Pole. *The Washington Post.* [Online] 7 05 2018. Https://Www.Washingtonpost.Com/News/Capital-Weather-Gang/Wp/2018/05/07/Another-Extreme-Heat-Wave-Strikes-The-North-Pole/?Utm_Term=.Eecc51296d2d.

31. Berardelli, Jeff. 100.4 Degree Arctic Temperature Record Confirmed As Study Suggests Earth Is Warmest In At Least 12,000 Years. *Cbs News.* [Online] Cbs News, 07 01, 2020. [Cited: 05 12, 2021.] Https://Www.Cbsnews.Com/News/Arctic-Temperature-Record-100-4-Degrees-Earth-Warmest-12000-Years/#:~:Text=Coronavirus%20pandemic-,100.4%20degree%20arctic%20temperature%20record%20confirmed%20as%20study%20 suggests%20earth,In%20at%20least%2012%2c000%20years&T.

32. Quinn, Eilís. The Barents Observer. *Record Breaking Temperatures Recorded In Arctic Russia.* [Online] The Barents Observer, 06 22, 2021. [Cited: 09 08, 2021.] Https://Thebarentsobserver.Com/En/Climate-Crisis/2021/06/Record-Breaking-Temperatures-Recorded-Arctic-Russia.

33. Harvey, Fiona. Heatwaves At Both Of Earth's Poles Alarm Climate Scientists . *The Guardian.* [Online] The Guardian, 03 20, 2022. [Cited: 03 21, 2022.] Https://Www.Theguardian.Com/Environment/2022/Mar/20/Heatwaves-At-Both-Of-Earth-Poles-Alarm-Climate-Scientists.

34. Woodward, Aylin. Greenland Is Approaching The Threshold Of An Irreversible Melt, And The Consequences For Coastal Cities Could Be Dire. [Online] Business Insider, 04 24, 2019. [Cited: 01 15, 2020.] Https://Www.Businessinsider.Com/Greenland-Approaching-Threshold-Of-Irreversible-Melting-2019-1.

35. Milman, Oliver. Greenland: Enough Ice Melted On Single Day To Cover Florida In Two Inches Of Water. *The Guardian.* [Online] The Guardian, 07 30, 2021. [Cited: 09 08, 2021.] Https://Www.Theguardian.Com/Environment/2021/Jul/30/Greenland-Ice-Sheet-Florida-Water-Climate-Crisis.

36. Harvey, Alison Rourke And Fiona. Photograph Lays Bare Reality Of Melting Greenland Sea Ice. [Online] The Guardian, 06 18, 2019. [Cited: 01 15, 2020.] Https://Www.Theguardian.Com/World/2019/Jun/18/Photograph-Melting-Greenland-Sea-Ice-Fjord-Dogs-Water.

37. Griffiths, James. Photo Of Sled Dogs Walking Through Water Shows Reality Of Greenland's Melting Ice Sheet. [Online] Cnn, 06 19, 2019. [Cited: 01 15, 2020.] Https://Edition.Cnn.Com/2019/06/17/Health/Greenland-Ice-Sheet-Intl-Hnk/Index.Html.

38. Pfeifer, Hazel. A 'Frozen Rainforest' Of Microscopic Life Is Melting Greenland's Ice Sheet. *Cnn World.* [Online] Cnn, 01 21, 2021. [Cited: 05 13, 2021.] Https://Edition.Cnn.Com/2020/12/14/World/Microscopic-Life-Melting-Greenland-Ice-Sheet-c2e-Spc-Intl/Index.Html#:~:Text=A%20'frozen%20rainforest'%20of%20microscopic%20life%20is%20melting%20greenland's%20ice%20sheet &Text=Greenland's%20ice%20sheet%20is%20the,.

39. National Snow And Ice Data Center. Greenland's 2020 Summer Melting: A New Normal? *National Snow And Ice Data Center.* [Online] Nasa, 08 12, 2020. [Cited: 05 13, 2021.] Http://Nsidc.Org/Greenland-Today/2020/08/Greenlands-2020-Summer-Melting-A-New-Normal/.

40. *Dynamic Ice Loss From The Greenland Ice Sheet Driven By Sustained Glacier Retreat.* Michalea D. King, Ian M. Howat, Salvatore G. Candela Et Al. 1, S.L : Communications Earth & Environment Volume, 2020, Vol. 1. Https://Doi.Org/10.1038/S43247-020-0001-2.

41. Lydon, Tim. Record-Breaking Heat In Alaska Wreaks Havoc On Communities And Ecosystems. *Smithsonian.Com.* [Online] 05 30, 2019. Https://Www.Smithsonianmag.Com/Science-Nature/Record-Breaking-Heat-Alaska-Wreaks-Havoc-Communities-And-Ecosystems-180972317/?Utm_Source=Facebook.Com&Utm_Medium=Socialmedia&Fbclid=Iwar3fpk7pxbj-9_Otrx6y7hszcvlxv6iyt9ap_Mwqssswvo6qbvb4yfoa1zc.

42. Milman, Oliver. Climate Crisis: Alaska Is Melting And It's Likely To Accelerate Global Heating. [Online] The Guardian, 06 14, 2019. [Cited: 01 15, 2020.] Https://Www.Theguardian.Com/Environment/2019/Jun/13/Climate-Crisis-Alaska-Is-Melting-And-Its-Likely-To-Accelerate-Global-Heating?.

43. Banguet, Laurent. Alaska Heat Wave Shatters City's Record, Disrupts Jobs And Lives. [Online] Phys.Org, 07 05, 2019. [Cited: 07 16, 2019.] Https://Phys.Org/News/2019-07-Alaska-Shatters-Temperature-Largest-City.Html.

44. Sinclare, Tracy. Record High Temperatures, Near Record Warmth And Green Up. *Alaska News Source.* [Online] Alaska News Source, 11 05 2020. [Cited: 13 05 2021.] Https://Www.Alaskanewssource.Com/Content/News/Record-High-Temperatures-Near-Record-Warmth-And-Green-Up-570355991.Html.

45. Alaska News Source. Anchorage Breaks Daily High Temperature Record. *Alaska News Source.* [Online] Alaska News Source, 08 16, 2020. [Cited: 05 13, 2021.] Https://Www.Alaskanewssource.Com/2020/08/16/Anchorage-Breaks-Daily-High-Temperature-Record/#:~:Text=Anchorage%2c%2oalaska%20(Ktuu)%20%2d,Temperature%2orecord%2ofor%20august%2015..

46. Lindsay, Bethany. Not Just A 'Lytton Problem'. *Cbc News.* [Online] Cbc News, 08 02, 2021. [Cited: 09 08, 2021.] Https://Newsinteractives.Cbc.Ca/Longform/Lytton-Feature.

47. Cagle, Susia. Heatwave Cooks Mussels In Their Shells On California Shore. *The Guardian.* [Online] The Guardian, 06 12, 2019. [Cited: 06 16, 2021.] Https://Www.Theguardian.Com/Environment/2019/Jun/28/California-Mussels-Cooked-Heat.

48. Nbc News. Sweltering Heat Is Shattering Records, Triggering Power Outages Across California. *Nbc News.* [Online] Nbc News, 08 15, 2020. [Cited: 05 13, 2021.] Https://Www.Cnbc.Com/2020/08/15/California-Heatwave-Triggers-Power-Outages.Html.

49. Livingston, Jason Samenow And Ian. Canada Sets New All-Time Heat Record Of 121 Degrees Amid Unprecedented Heat Wave. *The Washington Post.* [Online] The Washington Post, 06 29, 2021. [Cited: 09 08, 2021.] Https://Www.Washingtonpost.Com/Weather/2021/06/27/Heat-Records-Pacific-Northwest/.

50. *Chronic Kidney Disease And Associated Risk Factors In Two Salvadoran Farming Communities, 2012 .* Xavier F Vela 1, David O Henríquez, Susana M Zelaya, Delmy V Granados, Marcelo X Hernández, Carlos M Orantes. 2, Bethesda : National Center For Biotechnology Information, 2014, Vols. 55-60. Https://Pubmed.Ncbi.Nlm.Nih.Gov/24878650/.

51. Reuters. 'Another Hellish Day': South America Sizzles In Record Summer Temperatures. *The Guardian.* [Online] Reuters, 01 14, 2022. [Cited: 01 15, 2022.] Https://Www.Theguardian.Com/Environment/2022/Jan/14/South-America-Heat-Wave-Record-Summer-Temperatures.

52. Berwyn, Bob. This Summer's Heat Waves Could Be The Strongest Climate Signal Yet. *Inside Climate News.* [Online] 07 28, 2018. Https://Insideclimatenews.Org/News/27072018/Summer-2018-Heat-Wave-Wildfires-Climate-Change-Evidence-Crops-Flooding-Deaths-Records-Broken.

53. Samenow, Jason. Planet Is Entering 'New Climate Regime' With 'Extraordinary' Heat Waves Intensified By Global Warming, Study Says. [Online] Washington Post, 06 11, 2019. [Cited: 01 15, 2020.] Https://Www.Washingtonpost.Com/Weather/2019/06/11/Climate-Change-Intensified-Last-Summers-Northern-Hemisphere-Heat-Wave-It-May-Be-Starting-All-Over-Again/?Noredirect=On.

54. Watts, Jonathan. Welcome To The Fastest-Heating Place On Earth. [Online] The Guardian, 07 01, 2019. [Cited: 07 02, 2019.] Https://www.Theguardian.Com/Environment/Ng-Interactive/2019/Jul/01/Its-Getting-Warmer-Wetter-Wilder-The-Arctic-Town-Heating-Faster-Than-Anywhere?

55. Germanos, Andrea. Police Tear-Gas Climate Activists In Paris On 'Hottest Day In History Of France' . *Ecowatch.* [Online] Ecowatch, 07 01, 2019. [Cited: 03 24, 2022.] Https://Www.Ecowatch.Com/Extinction-Rebellion-Pepper-Sprayed-2639042841.Html.

56. Justin Huggler, Daniel Wittenberg And David Chazan. New Wildfires Rage In Europe As Punishing Heatwave Claims More Lives. [Online] The Telegraph, 06 29, 2019. [Cited: 07 01, 2019.] Https://Www.Telegraph.Co.Uk/News/2019/06/29/New-Wildfires-Rage-Europe-Punishing-Heatwave-Claims-Lives/?Utm_Campaign=Carbon%2obrief%2odaily%2obriefing&Utm_Medium=Email&Utm_Source=Revue%2onewsletter.

57. Forrest, Adam. Lake Discovered 11,000ft High In The Alps, In 'Truly Alarming' Sign Of Climate Change. *Independent.* [Online] Independent, 07 25, 2019. [Cited: 05 12, 2021.] Https://Www.Independent.Co.Uk/Climate-Change/News/Climate-Change-French-Alps-Lake-Snow-Melting-Global-Warming-Mount-Blanc-A9006916.Html.

58. Dellana, Alessio. Europe Heatwave: Tons Of Fish Found Dead In French Lake As Water Temperature Nears 29c. *Euronews.* [Online] Euronews, 08 14, 2020. [Cited: 05 13, 2021.] Https://Www.Euronews.Com/2020/08/14/Europe-Heatwave-Tons-Of-Fish-Found-Dead-In-French-Lake-As-Water-Temperature-Nears-29c.

59. Carrington, Damian. Climate Change Made European Heatwave At Least Five Times Likelier. [Online] The Guardian, 07 02, 2019. [Cited: 07 03, 2019.] Https://Www.Theguardian.Com/Science/2019/Jul/02/Climate-Change-European-Heatwave-Likelier?Cmp=Share_Androidapp_Facebook&Fbclid=Iwar2Giqzxpc_V_7tfdnntlv6mojiylob6d4nrcmaywy-Pveelzpqeycgsa1a.

60. High Heat: Spain Clocks Prelim Record Of 47.2 C (116.96 F). *U.S.News.* [Online] U.S.News, 08 14, 2021. [Cited: 09 08, 2021.] Https://Www.Usnews.Com/News/World/Articles/2021-08-14/Europe-Heat-Wave-Brings-Concern-For-Older-Adults-Homeless#:~:Text=By%20associated%20press-,Aug.,2021%2c%20at%203%3a10%20p.M.&Text=Madrid%20(Ap)%20%E2%80%94%20s pain%20set,Under%20a%20relentless%20summer.

61. Itv. Warnings Issued After At Least 31 Water Deaths Amid Uk Heatwave . *Itv.* [Online] Itv, 07 26, 2021. [Cited: 03 24, 2022.] Https://Www.Itv.Com/News/2021-07-23/Warnings-Issued-After-At-Least-20-Water-Deaths-Amid-Uk-Heatwave.

62. Beament, Emily. Heatwave Deaths Rise As Climate Increasingly Hits Health, Report Warns. *Belfast Telegraph.* [Online] Belfast Telegraph, 12 02, 2020. [Cited: 05 13, 2021.] Https://Www.Belfasttelegraph.Co.Uk/News/Uk/Heatwave-Deaths-Rise-As-Climate-Increasingly-Hits-Health-Report-Warns-39820378.Html.

63. Copernicus: 2020 Warmest Year On Record For Europe; Globally, 2020 Ties With 2016 For Warmest Year Recorded. *The Copernicus Programme.* [Online] Ecmwf As Part Of The Copernicus Programme, 01 08, 2021. [Cited: 05 13, 2021.] Https://Climate.Copernicus.Eu/2020-Warmest-Year-Record-Europe-Globally-2020-Ties-2016-Warmest-Year-Recorded.

64. Davidson, Jordan. 'Disturbing': Europe Is Warming Much Faster Than Science Predicted. *Ecowatch.* [Online] Ecowatch, 08 29, 2019. [Cited: 05 14, 2021.] Https://Www.Ecowatch.Com/Europe-Warming-Faster-Than-Predicted-2640103116.Html.

65. The Siberian Times Reporter. Massive Siberian Wildfires Affect North America With Smoke Via Jet Stream. *The Millenium Report.* [Online] 07 29, 2018. Http://Themillenniumreport.Com/2018/07/Massive-Siberian-Wildfires-Affect-North-America-With-Smoke-Via-Jet-Stream/.

66. Agence France-Presse. 2019 Was Hottest Year On Record For Russia. *Voa News.* [Online] Voa News, 12 30, 2019. [Cited: 05 14, 2021.] Https://Www.Voanews.Com/Europe/2019-Was-Hottest-Year-Record-Russia#:-:Text=He%20said%20moscow's%20average%20temperature,In%20russia%20as%20a%20whole..

67. Phys.Org. Russia Sees Record High Average Temperatures In 2020. *Phys.Org.* [Online] Phys.Org, 03 25, 2021. [Cited: 05 14, 2021.] Https://Phys.Org/News/2021-03-Russia-High-Average-Temperatures.Html.

68. Suh-Yoon, Lee. Korea's Electricity Consumption Set New Records In 2018 Amid Heat Wave. *The Korea Times.* [Online] 05 27, 2019. Http://Www.Koreatimes.Co.Kr/Www/Nation/2019/05/371_269569.Html.

69. All News. (Lead) Seoul Choked By Season's Highest Temperature. *All News.* [Online] All News, 06 22, 2020. [Cited: 05 14, 2021.] Https://En.Yna.Co.Kr/View/Aen20200622001951315.

70. Dunne, Daisy. Japan's Deadly 2018 Heatwave 'Could Not Have Happened Without Climate Change'. *Carbon Brief.* [Online] 05 30, 2019. Https://Www.Carbonbrief.Org/Japans-Deadly-2018-Heatwave-Could-Not-Have-Happened-Without-Climate-Change.

71. Staff Report. Two Die And Nearly 600 Taken To Hospitals As Heat Wave Roasts Much Of Japan. *Japan Times.* [Online] 05 26, 2019. Https://Www.Japantimes.Co.Jp/News/2019/05/26/National/Hokkaido-City-Temperature-Hits-May-Record-Heat-Wave-Envelops-Japan/#.Xpxadmgzauk.

72. Staff Report, Kyodo, Jiji . Japan's Weather Agency Issues Heatstroke Alert Amid Record-Setting Temperatures. *Thejapantimes.* [Online] Thejapantimes, 08 16, 2020. [Cited: 05 14, 2021.] Https://Www.Japantimes.Co.Jp/News/2020/08/16/National/Heatstroke-Alert-Japan/.

73. Yamagishi, Ryo. Eastern Japan Sweltered Under Record Heat During August. *The Asahi Shimbun.* [Online] The Asahi Shimbun, 09 02, 2020. [Cited: 05 14, 2021.] Http://Www.Asahi.Com/Ajw/Articles/13690327.

74. Marcus, Lilit. Quriyat In Oman Breaks World Temperature Record. *Cnn.* [Online] 06 29, 2018. Https://Edition.Cnn.Com/Travel/Article/Hottest-Low-Temperature-Quriyat-Oman-Wxc/Index.Html.

75. France 24. Think The Heat Wave In Europe And N. America Is Bad? Try Living In The Middle East . *France 24.* [Online] France 24, 07 26, 2019. [Cited: 05 14, 2021.] Https://Www.France24.Com/En/20190726-Heat-Wave-Europe-North-America-Worse-Middle-East-Extreme-Temperatures-Climate-Change.

76. Wallace, Paul. Record Heat Sets Off A Cascade Of Suffering In Baghdad . *Bloomberg Green.* [Online] Bloomberg Green, 08 04, 2020. [Cited: 05 14, 2021.] Https://Www.Bloomberg.Com/News/Articles/2020-08-03/Record-Heat-Sets-Off-A-Cascade-Of-Suffering-In-Baghdad.

77. Peltier, Falih Hassan And Elian. Scorching Temperatures Bake Middle East Amid Eid Al-Adha Celebrations. *The New York Times.* [Online] The New York Times, 07 31, 2020. [Cited: 05 12, 2021.] Https://Www.Nytimes.Com/2020/07/31/World/Middleeast/Middle-East-Heat-Wave.Html.

78. News.Com.Au. India Weather: Temperature Passes 50c Celsius In Northern India. *News.Com.Au.* [Online] 06 03, 2019. Https://Www.News.Com.Au/Technology/Environment/Climate-Change/India-Weather-Temperature-Passes-50c-Celsius-In-Northern-India-News-Story/0985e58e9ded4524fe586312b5154aef?Fbclid=Iwar1aseqlgxnna4m9kstxdtkwmxtcjsfvdkkloihtggcajh4dscruffvcyok.

79. Mishra, Himanshu Shekhar. Delhi At 48 Degrees, Highest Ever In June As Heat Wave Sweeps North India. [Online] Ndtv, 06 10, 2019. [Cited: 01 15, 2020.] Https://Www.Ndtv.Com/Delhi-News/Delhi-Weather-Delhi-At-48-Degrees-Highest-Ever-In-June-Says-Weather-Agency-Skymet-2051014?Fbclid=Iwar2bbfwqwrqecy5moroyr7zhnluwf-Wbkzp9ehd2pd5ezc0-Ghbhko_Beku.

80. Rosane, Olivia. All-Time Temperature. [Online] Ecowatch, 06 13, 2019. [Cited: 01 15, 2020.] Https://Www.Ecowatch.Com/India-Heat-Wave-Deaths-2638801311.Html.

81. Earth Observing System. Soil Temperature As A Factor Of Crops Development. *Earth Observing System.* [Online] Earth Observing System, 12 31, 2020. [Cited: 05 05, 2022.] Https://Eos.Com/Blog/Soil-Temperature/.

82. Baloch, Hannah Ellis-Peterson And Shar Meer. 'We Are Living In Hell': Pakistan And India Suffer Extreme Spring Heatwaves. *The Guardian.* [Online] The Guardian, 05 02, 2022. [Cited: 05 04, 2022.] Https://Www.Theguardian.Com/World/2022/May/02/Pakistan-India-Heatwaves-Water-Electricity-Shortages.

83. Braganza, Karl. Australia's 2018 In Weather: Drought, Heat And Fire. *The Conversation.* [Online] 01 10, 2019. Https://Theconversation.Com/Australias-2018-In-Weather-Drought-Heat-And-Fire-109575.

84. Associated Press. Australia Sees Highest Temperature Ever Recorded In A Major City With Adelaide Reaching 46.6 Degrees. *National Post.* [Online] 01 24, 2019. Https://Nationalpost.Com/News/World/Australian-City-Adelaide-Sets-New-National-Heat-Record.

85. Pappas, Stephanie. Hundreds Of 'Boiled' Bats Fall From Sky In Australian Heat Wave. *Live Science.* [Online] 01 08, 2018. Https://Www.Livescience.Com/61372-Boiled-Bats-Australia.Html.

86. Mcguirk, Rod. Australia Sweltered Through Its 4th-Hottest Year In 2020. *Abc News.* [Online] Abc News, 01 08, 2021. [Cited: 05 14, 2021.] Https://Abcnews.Go.Com/Technology/Wirestory/Australia-Sweltered-4th-Hottest-Year-2020-75122797#:~:Text=The%20hottest%20temperature%20in%20australia,Years%20were%202013%20and%202005..

87. The Herald's View. Nationalbushfires This Was Published 5 Months Ago Editorial Heatwave A Stark Reminder Of Potential Dangers For Australia. *The Sydney Morning Herald.* [Online] The Sydney Morning Herald, 12 03, 2020. [Cited: 05 14, 2021.] Https://Www.Smh.Com.Au/National/Heatwave-A-Stark-Reminder-Of-Potential-Dangers-For-Australia-20201203-P56k71.Html?Utm_Campaign=Carbon%20brief%20daily%20briefing&Utm_Content=20201203&Utm_Medium=Email&Utm_Source=Revue%20daily.

88. Masters, Jeff. Earth Had Its Second Warmest Year In Recorded History In 2019. *Scientific American.* [Online] Scientific American, 01 15, 2020. [Cited: 05 14, 2021.] Https://Blogs.Scientificamerican.Com/Eye-Of-The-Storm/Earth-Had-Its-Second-Warmest-Year-In-Recorded-History-In-2019/.

89. Hannam, Peter. Australia Matches Its Hottest Day On Record As Western Australia Town Hits 50.7c. *The Guardian.* [Online] The Guardian, 01 13, 2022. [Cited: 01 15, 2022.] Https://Www.Theguardian.Com/Australia-News/2022/Jan/13/Hottest-Day-On-Record-In-Parts-Of-Western-Australia-As-Temperature-Reaches-50c.

90. Redfearn, Graham. Antarctica Logs Hottest Temperature On Record With A Reading Of 18.3c. *The Guardian.* [Online] The Guardian, 02 07, 2020. [Cited: 05 14, 2021.] Https://Www.Theguardian.Com/World/2020/Feb/07/Antarctica-Logs-Hottest-Temperature-On-Record-With-A-Reading-Of-183c?Utm_Campaign=Carbon%20brief%20daily%20briefing&Utm_Medium=Email&Utm_Source=Revue%20newsletter.

91. Carrington, Damian. 'Precipitous' Fall In Antarctic Sea Ice Since 2014 Revealed. [Online] The Guardian, 07 01, 2019. [Cited: 07 02, 2019.] Https://Www.Theguardian.Com/World/2019/Jul/01/Precipitous-Fall-In-Antarctic-Sea-Ice-Revealed?.

92. Datta, Anusaya. It's Official Now: 2020 Is The Warmest Year On Record Along With 2016. *Geospatial World:.* [Online] Geospatial Media And Communications, 01 14, 2021. [Cited: 05 14, 2021.] Https://Www.Geospatialworld.Net/Blogs/Its-Official-Now-2020-Is-The-Warmest-Year-On-Record-Along-With-2016/.

93. Neslen, Arthur. Flooding And Heavy Rains Rise 50% Worldwide In A Decade, Figures Show. *The Guardian.* [Online] 03 21, 2018. Https://Www.Theguardian.Com/Environment/2018/Mar/21/Flooding-And-Heavy-Rains-Rise-50-Worldwide-In-A-Decade-Figures-Show.

94. Samenow, Jason. Planet Is Entering 'New Climate Regime' With 'Extraordinary' Heat Waves Intensified By Global Warming, Study Says. [Online] The Washington Post, 06 11, 2019. [Cited: 01 15, 2020.] Https://Www.Washingtonpost.Com/Weather/2019/06/11/Climate-Change-Intensified-Last-Summers-Northern-Hemisphere-Heat-Wave-It-May-Be-Starting-All-Over-Again/?Noredirect=On.

95. Marshall, Michael. Record-Breaking Hot Years Look Set To Continue Through The Next Decade. *New Scientist.* [Online] New Scientist, 02 10, 2020. [Cited: 05 14, 2021.] Https://Www.Newscientist.Com/Article/2232998-Record-Breaking-Hot-Years-Look-Set-To-Continue-Through-The-Next-Decade/?Fbclid=Iwar1dwhtmzwmhgx3xfyvks97ftmzgercmevqx_Jnn5nzg-4z0zxju85bqz0a.

96. Hausfather, Zeke. Analysis: When Might The World Exceed 1.5c And 2c Of Global Warming? *Carbonbrief.* [Online] Carbonbrief, 12 04, 2020. [Cited: 09 08, 2021.] Https://Www.Carbonbrief.Org/Analysis-When-Might-The-World-Exceed-1-5c-And-2c-Of-Global-Warming?Utm_Campaign=Carbon%20brief%20daily%20briefing&Utm_Content=20201207&Utm_Medium=Email&Utm_Source=Revue%20daily.

97. Tollefson, Jeff. How Hot Will Earth Get By 2100? *Nature.* [Online] Nature, 04 22, 2020. [Cited: 09 08, 2021.] Https://Www.Nature.Com/Articles/D41586-020-01125-X.

98. Vince, Gaia. The Heat Is On Over The Climate Crisis. Only Radical Measures Will Work. [Online] The Guardian, 05 18, 2019. [Cited: 01 15, 2020.] Https://Www.Theguardian.Com/Environment/2019/May/18/Climate-Crisis-Heat-Is-On-Global-Heating-Four-Degrees-2100-Change-Way-We-Live.

99. Carbonbrief. The Impacts Of Climate Change At 1.5c, 2c And Beyond. [Online] Carbonbrief. [Cited: 01 15, 2020.] Https://Interactive.Carbonbrief.Org/Impacts-Climate-Change-One-Point-Five-Degrees-Two-Degrees/?Utm_Source=Web&Utm_Campaign=Redirect.

100. Rice, Doyle. 'Breaking' The Heat Index: Us Heat Waves To Skyrocket As Globe Warms, Study Suggests. *Usa Today.* [Online] Usa Today, 07 16, 2019. [Cited: 05 14, 2021.] Https://Www.Usatoday.Com/Story/News/Nation/2019/07/16/Heat-Waves-Worsen-Because-Global-Warming-Study-Says/

101. Eltahir, Jeremy S. Pal And Elfatih A. B. Future Temperature In Southwest Asia Projected To Exceed A Threshold For Human Adaptability. [Online] Nature, 10 26, 2015. [Cited: 01 15, 2020.] Https://Www.Nature.Com/Articles/Nclimate2833.

102. Rees, Prof. William E. Yes, The Climate Crisis May Wipe Out Six Billion People. *The Tyee.* [Online] The Tyee, 09 18, 2019. [Cited: 11 08, 2021.] Https://Thetyee.Ca/Analysis/2019/09/18/Climate-Crisis-Wipe-Out/.

103. The Economist. The Likelihood Of Floods Is Changing With The Climate. *The Economist.* [Online] 08 31, 2017. Https://Www.Economist.Com/Briefing/2017/08/31/The-Likelihood-Of-Floods-Is-Changing-With-The-Climate.

104. Mcgreal, Chris. 'So Much Land Under So Much Water': Extreme Flooding Is Drowning Parts Of The Midwest. *The Guardian.* [Online] 06 03, 2019. Https://Www.Theguardian.Com/Us-News/2019/Jun/03/So-Much-Land-Under-So-Much-Water-Extreme-Flooding-Is-Drowning-Parts-Of-The-Midwest.

105. Waldrop, Theresa. Hurricane Ida Took Down More Power Poles In 2 States Than Katrina, Ike, Delta And Zeta Combined, Power Company Says. *Cnn.* [Online] Cnn, 09 07, 2021. [Cited: 09 08, 2021.] Https://Edition.Cnn.Com/2021/09/07/Us/Hurricane-Ida-Aftermath-Louisiana-Tuesday/Index.Html.

106. Macaya, Melissa Mahtani And Melissa. Biden Surveys Ida Aftermath In New York And New Jersey. *Cnn.* [Online] Cnn, 09 07, 2021. [Cited: 09 08, 2021.] Https://Edition.Cnn.Com/Us/Live-News/Biden-Visits-Ny-Nj-Hurricane-Ida-Aftermath-09-07-21/Index.Html.

107. Hardin, Kat So And Sally. Extreme Weather Cost U.S. Taxpayers $99 Billion Last Year, And It Is Getting Worse. *Center For American Progress.* [Online] Center For American Progress, 09 01, 2021. [Cited: 09 08, 2021.] Https://Www.Americanprogress.Org/Issues/Green/Reports/2021/09/01/503251/Extreme-Weather-Cost-U-S-Taxpayers-99-Billion-Last-Year-Getting-Worse/.

108. Stephenson, Amanda. 'An Absolutely Brutal Year': Alberta Farm Incomes Plummet 70 Per Cent In 2018. *Edmonton Sun.* [Online] 05 29, 2019. Https://Edmontonsun.Com/Business/Local-Business/An-Absolutely-Brutal-Year-Alberta-Farm-Incomes-Plummet-70-Per-Cent-In-2018/

109. Gibbens, Sarah. 2018's Deadly Hurricane Season, Visualized. *National Geographic.* [Online] 20 December 2018. Https://Www.Nationalgeographic.Com/Environment/2018/12/Hurricane-Season-Explained-Maps-Photos/.

110. Poon, Linda. Climate Change Is Testing Asia's Megacities. *Citylab.* [Online] Citylab, 10 09, 2018. [Cited: 01 15, 2020.] Https://Www.Citylab.Com/Environment/2018/10/Asian-Megacities-Vs-Tomorrows-Typhoons/572062/.

111. 2018 Pacific Typhoon Season. *Wikipedia.* [Online] 25 May 2019. Https://En.Wikipedia.Org/Wiki/2018_Pacific_Typhoon_Season#Typhoon_Jebi_(Maymay).

112. Berlinger, Junko Ogura And Joshua. Japan Floods: Death Toll Rises To 200 As Un Offers Assistance. *Cnn.* [Online] Cnn, 07 12, 2018. [Cited: 05 12, 2021.] Https://Edition.Cnn.Com/2018/07/10/Asia/Japan-Floods-Intl/Index.Html.

113. Phys.Org. Floods In Indonesia Kill 29, Dozen Missing. *Phys.Org.* [Online] Phys.Org, 04 29, 2019. [Cited: 05 12, 2021.] Https://Phys.Org/News/2019-04-Indonesia-Dozen.Html#:-:Text=On%20monday%2c%20indonesia's%20disaster%20agency,Killed%20a%20family%20of%20six..

114. Simon, Matt. The Sea Is Consuming Jakarta, And Its People Aren't Insured. *Wired.* [Online] Wired, 07 19, 2019. [Cited: 05 12, 2021.] Https://Www.Wired.Com/Story/Jakarta-Insurance/?Fbclid=Iwar15217e9fdwrg6hrlnielcjlubcfr2-s203laxzutdit5htqn3a2wpymhc&Verso=True.

115. Everington, Keoni. Video Shows Tornado Strike S. Taiwan. [Online] Taiwan News, 07 02, 2019. [Cited: 07 04, 2019.] Https://Www.Taiwannews.Com.Tw/En/News/3736427?Fbclid=Iwaroglguh6yafhs3vlmyq7znkmlsblifximv4hmg8s1sl2dml2qjhqzzy20a.

116. Reuters. Tornadoes Hit Two Chinese Provinces, Killing 12, Injuring Hundreds. *Reuters.* [Online] Reuters, 05 15, 2021. [Cited: 05 16, 2021.] Https://Www.Reuters.Com/Business/Environment/Many-Injured-Tornado-Strikes-Wuhan-China-Xinhua-2021-05-14/.

117. —. China Braces For Summer Floods As 71 Rivers Exceed Warning Levels. *Reauters.* [Online] Reauters, 05 26, 2021. [Cited: 06 16, 2021.] Https://Www.Reuters.Com/World/China/China-Braces-For-Summer-Floods-71-Rivers-Exceed-Warning-Levels-2021-05-26/.

118. Aljazeera. Worst Floods In Years 'Submerge' Bangladesh Villages. *Aljazeera.* [Online] Aljazeera, 07 19, 2019. [Cited: 05 12, 2021.] Https://Www.Aljazeera.Com/News/2019/07/19/Worst-Floods-In-Years-Submerge-Bangladesh-Villages/?Fbclid=Iwaroqsqfqobnjdkzojovv7uuxyftbi_Ctmt2zfjzmrhdb6bez31ht2ezfcbu.

119. Vallangi, Neelima. Rapid Heating Of Indian Ocean Worsening Cyclones, Say Scientists. *The Guardian.* [Online] The Guardian, 05 27, 2021. [Cited: 06 16, 2021.] Https://Www.Theguardian.Com/Environment/2021/May/27/Rapid-Heating-Of-Indian-Ocean-Worsening-Cyclones-Say-Scientists.

120. Page, Michael Le. Extreme Flooding Leads To Deaths In Indonesia And Mozambique. *New Scientist.* [Online] 04 30, 2019. Https://Www.Newscientist.Com/Article/2201224-Extreme-Flooding-Leads-To-Deaths-In-Indonesia-And-Mozambique/.

121. Agence France-Presse. South Africa Floods: Deadliest Storm On Record Kills Over 300 People. *The Guardian.* [Online] The Guardian, 04 13, 2022. [Cited: 04 14, 2022.] Https://Www.Theguardian.Com/World/2022/Apr/13/South-Africa-Floods-Deadliest-Storm-On-Record-Kills-Over-250-People.

122. Willsher, Kim. France To Declare Natural Disaster After Storms Rip Through Crops. [Online] Guardian, 06 16, 2019. [Cited: 01 15, 2020.] Https://Www.Theguardian.Com/World/2019/Jun/16/France-To-Declare-Natural-Disaster-After-Storms-Rip-Through-Crops.

123. Bbc News. Mexico Hail: Ice 1.5m Thick Carpets Guadalajara. [Online] Bbc News, 07 01, 2019. [Cited: 07 01, 2019.] Https://Www.Bbc.Com/News/World-Latin-America-48821306.

124. Calahan, Rick. After Several Quiet Years, Tornadoes Are Erupting Across The Us In What Experts Are Calling An 'Outbreak'. *Business Insider.* [Online] Business Insider, 05 29, 2019. [Cited: 04 25, 2022.] Https://Www.Businessinsider.Com/After-Several-Quiet-Years-Tornadoes-Erupt-In-United-States-2019-5.

125. Niccitelli, Jeff Masters And Dana. The Top 10 Extreme Weather And Climate Events Of 2020. *Ecowatch.* [Online] Yale Climate Connections, 12 23, 2020. [Cited: 05 12, 2021.] Https://Www.Ecowatch.Com/Extreme-Weather-Climate-2020-2649628910.Html.

126. Sciencedaily. New Data Confirm Increased Frequency Of Extreme Weather Events. [Online] Sciencedaily, 03 21, 2018. [Cited: 01 15, 2020.] Https://Www.Sciencedaily.Com/Releases/2018/03/180321130859.Htm.

127. Connor, Steve. Gulf Stream Is Slowing Down Faster Than Ever, Scientists Say. [Online] Independent, 03 23, 2015. [Cited: 01 15, 2020.] Https://Www.Independent.Co.Uk/Environment/Gulf-Stream-Is-Slowing-Down-Faster-Than-Ever-Scientists-Say-10128700.Html.

128. Carrington, Damian. Climate Crisis: Scientists Spot Warning Signs Of Gulf Stream Collapse This Article Is More Than 1 Mo. *The Guardian.* [Online] The Guardian, 08 05, 2021. [Cited: 09 08, 2021.] Https://Www.Theguardian.Com/Environment/2021/Aug/05/Climate-Crisis-Scientists-Spot-Warning-Signs-Of-Gulf-Stream-Collapse.

129. Gabbatiss, Josh. 'Climate Chaos': Melting Ice Sheets Will Trigger Extreme Weather Across The World, Warn Scientists. [Online] Independent, 02 06, 2019. [Cited: 01 15, 2020.] Https://Www.Independent.Co.Uk/Environment/Climate-Change-Melting-Ice-Antarctica-Greenland-Arctic-Global-Warming-Gulf-Stream-Amoc-A8766521.Html.

130. Robb, Trevor. 'We Have To Learn To Live With Fire': Q And A With University Of Alberta Climate Change Expert. [Online] Edmonton Journal, 05 31, 2019. [Cited: 01 15, 2020.] Https://Edmontonjournal.Com/News/Local-News/We-Have-To-Learn-To-Live-With-Fire-Q-And-A-With-University-Of-Alberta-Climate-Change-Expert.

131. *Anthropogenic Influences On Major Tropical Cyclone Events.* Christina M. Patricola & Michael F. Wehner. S.L. : Nature, 11 14, 2018, Vol. 563, Pp. 339-346.

132. Agence France-Presse. Asia Is Home To 99 Of World's 100 Most Vulnerable Cities. *The Guardian.* [Online] The Guardian, 05 13, 2021. [Cited: 05 14, 2021.] Https://Www.Theguardian.Com/Cities/2021/May/13/Asia-Is-Home-To-99-Of-Worlds-100-Most-Vulnerable-Cities?Cmp=Share_Iosapp_Other&Fbclid=Iwar27bycofaixar_Dscb86gscanxcuv6q7aeqryoqb7beagttirvqsy8cszo.

133. Statista Research Department. Natural Rate Of Population Growth By Continent In 2014. [Online] Statista, 04 29, 2019. [Cited: 01 15, 2020.] Https://Www.Statista.Com/Statistics/270859/Natural-Rate-Of-Population-Growth-By-Continent/.

134. Mckie, Robin. Global Heating To Inflict More Droughts On Africa As Well As Floods. [Online] Guardian, 06 16, 2019. [Cited: 01 15, 2020.] Https://Www.Theguardian.Com/Science/2019/Jun/14/Africa-Global-Heating-More-Droughts-And-Flooding-Threat?Cmp=Share_Androidapp_Facebook&Fbclid=Iwar2gusokky2pqweecceulodpbwjwez3zpb6iei2ifzo6bprpyfzobjrn5aa.

135. Harvey, Fiona. One In Four Cities Cannot Afford Climate Crisis Protection Measures – Study. *The Guardian.* [Online] The Guardian, 05 12, 2021. [Cited: 05 13, 2021.] Https://Www.Theguardian.Com/Environment/2021/May/12/One-In-Four-Cities-Cannot-Afford-Climate-Crisis-Protection-Measures-Study#:-:Text=A%20survey%20of%20800%20cities,Adapt%20to%20the%20climate%20crisis..

136. Cho, Renee. What Do Wildfires Have To Do With Climate Change? [Online] Columbia University, 10 13, 2014. [Cited: 01 15, 2020.] Https://Blogs.Ei.Columbia.Edu/2014/10/13/What-Do-Wildfires-Have-To-Do-With-Climate-Change/.

137. *Economic Footprint Of California Wildfires In 2018.* Daoping Wang, Dabo Guan And Shupeng Zhu Et Al. S.L. : Nature Sustainability , 2020, Vol. 4. Https://Doi.Org/10.1038/S41893-020-00646-7.

138. Meyer, Robinson. California's Wildfires Are 500 Percent Larger Due To Climate Change. *The Atlantic.* [Online] The Atlantic, 07 16, 2019. [Cited: 05 12, 2021.] Https://Www.Theatlantic.Com/Science/Archive/2019/07/Climate-Change-500-Percent-Increase-California-Wildfires/594016/?Utm_Campaign=Carbon%20brief%20daily%20briefing&Utm_Medium=Email&Utm_Source=Revue%20newsletter.

139. Hackney, Amir Vera And Deanna. 91 Wildfires Are Now Burning Across The Us, With Oregon's Bootleg Fire Growing To Over 400,000 Acres. *Cnn.* [Online] Cnn, 08 02, 2021. [Cited: 09 08, 2021.] Https://Edition.Cnn.Com/2021/08/01/Us/Us-Western-Wildfires-Sunday/Index.Html.

140. Irfan, Umair. California's Wildfires Are Hardly "Natural" — Humans Made Them Worse At Every Step. Vox. [Online] 19 11 2018. Https://Www.Vox.Com/2018/8/7/17661096/California-Wildfires-2018-Camp-Woolsey-Climate-Change.

141. Thompson, Andrea. Lightning May Increase With Global Warming. [Online] Scientific American, 11 13, 2014. [Cited: 01 15, 2020.] Https://Www.Scientificamerican.Com/Article/Lightning-May-Increase-With-Global-Warming/.

142. Climate Reality Project. How Does Climate Change Affect Forest Fires? [Online] Climate Reality Project, 05 24, 2017. [Cited: 01 15, 2020.] Https://Www.Climaterealityproject.Org/Blog/How-Does-Climate-Change-Cause-Forest-Fires.

143. Berwyn, Bob. The Future Of Fighting Wildfires In The Era Of Climate Change. [Online] Pacific Standard, 08 04, 2017. [Cited: 01 15, 2020.] Https://Psmag.Com/News/The-Future-Of-Fighting-Wildfires-In-The-Era-Of-Climate-Change.

144. *Climate Change Lengthens Southeastern Usa Lightning-Ignited Fire Seasons.* Corey N. Davis, Raelene M. Crandall. 1, S.L. : Global Change Biology, 2019, Global Change Biology, Vol. 11, P. 07. Https://Www.Researchgate.Net/Publication/334424539_Climate_Change_Lengthens_Southeastern_Usa_Lightning-Ignited_Fire_Seasons.

145. Bbc News. *Bbc.* [Online] 24 07 2018. Https:// Https://Www.Bbc.Com/News/World-Europe-44932366.

146. Burgen, Stephen. Firefighters Battle Forest Blaze In Central Spain. [Online] The Guardian, 06 30, 2019. [Cited: 07 01, 2019.] Https://Www.Theguardian.Com/World/2019/Jun/30/Firefighters-Battle-Forest-Blaze-In-Central-Spain-Madrid?

147. Cnn World. In Photos: Europe Battles Wildfires Amid Scorching Heat Waves. *Cnn World.* [Online] Cnn World, 08 17, 2021. [Cited: 09 08, 2021.] Https://Edition.Cnn.Com/2021/08/06/World/Gallery/Europe-Extreme-Summer-Weather-2021/Index.Html.

148. Brown, Mark. Huge Wildfire Sweeps Across Canford Heath Nature Reserve In Dorset. *The Guardian.* [Online] The Guardian, 04 24, 2022. [Cited: 04 27, 2022.] Https://Www.Theguardian.Com/World/2022/Apr/24/Wildfire-Canford-Heath-Nature-Reserve-Dorset-Fire.

149. Ross Bradstock, Hamish Clarke And Luke Collins Et Al. A Staggering 1.8 Million Hectares Burned In 'High-Severity' Fires During Australia's Black Summer. *The Conversation.* [Online] The Conversation, 03 30, 2021. [Cited: 05 12, 2021.] Https://Theconversation.Com/A-Staggering-1-8-Million-Hectares-Burned-In-High-Severity-Fires-During-Australias-Black-Summer-157883.

150. Upadhyay, Vineet. Second-Worst In 16 Yrs, Fires Destroy 2,521ha Forest Cover I .. [Online] Times Of India, 06 07, 2019. [Cited: 01 15, 2020.] Https://Timesofindia.Indiatimes.Com/India/Forest-Fires-Rage-In-Ukhand-Despite-Crores-Spent-On-Mitigation/Articleshow/

151. Dw. Massive Wildfires Cause 'Greatest Forest Disaster' In Chile's History. [Online] Dw, 01 24, 2017. [Cited: 01 15, 2020.] Https://Www.Dw.Com/En/Massive-Wildfires-Cause-Greatest-Forest-Disaster-In-Chiles-History/A-37246386.

152. Watts, Jonathan. The Swedish Town On The Frontline Of The Arctic Wildfires. *The Guardian.* [Online] 30 07 2018. Https://Www.Theguardian.Com/World/2018/Jul/30/The-Swedish-Town-On-The-Frontline-Of-The-Arctic-Wildfires.

153. Miles, Tom. Arctic Wildfires In June Equivalent To Sweden's Annual Emissions - U.N. [Online] Reuters, 07 12, 2019. [Cited: 07 16, 2019.] Https://Uk.Reuters.Com/Article/Us-Weather-Wildfires/Arctic-Wildfires-In-June-Equivalent-To-Swedens-Annual-Emissions-U-N-I

154. Reevell, Patrick. Siberian Wildfires Now Bigger Than All Other Fires In World Combined. *Abc News.* [Online] Abc News, 08 13, 2021. [Cited: 09 08, 2021.] Https://Abcnews.Go.Com/International/Siberian-Wildfires-Now-Bigger-Fires-World-Combined/Story?Id=79422602.

155. Bbc Newsround. What Are 'Zombie Fires' And Why Is The Arctic Circle On Fire? *Bbc Newsround.* [Online] Bbc Newsround, 05 20, 2021. [Cited: 09 08, 2021.] Https://Www.Bbc.Co.Uk/Newsround/57173570.

156. Berwyn, Bob. How Wildfires Can Affect Climate Change (And Vice Versa). [Online] Inside Climate News, 08 23, 2018. [Cited: 01 15, 2020.] Https://Insideclimatenews.Org/News/23082018/Extreme-Wildfires-Climate-Change-Global-Warming-Air-Pollution-Fire-Management-Black-Carbon-Co2.

157. Parry, Wynne. 20 Years After Pinatubo: How Volcanoes Could Alter Climate. [Online] Live Science, 06 15, 2011. [Cited: 01 15, 2020.] Https://Www.Livescience.Com/14513-Pinatubo-Volcano-Future-Climate-Change-Eruption.Html.

158. Howard, Peter. Flammable Planet: Wildfires And The Social Cost Of Carbon. [Online] Cost Of Carbon, 09 2014. [Cited: 01 15, 2020.] Https://Costofcarbon.Org/Files/Flammable_Planet__Wildfires_And_Social_Cost_Of_Carbon.Pdf.

159. Brändlin, Anne-Sophie. How Climate Change Is Increasing Forest Fires Around The World. [Online] Dw, 06 19, 2017. [Cited: 01 15, 2020.] Https://Www.Dw.Com/En/How-Climate-Change-Is-Increasing-Forest-Fires-Around-The-World/A-19465490.

160. Gee, Alastair. Is This The End Of Forests As We've Known Them? *The Guardian.* [Online] The Guardian, 03 10, 2021. [Cited: 05 12, 2021.] Https://Www.Theguardian.Com/Environment/2021/Mar/10/Is-This-The-End-Of-Forests-As-Weve-Known-Them.

161. Williams, Terri-Ann. Food Prices Are Set To Rise By £7.15 A Week As Heatwave Hits Britain's Struggling Farmers With Crops Of Potatoes, Onions And Carrots Slumping By As Much As One-Fifth. *Daily Mail Online.* [Online] 04 09 2018. Https://Www.Dailymail.Co.Uk/News/Article-6129469/Heatwave-Hits-Britains-Struggling-Farmers-Crops-Potatoes-Onions-Carrots-Slump.Html.

162. Harris, Chris. Heat, Hardship And Horrible Harvests: Europe's Drought Explained. *Euronews.* [Online] 12 08 2018. Https://Www.Euronews.Com/2018/08/10/Explained-Europe-S-Devastating-Drought-And-The-Countries-Worst-Hit.

163. Helsel, Phil. Drought & Deluge In 2018. *Nbc News.* [Online] Nbc News, 15 03 2019. [Cited: 29 04 2022.] Https://Www.Nbcnews.Com/Storyline/California-Drought/California-Drought-Officially-Over-After-More-Seven-Years-N983461.

164. Yulsman, Tom. Drought In The Western United States Sets A 122-Year Record. *Discover.* [Online] Discover, 07 23, 2021. [Cited: 09 08, 2021.] Https://Www.Discovermagazine.Com/Environment/Drought-In-The-Western-United-States-Sets-A-122-Year-Record.

165. Borunda, Alejanda. The Drought In The Western U.S. Could Last Until 2030. *National Geographic.* [Online] National Geographic, 02 15, 2022. [Cited: 04 29, 2022.] Https://Www.Nationalgeographic.Com/Environment/Article/The-Drought-In-The-Western-Us-Could-Last-Until-2030.

166. Patel, Kasha. After Two Years Of Drought Ends, Southeastern Australia Turns Green. *Scitechdaily.* [Online] Nasa Earth Observatory, 06 21, 2020. [Cited: 05 12, 2021.] Https://Scitechdaily.Com/After-Two-Years-Of-Drought-Ends-Southeastern-Australia-Turns-Green/

167. Nick Evershed, Andy Ball, Gabrielle Chan, And Mike Bowers. The New Normal? How Climate Change Is Making Droughts Worse. *The Guardian.* [Online] 03 10 2018. Https://Www.Theguardian.Com/Environment/Ng-Interactive/2018/Oct/03/The-New-Normal-How-Climate-Change-Is-Making-Droughts-Worse.

168. Fountain, Henry. Researchers Link Syrian Conflict To A Drought Made Worse By Climate Change. *The New York Times.* [Online] 03 02, 2015. Https://Www.Nytimes.Com/2015/03/03/Science/Earth/Study-Links-Syria-Conflict-To-Drought-Caused-By-Climate-Change.Html.

169. Aljazeera. Water Crisis And Drought Threaten 12 Million In Syria, Iraq. *Aljazeera.* [Online] Aljazeera, 08 23, 2021. [Cited: 09 08, 2021.] Https://Www.Aljazeera.Com/News/2021/8/23/Water-Crisis-And-Drought-Threaten-12-Million-In-Syria-Iraq.

170. Steffens, Gena. Changing Climate Forces Desperate Guatemalans To Migrate. *National Georgraphic.* [Online] 10 23, 2018. Https://Www.Nationalgeographic.Com/Environment/2018/10/Drought-Climate-Change-Force-Guatemalans-Migrate-To-Us/.

171. Datta, Anusuya. Satellite Data Shows Shrinking Reservoirs That May Spark Major Water Crisis Globally. [Online] Geospatial World, 06 30, 2019. [Cited: 07 02, 2019.] Https://Www.Geospatialworld.Net/Blogs/Satellite-Data-Show-Shrinking-Reservoirs-That-May-Spark-Major-Water-Crisis-Globally/?Fbclid=Iwar3jwobovonfszwejoyzkpipkadjshowjp_4havixuasaoiipuakrkjntvw.

172. Hamaide, Bate Felix And Sybille De La. Latest Hot Spell Set To Deepen Drought Pain In France. *Reuters.* [Online] Reuters, 07 17, 2019. [Cited: 05 12, 2021.] Https://Www.Reuters.Com/Article/Us-France-Drought/Latest-Hot-Spell-Set-To-Deepen-Drought-Pain-In-France

173. Connexion Journalist. France Imposes Water Restrictions In 55 Departments. *Connexionfrance.* [Online] Connexionfrance, 07 23, 2020. [Cited: 05 12, 2021.] Https://Www.Connexionfrance.Com/French-News/France-Imposes-Water-Restrictions-In-55-Departments-As-Alert-Levels-Rise-In-Dry-July-2020

174. Davidson, Helen. Parched Taiwan Prays For Rain As Sun Moon Lake Is Hit By Drought. *The Guardian.* [Online] The Guardian, 05 09, 2021. [Cited: 05 12, 2021.] Https://Www.Theguardian.Com/Environment/2021/May/09/Parched-Taiwan-Prays-For-Rain-As-Sun-Moon-Lake-Is-Hit-By-Drought.

175. Barbolini, Natasha. Central Asia Risks Becoming A Hyperarid Desert In The Near Future. *The Conversation*. [Online] The Conversation, 10 30, 2020. [Cited: 05 12, 2021.] Https://Theconversation.Com/Central-Asia-Risks-Becoming-A-Hyperarid-Desert-In-The-Near-Future-

176. Ratcliffe, Rebecca. Two Million People At Risk Of Starvation As Drought Returns To Somalia. *The Guardian*. [Online] 06 06, 2019. Https://Www.Theguardian.Com/Global-Development/2019/Jun/06/Two-Million-People-At-Risk-Of-Starvation-As-Drought-Returns-To-Somalia.

177. Tahir, Mustapha. In Somalia, The Climate Emergency Is Already Here. The World Cannot Ignore It. [Online] The Guardian, 07 08, 2019. [Cited: 07 16, 2019.] Https://Www.Theguardian.Com/Commentisfree/2019/Jul/08/Somalia-Climate-Emergency-World-Drought-Somalis.

178. Schulman, Susan. Shortages Of Food, Water And Electricity: How Djibouti Has Been Destroyed By Climate Change. [Online] The Telegraph, 06 18, 2019. [Cited: 01 15, 2020.] Https://Www.Telegraph.Co.Uk/Global-Health/Climate-And-People/Shortages-Food-Water-Electricity-Djibouti-Has-Been-Destroyed-Climate/.

179. Afp. Drought Forces Namibia To Auction 1,000 Wild Animals. [Online] The Guardian, 06 15, 2019. [Cited: 01 15, 2020.] Https://Guardian.Ng/News/Drought-Forces-Namibia-To-Auction-1000-Wild-Animals/.

180. Reuters. Severe Drought To Hit Namibia's Meat Industry After Cattle Herds Decimated. *Reuters*. [Online] Reuters, 12 11, 2020. [Cited: 05 12, 2021.] Https://Www.Reuters.Com/Article/Ozatp-Uk-Namibia-Meat-Idafkbn28m09m-Ozatp.

181. Mbugua, Sophie. Two Million In Zimbabwe's Capital Have No Water As City Turns Off Taps. *Climate Home News*. [Online] Climate Home News, 07 15, 2019. [Cited: 05 12, 2021.] Https://Www.Climatechangenews.Com/2019/07/15/Two-Million-Zimbabwes-Capital-No-Water-City-Turns-Off-Taps/?Utm_Campaign=Carbon%20brief%20daily%20briefing&Utm_Medium=Email&Utm_Source=Revue%20newsletter.

182. Reliefweb. Kenya: Drought - 2014-2022. *Reliefweb*. [Online] Reliefweb, 09 30, 2021. [Cited: 01 23, 2022.] Https://Reliefweb.Int/Disaster/Dr-2014-000131-Ken.

183. Newsday. Drought Cuts Across 13 African Countries. [Online] Newsday, 06 15, 2019. [Cited: 01 15, 2020.] Https://Www.Newsday.Co.Zw/2019/06/Drought-Cuts-Across-13-African-Countries/.

184. Ngonou, Boris. Madagascar: Drought Drives 730,000 People To Food Insecurity. *Afrik21*. [Online] Afrik21, 02 13, 2020. [Cited: 05 14, 2021.] Https://Www.Afrik21.Africa/En/Madagascar-Drought-Drives-730000-People-To-Food-Insecurity/

185. Relph, Sam. Indian Villages Lie Empty As Drought Forces Thousands To Flee. [Online] The Guardian, 06 12, 2019. [Cited: 06 24, 2019.] Https://Www.Theguardian.Com/World/2019/Jun/12/Indian-Villages-Lie-Empty-As-Drought-Forces-Thousands-To-Flee?

186. Bbc. Chennai Water: How India's Sixth Biggest City Is Coping With Shortages. [Online] Bbc, 06 20, 2019. [Cited: 01 15, 2020.] Https://Www.Bbc.Com/News/World-Asia-India-48703464.

187. Climate Home News. Trains Deliver Emergency Water To Drought-Hit Chennai. [Online] Climate Home News, 07 12, 2019. [Cited: 07 16, 2019.] Https://Www.Climatechangenews.Com/2019/07/12/Trains-Deliver-Emergency-Water-Drought-Hit-Chennai/?Utm_Campaign=Carbon%20brief%20daily%20briefing&Utm_Medium=Email&Utm_Source=Revue%20newsletter.

188. Shaikh, Saleem. In Rural Pakistan, 'Worst Drought In Years' Drives Displacement And Hunger. [Online] The New Humanitarian, 02 06, 2019. [Cited: 01 15, 2020.] Https://Www.Thenewhumanitarian.Org/News/2019/02/06/Rural-Pakistan-Worst-Drought-Years-Drives-Displacement-And-Hunger.

189. Weiss, Kenneth R. Some Of The World's Biggest Lakes Are Drying Up. Here's Why. *National Geographic*. [Online] 03 2018. Https://Www.Nationalgeographic.Com/Magazine/2018/03/Drying-Lakes-Climate-Change-Global-Warming-Drought/.

190. Climate Reality Project. The Facts About Climate Change And Drought. [Online] Climate Reality Project, 06 15, 2016. [Cited: 06 15, 2020.] Https://Www.Climaterealityproject.Org/Blog/Facts-About-Climate-Change-And-Drought.

191. Hodgson, Jonathan. Freshwater Crisis. [Online] National Geographic. [Cited: 01 15, 2020.] Https://Www.Nationalgeographic.Com/Environment/Freshwater/Freshwater-Crisis/.

192. Mann, Charles C. *The Wizard And The Prophet*. London : Picador, 2018. 978-1-5098-8416-2.

193. Aarhus University. Worldwide Water Shortage By 2040. [Online] Science Daily, 06 29, 2014. [Cited: 01 15, 2020.] Https://Www.Sciencedaily.Com/Releases/2014/07/140729093112.Htm.

194. *Keeping Global Warming Within 1.5 °C Constrains Emergence Of Aridification*. Chang-Eui Park, Su-Jong Jeong, Manoj Joshi. 2018, Nature Climate Change, Vol. 8, Pp. 70-74.

195. *The Water Footprint Of Poultry, Pork And Beef: A Comparative Study In Different Countries And Production Systems*. Gerbens-Leenes, Mekonnen, Hoekstra. Water Footprint Assessment (Wfa) For Better Water Governance And Sustainable Development, 2013, Water Resources And Industry, Vols. 1-2, Pp. 25-36.

196. *A Global Assessment Of The Water Footprint Of Farm Animal Products*. Hoekstra, Mesfin M. Mekonnen & Arjen Y. 1, S.L. : Springer, 2012, Ecosystems, Vol. 15, Pp. 401-415.

197. U.N. Environment Programme. Half The World To Face Severe Water Stress By 2030 Unless Water Use Is "Decoupled" From Economic Growth, Says International Resource Panel. *U.N. Environment Programme*. [Online] U.N. Environment Programme, 03 21, 2016. [Cited: 06 15, 2022.] Https://Www.Theatlantic.Com/International/Archive/2012/05/The-Coming-Global-Water-Crisis/256896/.

198. Cook, Dr. Benjamin. Guest Post: Climate Change Is Already Making Droughts Worse. [Online] Carbon Brief, 14 05, 2018. [Cited: 06 24, 2019.] Https://Www.Carbonbrief.Org/Guest-Post-Climate-Change-Is-Already-Making-Droughts-Worse.

199. The Straits Times. Africa Is Running Out Of Water, Plagued By Drought, Population Growth. [Online] The Straits Times, 03 19, 2019. [Cited: 06 25, 2019.] Https://Www.Straitstimes.Com/World/Africa/Africa-Is-Running-Out-Of-Water-Plagued-By-Drought-Population-Growth.

200. *Unprecedented Drought Over Tropical South America In 2016: Significantly Under-Predicted By Tropical Sst*. Amir Erfanian, Guiling Wong And Lori Fomenko. 7, 2017, Scientific Reports, Vol. 19, P. 07.

201. Schubert, Cecilia. A Look At How A Changing Climate Will Hit South And Central America. [Online] Climate Change, Agriculture And Food Security (Ccafs), 09 05, 2014. [Cited: 06 25, 2019.] Https://Ccafs.Cgiar.Org/Research-Highlight/Look-How-Changing-Climate-Will-Hit-South-And-Central-America#.Xrbziegzauk.
202. Food & Water Watch. Ocean Desalination No Solution To Water Shortages. [Online] Food & Water Watch, 04 02, 2009. [Cited: 06 23, 2019.] Https://Www.Foodandwaterwatch.Org/News/Ocean-Desalination-No-Solution-Water-Shortages.
203. Heather Cooley, Newsha Ajami, And Matthew Heberger. Key Issues In Seawater. Oakland : Pacific Institute, 2013.
204. Why Desalination Is Not The Answer To The World's Water Issues. [Online] Hydrofinity, 10 16, 2018. [Cited: 06 23, 2019.] Https://Www.Hydrofinity.Com/Blog/Why-Desalination-Is-Not-The-Answer-To-The-Worlds-Water-Issues.
205. World Hunger Education. 2018 World Hunger And Poverty Facts And Statistics. [Online] World Hunger Education, 2019. [Cited: 06 26, 2019.] Https://Www.Worldhunger.Org/World-Hunger-And-Poverty-Facts-And-Statistics/#Hunger-Number.
206. Reset. Global Food Waste And Its Environmental Impact. Reset. [Online] Reset. [Cited: 05 14, 2021.] Https://En.Reset.Org/Knowledge/Global-Food-Waste-And-Its-Environmental-Impact-09122018.
207. Eating Meat: Evolution, Patterns, And Consequences. . Smil, Vaclav. Https://Doi.Org/10.1111/J.1728-4457.2002.00599.X, S.L. : Wiley Online Library, 01 27, 2004, Population And Development Review, Vol. 28, Pp. 599-639.
208. Wallace-Wells, David. The Uninhabitable Earth: Life After Warming. New York : Tim Duggan Books, 2019.
209. Eclac (Economic Commission For Latin America And The Caribbean). Climate Change In Central America: Potential Impacts And Public Policy Options. Mexico City : United Nations , 2018.
210. Sethi, Nitin. Cereal Prices May Rise 29% By 2050 Because Of Climate Change: Leaked Un Report. Science: The Wire. [Online] Science: The Wire, 07 16, 2019. [Cited: 05 15, 2021.] Https://Science.Thewire.In/Environment/Climate-Change-Cereal-Prices-Un-Report/.
211. Bounoua, Lahouari. Climate Change Is Hitting African Farmers The Hardest Of All. [Online] The Conversation, 05 12, 2015. [Cited: 06 24, 2019.] Https://Theconversation.Com/Climate-Change-Is-Hitting-African-Farmers-The-Hardest-Of-All-40845.
212. Stone, Madelein. A Plague Of Locusts Has Descended On East Africa. Climate Change May Be To Blame. National Geographic. [Online] National Geographic, 02 15, 2020. [Cited: 05 15, 2020.] Https://Www.Nationalgeographic.Com/Science/Article/Locust-Plague-Climate-Science-East-Africa.
213. Wright, Oliver. Britain's Food Self-Sufficiency At Risk From Reliance On Overseas Imports Of Fruit And Vegetables That Could Be Produced At Home. [Online] The Independent, 07 01, 2014. [Cited: 07 16, 2019.] Https://Www.Independent.Co.Uk/News/Uk/Politics/Britains-Food-Self-Sufficiency-At-Risk-From-Reliance-On-Overseas-Imports-Of-Fruit-And-Vegetables-9574238.Html.
214. Lama, Pravhat. Japan's Food Security Problem: Increasing Self-Sufficiency In Traditional Food. [Online] Medium, 07 27, 2017. [Cited: 07 16, 2019.] Https://Medium.Com/Indrastra/Japans-Food-Security-Problem-Increasing-Self-Sufficiency-In-Traditional-Food-F48937A757C5.
215. Gustin, Georgina. Climate Change Could Lead To Major Crop Failures In World's Biggest Corn Regions. [Online] Inside Climate News, 06 11, 2018. [Cited: 01 30, 2020.] Https://Insideclimatenews.Org/News/11062018/Climate-Change-Research-Food-Security-Agriculture-Impacts-Corn-Vegetables-Crop-Prices.
216. Butler, Hilary Osborne And Sarah. From Milk To Crisps: Why The Price Of Basic Food Items Is Rising. The Guardian. [Online] The Guardian, 01 29, 2022. [Cited: 02 05, 2022.] Https://Www.Theguardian.Com/Business/2022/Jan/29/From-Milk-To-Crisps-Why-The-Price-Of-Basic-Food-Items-Is-Rising.
217. Clemens Breisinger, David Laborde Debucquet, Joseph Glauber Et Al. Russia-Ukraine Conflict Is Driving Up Wheat Prices: This Could Fuel Instability In Sudan. The Conversation. [Online] The Conversation, 04 13, 2022. [Cited: 04 30, 2022.] Https://Thewire.In/Agriculture/Wheat-Harvest-Punjab-Haryana-Export.
218. Ap. Wheat, Corn Prices Surge Deepening Consumer Pain . Al Jazeera. [Online] Al Jazeera, 03 02, 2022. [Cited: 04 30, 2022.] Https://Www.Aljazeera.Com/Economy/2022/3/3/Wheat-Corn-Prices-Surge-As-Consumer-Pain-Mounts.
219. Ghosal, Aniruddha. Heat Wave Scorches India's Wheat Crop, Snags Export Plans . Ap News. [Online] Ap News, 04 30, 2022. [Cited: 04 30, 2022.] Https://Apnews.Com/Article/Russia-Ukraine-Science-Business-India-Global-Trade-4d32889d982bf0a60396ff4ba817ca16.
220. Brendan Rice, Manuel A. Hernández, Joseph Glauber And Rob Vos. The Russia-Ukraine War Is Exacerbating International Food Price Volatility. International Food Policy Research Institute . [Online] International Food Policy Research Institute , 03 30, 2022. [Cited: 04 30, 2022.] Https://Www.Ifpri.Org/Blog/Russia-Ukraine-War-Exacerbating-International-Food-Price-Volatility.
221. Roberts, Freya. What The New Ipcc Report Says About Sea Level Rise. [Online] Carbonbrief, 10 03, 2013. [Cited: 01 15, 2020.] Https://Www.Carbonbrief.Org/What-The-New-Ipcc-Report-Says-About-Sea-Level-Rise.
222. Liberatore, Stacy. New York And London Could Be Underwater Within Decades: Scientists Say Devastating Climate Change Will Take Place Sooner Than Thought. [Online] Daily Mail, 03 22, 2016. [Cited: 01 15, 2020.] Https://Www.Dailymail.Co.Uk/Sciencetech/Article-3504667/New-York-London-Underwater-Decades-Scientists-Say-Devastating-Climate-Change-Place-Sooner-Thought.Html.
223. Holden, Emily. Flooding Will Affect Double The Number Of People Worldwide By 2030. The Guardian. [Online] The Guardian, 04 23, 2020. [Cited: 05 15, 2021.] Https://Www.Theguardian.Com/Environment/2020/Apr/23/Flooding-Double-Number-People-Worldwide-2030.
224. Millman, Oliver. Us To Have Major Floods On Daily Basis Unless Sea-Level Rise Is Curbed – Study. The Guardian. [Online] The Guardian, 04 16, 2020. [Cited: 05 15, 2021.] Https://Www.Theguardian.Com/Environment/2020/Apr/16/Us-Climate-Change-Floods-Sea-Level-Rise.

225. Josh Holder, Niko Kommenda And Jonathan Watts. The Three-Degree World: The Cities That Will Be Drowned By Global Warming. [Online] The Guardian, 11 03, 2017. [Cited: 01 15, 2020.] Https://Www.Theguardian.Com/Cities/Ng-Interactive/2017/Nov/03/Three-Degree-World-Cities-Drowned-Global-Warming.

226. Voosen, Paul. At Many River Deltas, Scientists Are Missing A Major Source Of Sea Level Rise. [Online] Science, 01 30, 2019. [Cited: 01 15, 2020.] Https://Www.Sciencemag.Org/News/2019/01/Many-River-Deltas-Scientists-Are-Missing-Major-Source-Sea-Level-Rise.

227. National Snow And Ice Data Center. Quick Facts On Ice Sheets. [Online] National Snow And Ice Data Center. [Cited: 01 15, 2020.] Https://Nsidc.Org/Cryosphere/Quickfacts/Icesheets.Html.

228. Pa. Media. Third Of Antarctic Ice Shelves 'Will Collapse Amid 4c Global Heating'. The Guardian. [Online] The Guardian, 04 08, 2021. [Cited: 05 15, 2021.] Https://Www.Theguardian.Com/World/2021/Apr/08/Third-Of-Antarctic-Ice-Shelves-Will-Collapse-Amid-4c-Global-Heating?Fbclid=Iwar1t5n93mip9krrj-Oacndr61ssptiumbxao2ptch9rn-2xvxuyxo5oipvc.

229. Arenschield, Laura. Warming Greenland Ice Sheet Passes Point Of No Return. Phys.Org. [Online] Phys.Org, 08 13, 2020. [Cited: 05 15, 2021.] Https://Phys.Org/News/2020-08-Greenland-Ice-Sheet.Html?Fbclid=Iwar0bg7weg8fholoxifgfia6_Gnmchatukblmtg_Bsspcsa-b2jn1-j7sfbw.

230. Milman, Oliver. Climate Guru James Hansen Warns Of Much Worse Than Expected Sea Level Rise. [Online] The Guardian, 03 22, 2016. [Cited: 01 15, 2020.] Https://Www.Theguardian.Com/Science/2016/Mar/22/Sea-Level-Rise-James-Hansen-Climate-Change-Scientist.

231. Climate Central. Land Projected To Be Below Annual Flood Level In 2030. Climate Central. [Online] Climate Central. [Cited: 05 15, 2021.] Https://Coastal.Climatecentral.Org/Map.

232. World Health Organization. Air Pollution. [Online] World Health Organization, 2019. [Cited: 06 26, 2019.] Https://Www.Who.Int/Airpollution/En/.

233. Carrington, Damien. Action On Air Pollution Works But Far More Is Needed, Study Shows. [Online] The Guardian, 06 26, 2019. [Cited: 06 28, 2019.] Https://Www.Theguardian.Com/Environment/2019/Jun/26/Action-On-Air-Pollution-Works-But-Far-More-Is-Needed-Study-Shows.

234. Roser, Hannah Ritchie And Max. Air Pollution. [Online] Our World In Data, 04 2019. [Cited: 06 27, 2019.] Https://Ourworldindata.Org/Air-Pollution.

235. Wu, Huizhong. Breathing In Delhi Air Equivalent To Smoking 44 Cigarettes A Day. [Online] Cnn, 11 10, 2017. [Cited: 06 28, 2019.] Https://Edition.Cnn.Com/2017/11/10/Health/Delhi-Pollution-Equivalent-Cigarettes-A-Day/Index.Html.

236. Forrest, Adam. Climate Change Likely To Make Us More Stupid, Study Finds. [Online] The Independent, 12 9, 2018. [Cited: 06 28, 2019.] Https://Www.Independent.Co.Uk/Environment/Climate-Change-Carbon-Dioxide-Intelligence-Greenhouse-Gas-More-Stupid-Ucl-Study-A8674706.Html.

237. Bridle, James. Air Pollution Rots Our Brains. Is That Why We Don't Do Anything About It? [Online] The Guardian, 09 24, 2018. [Cited: 06 28, 2019.] Https://Www.Theguardian.Com/Commentisfree/2018/Sep/24/Air-Pollution-Cognitive-Improvement-Environment.

238. Tatchell, Peter. The Oxygen Crisis. [Online] The Guardian, 08 13, 2008. [Cited: 06 28, 2019.] Https://Www.Theguardian.Com/Commentisfree/2008/Aug/13/Carbonemissions.Climatechange.

239. Ramsey, Lydia. These Are The Top 15 Deadliest Animals On Earth. [Online] Science Alert, 02 23, 2018. [Cited: 07 01, 2019.] Https://Www.Sciencealert.Com/What-Are-The-Worlds-15-Deadliest-Animals.

240. U.N. What Is Malaria? U.N. [Online] U.N., 05 23, 2022. [Cited: 05 28, 2022.] Https://Www.Un.Org/Ar/Node/178575#:-:Text=According%20to%20a%20world%20health%20organizations's,More%20deaths%20co mpared%20to%202020..

241. Jordan, Rob. Stanford Researchers Explore The Effects Of Climate Change On Disease. [Online] Stanford News, 03 15, 2019. [Cited: 07 01, 2009.] Https://News.Stanford.Edu/2019/03/15/Effect-Climate-Change-Disease/.

242. Pappas, Stephanie. 5 Deadly Diseases Emerging From Global Warming. [Online] Live Science, 08 03, 2016. [Cited: 07 01, 2019.] Https://Www.Livescience.Com/55632-Deadly-Diseases-Emerge-From-Global-Warming.Html.

243. Barkham, Patrick. Study Finds Ticks Choose Humans Over Dogs When Temperature Rises. The Guardian. [Online] The Guardian, 11 16, 2020. [Cited: 05 15, 2021.] Https://Www.Theguardian.Com/Science/2020/Nov/16/Study-Finds-Ticks-Choose-Humans-Over-Dogs-When-Temperature-Rises?

244. Fox-Skelly, Jasmin. There Are Diseases Hidden In Ice, And They Are Waking Up. [Online] Bbc, 05 04, 2017. [Cited: 07 01, 2019.] Http://Www.Bbc.Com/Earth/Story/20170504-There-Are-Diseases-Hidden-In-Ice-And-They-Are-Waking-Up.

245. Harari, Yuval Noah. Sapiens : A Brief History Of Humankind. Toronto : Mcclelland & Stewart, 2014.

246. Donnelly, Jim. The Irish Famine. [Online] Bbc History, 02 17, 2011. [Cited: 07 02, 2019.] Http://Www.Bbc.Co.Uk/History/British/Victorians/Famine_01.Shtml.

247. Iom. World Migration Report 2022. Geneva : U.N. Migration, 2022. Https://Publications.Iom.Int/Books/World-Migration-Report-2022.

248. Amnesty International. Syria's Refugee Crisis In Numbers. [Online] Amnesty International, 02 03, 2016. [Cited: 07 02, 2019.] Https://Www.Amnesty.Org/En/Latest/News/2016/02/Syrias-Refugee-Crisis-In-Numbers/.

249. Baker, Aryn. How Climate Change Is Behind The Surge Of Migrants To Europe. [Online] Time, 09 07, 2015. [Cited: 07 02, 2019.] Https://Time.Com/4024210/Climate-Change-Migrants/.

250. U.S Customs And Border Protection. Southwest Border Migration Fy 2019. [Online] U.S Customs And Border Protection, 06 07, 2019. [Cited: 07 02, 2019.] Https://Www.Cbp.Gov/Newsroom/Stats/Sw-Border-Migration.

251. Raff, Jeremy. What A Pediatrician Saw Inside A Border Patrol Warehouse. [Online] The Atlantic, 07 03, 2019. [Cited: 07 04, 2019.] Https://Www.Theatlantic.Com/Politics/Archive/2019/07/Border-Patrols-Oversight-Sick-Migrant-Children/593224/.

252. The Star. On The Fourth Of July, America Confronts A Moral Crisis On Its Border. [Online] The Star, 07 03, 2019. [Cited: 07 04, 2019.] Https://Www.Thestar.Com/Opinion/Editorials/2019/07/03/On-The-Fourth-Of-July-America-Confronts-A-Moral-Crisis-On-Its-Border.Html.
253. Withers, Rachel. Trump Is Right That "Much Can Be Learned" From Australia's Immigration Policies. [Online] Slate, 07 03, 2019. [Cited: 07 04, 2019.] Https://Slate.Com/News-And-Politics/2019/07/Trump-Morrison-Australia-Immigration-Manus-Nauru.Html.
254. Wikipedia. List Of Countries By Refugee Population. [Online] Wikipedia, 2019. [Cited: 07 04, 2019.] Https://En.Wikipedia.Org/Wiki/List_Of_Countries_By_Refugee_Population.
255. Environmental Justice Foundation. Climate Displacement In Bangladesh. *Environmental Justice Foundation.* [Online] Environmental Justice Foundation. [Cited: 05 15, 2021.] Https://Ejfoundation.Org/What-We-Do/On-The-Front-Line-Of-Climate-Change.
256. Fosco, Molly. Why India Is Building The Longest Border Wall In The World. *Seeker.* [Online] Seeker, 04 18, 2015. [Cited: 05 15, 2021.] Https://Www.Seeker.Com/Why-India-Is-Building-The-Longest-Border-Wall-In-The-World-1501495427.Html.
257. World Bank Group. Climate Change Could Force Over 140 Million To Migrate Within Countries By 2050: World Bank Report. [Online] World Bank Group, 03 19, 2018. [Cited: 07 02, 2019.] Https://Www.Worldbank.Org/En/News/Press-Release/2018/03/19/Climate-Change-Could-Force-Over-140-Million-To-Migrate-Within-Countries-By-2050-World-Bank-Report.
258. Adamo, Elizabeth Warn And Susana B. The Impact Of Climate Change: Migration And Cities In South America. [Online] World Meterological Organization, 2014. [Cited: 07 03, 2019.] Https://Public.Wmo.Int/En/Resources/Bulletin/Impact-Of-Climate-Change-Migration-And-Cities-South-America.
259. Al-Arian, Ruth Sherlock And Lama. Migrants Captured In Libya Say They End Up Sold As Slaves. [Online] National Public Radio, 03 21, 2018. [Cited: 07 02, 2019.] Https://Www.Npr.Org/Sections/Parallels/2018/03/21/595497429/Migrants-Passing-Through-Libya-Could-End-Up-Being-Sold-As-Slaves?
260. Cbs News. Libya Airstrike Hits Migrant Detention Center In Tripoli, Killing Scores. [Online] Cbs News, 07 03, 2019. [Cited: 07 03, 2019.] Https://Www.Cbsnews.Com/News/Libya-Airstrike-Tripoli-Migrant-Detention-Center-Today-Kills-Scores-Khalifa-Haftar-Lna-2019-07-03/.
261. Taylor, Matthew. Climate Change 'Will Create World's Biggest Refugee Crisis'. [Online] The Guardian, 11 02, 2017. [Cited: 07 03, 2019.] Https://Www.Theguardian.Com/Environment/2017/Nov/02/Climate-Change-Will-Create-Worlds-Biggest-Refugee-Crisis.
262. Hagel, Chuck. Department Of Defence. [Online] 10 13, 2014. [Cited: 07 22, 2019.] Https://Dod.Defense.Gov/News/Speeches/Speech-View/Article/605617/.
263. Bloomberg Nef. Ex-Military Brass Back Tillerson, Mattis In Climate Fight. *Bloomberg Nef.* [Online] Bloomberg Nef, 05 08, 2017. [Cited: 05 16, 2021.] Https://About.Bnef.Com/Blog/Ex-Military-Brass-Back-Tillerson-Mattis-In-Climate-Change-Fight/.
264. Bergengruen, Vera. Trump May Doubt Climate Change, Pentagon Sees It As Threat Multiplier. [Online] Military, 06 02, 2017. [Cited: 07 24, 2019.] Https://Www.Military.Com/Daily-News/2017/06/02/Trump-May-Doubt-Climate-Change-Pentagon-Sees-It-Looming-Threat.Html.
265. Stanmeyer, John. National Geographic. [Online] 03 02, 2015. [Cited: 07 23, 2019.] Https://News.Nationalgeographic.Com/News/2015/03/150302-Syria-War-Climate-Change-Drought/.
266. Borger, Julian. The Guardian. [Online] Darfur Conflict Heralds Era Of Wars Triggered By Climate Change, Un Report Warns, 06 23, 2007. [Cited: 07 23, 2019.] Https://Www.Theguardian.Com/Environment/2007/Jun/23/Sudan.Climatechange.
267. Stanford News. Global Warming Increases Risk Of Civil War In Africa, Stanford Researchers Say. *Stanford News.* [Online] Stanford News, 11 23, 2019. [Cited: 05 30, 2022.] Https://News.Stanford.Edu/News/2009/November23/Climate-Civil-Wars-112309.Html.
268. *Warming Climate Drives Human Conflict.* Morello, Lauren. S.L. : Nature, 08 01, 2013. Https://Doi.Org/10.1038/Nature.2013.13464.
269. Illing, Sean. How Climate Change Could Lead To More Wars In The 21st Century. [Online] Vox, 11 14, 2017. [Cited: 07 23, 2019.] Https://Www.Vox.Com/World/2017/11/14/16589878/Global-Climate-Change-Conflict-Environment.
270. Wendle, John. When Climate Change Starts Wars. [Online] Nautius, 02 16, 2017. [Cited: 07 23, 2019.] Http://Nautil.Us/Issue/45/Power/When-Climate-Change-Starts-Wars.
271. Committee On Foreign Relations. 2011. Washington : 22, 02.
272. Arsenault, Chris. Risk Of Water Wars Rises With Scarcity. [Online] Aljazeera, 07 26, 2012. [Cited: 07 24, 019.] Https://Www.Aljazeera.Com/Indepth/Features/2011/06/2011622193147231653.Html.
273. Caitlin Werrell, Francesco Femia. A Storm Without Rain: Yemen, Water, Climate Change, And Conflict. [Online] Home, 08 03, 2016. [Cited: 07 24, 2019.] Https://Climateandsecurity.Org/2016/08/03/A-Storm-Without-Rain-Yemen-Water-Climate-Change-And-Conflict/.
274. Carrington, Damian. Tackle Climate Or Face Financial Crash, Say World's Biggest Investors. [Online] The Guardian, 12 10, 2018. [Cited: 12 12, 2019.] Https://Www.Theguardian.Com/Environment/2018/Dec/10/Tackle-Climate-Or-Face-Financial-Crash-Say-Worlds-Biggest-Investors.
275. Costa, Pedro Nicolaci Da. How Climate Change Could Trigger The Next Financial Crisis. [Online] Market Watch, 04 22, 2019. [Cited: 12 12, 2019.] Https://Www.Marketwatch.Com/Story/How-Climate-Change-Could-Trigger-The-Next-Financial-Crisis-2019-04-18.
276. Kurmelovs, Royce. Fire And Flood: 'Whole Areas Of Australia Will Be Uninsurable'. *The Guardian.* [Online] The Guardian, 04 01, 2021. [Cited: 05 16, 2021.] Https://Www.Theguardian.Com/Australia-News/2021/Apr/02/Fire-And-Flood-Whole-Areas-Of-Australia-Will-Be-Uninsurable.

277. Brown, Courtenay. Climate Change Rains Insurance Misery On Homeowners. [Online] Axios, 11 26, 2019. [Cited: 12 18, 2019.] Https://Www.Axios.Com/Climate-Change-Home-Insurance-F07655b5-969f-4523-8042-c546ca997801.Html.

278. Marcacci, Silvio. The Global Insurance Industry's $6 Billion Existential Threat: Coal Power. [Online] Forbes, 05 22, 2019. [Cited: 12 18, 2019.] Https://Www.Forbes.Com/Sites/Energyinnovation/2019/05/22/The-Global-Insurance-Industrys-6-Billion-Existential-Threat-Coal-Power/#78184cf963c1.

279. Quiggin, John. Blackrock Is The Canary In The Coalmine. Its Decision To Dump Coal Signals What's Next. The Conversation. [Online] The Conversation, 01 16, 2020. [Cited: 05 16, 2020.] Https://Theconversation.Com/Blackrock-Is-The-Canary-In-The-Coalmine-Its-Decision-To-Dump-Coal-Signals-Whats-Next-129972?Fbclid=Iwar3_Xsxnp_z0fcrgoanwrrothywodlbhub7636fniewb6aiimjmklc8b-Cu.

280. Colman, Zack. How Climate Change Could Spark The Next Home Mortgage Disaster. Politico. [Online] Politico, 11 30, 2020. [Cited: 05 16, 2021.] Https://Www.Politico.Com/News/2020/11/30/Climate-Change-Mortgage-Housing-Environment-433721?Utm_Campaign=Carbon%20brief%20daily%20briefing&Utm_Content=20201201&Utm_Medium=Email&Utm_Source=Re vue%20daily.

281. Daily Mail City & Finance Reporter. Investors Face £33trn Climate Change Threat: Fund Manager Warns Shares Could Fall 30% If Global Warming Is Not Controlled. This Is Money. [Online] This Is Money, 12 19, 2019. [Cited: 05 16, 2021.] Https://Www.Thisismoney.Co.Uk/Money/Markets/Article-7838217/Investors-Face-33trn-Climate-Change-Threat-Shares-Fall-30.Html.

282. Inman, Fiona Harvey And Phillip. Global Economic Growth Will Take Big Hit Due To Loss Of Nature. The Guardian. [Online] The Guardian, 02 12, 2020. [Cited: 05 16, 2021.] Https://Www.Theguardian.Com/Environment/2020/Feb/12/Global-Economic-Growth-Will-Take-Big-Hit-Due-To-Loss-Of-Nature?

283. Kemp, Luke. Are We On The Road To Civilisation Collapse? [Online] Bbc Future, 02 19, 2019. [Cited: 12 18, 2019.] Https://Www.Bbc.Com/Future/Article/20190218-Are-We-On-The-Road-To-Civilisation-Collapse.

284. Diamond, Jared. Collapse: How Societies Choose To Fail Or Succeed. New York : Penguin Books, 2005. 0-670-03337-5.

285. Ahmed, Nafeez. U.S. Military Could Collapse Within 20 Years Due To Climate Change, Report Commissioned By Pentagon Says. [Online] Vice, 10 24, 2019. [Cited: 12 17, 2019.] Https://Www.Vice.Com/En_Us/Article/Mbmkz8/Us-Military-Could-Collapse-Within-20-Years-Due-To-Climate-Change-Report-Commissioned-By-Pentagon-Says.

286. Doré, Louis. Society Will Collapse By 2040 Due To Catastrophic Food Shortages, Says Study. [Online] The Independent, 06 22, 2015. [Cited: 12 17, 2019.] Https://Www.Independent.Co.Uk/Environment/Climate-Change/Society-Will-Collapse-By-2040-Due-To-Catastrophic-Food-Shortages-Says-Study-10335606.Html.

287. David Spratt, Ian Dunlop. Existential Climate-Related Security Risk: A Scenario Approach. [Online] Breakthrough - National Centre For Climate Restoration, 05 2019. [Cited: 12 17, 2019.] Https://Docs.Wixstatic.Com/Ugd/148cbo_b2c0c79dc4344b279bcf2365336ff23b.Pdf.

288. Turner, Dr Graham M. Is Global Collapse Imminent? An Updated Comparison Of The Limits To Growth With Historical Data. Melbourne : Melbourne Sustainable Society Institute, 2014. 978 0 7340 4940 7.

289. Harvey, Fiona. Humanity Under Threat From Perfect Storm Of Crises – Study. The Guardian. [Online] The Guardian, 02 06, 2020. [Cited: 05 16, 2021.] Https://Www.Theguardian.Com/Environment/2020/Feb/06/Humanity-Under-Threat-Perfect-Storm-Crises-Study-Environment?Utm_Campaign=Carbon%20brief%20daily%20briefing&Utm_Medium=Email&Utm_Source=Revue%20newsletter.

290. Future Earth. Risks Perceptions Report 2020. S.L. : Future Earth, 2020. Https://Futureearth.Org/Wp-Content/Uploads/2020/02/Rpr_2020_Report.Pdf.

291. Ahmed, Nafeez. The Collapse Of Civilization May Have Already Begun. Vice. [Online] Vice, 11 22, 2019. [Cited: 12 17, 2019.] Https://Www.Vice.Com/En_Us/Article/8xwygg/The-Collapse-Of-Civilization-May-Have-Already-Begun.

292. Bendell, Jim. Deep Adaptation: A Map For Navigating Climate Tragedy. S.L. : Www.Iflas.Info, 2018. P. 07. Https://Www.Lifeworth.Com/Deepadaptation.Pdf.

293. Jatan. Japan's Tropical Timber Imports-An Overview. [Online] Jatan, 1999. [Cited: 12 18, 2019.] Http://Jatan.Org/Eng/Tropicaltimber.Html.

294. Kubokawa, Kenta. Hardwood Report 2018 Japan. 28 : 09, 2018.

295. Environmental Investigation Agency. Japan. [Online] Environmental Investigation Agency, 09 09, 2018. [Cited: 12 18, 2019.] Https://Eia-Global.Org/Subinitiatives/Japan-Forests.

296. Mcdonald, Joshua. The Island With No Water: How Foreign Mining Destroyed Banaba. The Guardian. [Online] The Guardian, 06 08, 2021. [Cited: 06 09, 2021.] Https://Www.Theguardian.Com/World/2021/Jun/09/The-Island-With-No-Water-How-Foreign-Mining-Destroyed-Banaba.

297. Wikipedia. Electric Energy Consumption. Wikipedia. [Online] Wikipedia. [Cited: 05 31, 2022.] Https://En.Wikipedia.Org/Wiki/Electric_Energy_Consumption.

298. World Nuclear Association. Nuclear Power In The World Today. World Nuclear Association. [Online] World Nuclear Association, 05 2022. [Cited: 05 31, 2022.] Https://World-Nuclear.Org/Information-Library/Current-And-Future-Generation/Nuclear-Power-In-The-World-Today.Aspx#:-:Text=Around%2010%25%20of%20the%20world's,From%202657%20twh%20in%202019..

299. Wikipedia. Nuclear And Radiation Accidents And Incidents. [Online] Wikipedia, 10 28, 2019. [Cited: 10 28, 2019.] Https://En.Wikipedia.Org/Wiki/Nuclear_And_Radiation_Accidents_And_Incidents.

300. Chen, Christina. Nuclear Vs. Climate Change: Rising Seas. [Online] Natural Resources Defense Council, 11 16, 2019. [Cited: 12 10, 2019.] Https://Www.Nrdc.Org/Experts/Christina-Chen/Nuclear-Vs-Climate-Change-Rising-Seas.

301. Vidal, John. What Are Nuclear Power Plants Doing To Address Climate Threats? [Online] Beyond Nuclear International, 09 16, 2018. [Cited: 12 10, 2019.] Https://Beyondnuclearinternational.Org/2018/09/16/What-Are-Nuclear-Power-Plants-Doing-To-Address-Climate-Threats/.

302. Perkins, Robert. Fukushima Disaster Was Preventable, New Study Finds. [Online] University Of Southern California, 09 21, 2015. [Cited: 12 11, 2019.] Https://News.Usc.Edu/86362/Fukushima-Disaster-Was-Preventable-New-Study-Finds/.

303. Goodell, Jeff. The Water Will Come. New York : Little Brown, 2017.

304. Mccurry, Justin. Fukushima 50: 'We Felt Like Kamikaze Pilots Ready To Sacrifice Everything'. [Online] The Guardian, 01 11, 2013. [Cited: 12 12, 2019.] Https://Www.Theguardian.Com/Environment/2013/Jan/11/Fukushima-50-Kamikaze-Pilots-Sacrifice.

305. Weyler, Rex. Nuclear Power And The Collapse Of Society. [Online] Rex Weyler, 05 05, 2017. [Cited: 12 12, 2019.] Https://Www.Rexweyler.Ca/Ecologue/2017/9/11/Nuclear-Power-And-The-Collapse-Of-Society.

306. Woodward, Aylin. More Than 50% Of Insects Have Disappeared Since 1970, An Ecologist Warns — Even More Evidence Of An 'Insect Apocalypse'. [Online] Business Insider, 11 15, 2019. [Cited: 01 09, 2020.] Https://Www.Businessinsider.Com/Insect-Apocalypse-Ecosystem-Collapse-Food-Insecurity-2019-11.

307. Carrington, Damian. 'Insect Apocalypse' Poses Risk To All Life On Earth, Conservationists Warn. [Online] Guardian, 11 13, 2019. [Cited: 12 20, 2019.] Https://Www.Theguardian.Com/Environment/2019/Nov/13/Insect-Apocalypse-Poses-Risk-To-All-Life-On-Earth-Conservationists-Warn.

308. —. Plummeting Insect Numbers 'Threaten Collapse Of Nature'. [Online] Guarian, 02 10, 2019. [Cited: 12 20, 2019.] Https://Www.Theguardian.Com/Environment/2019/Feb/10/Plummeting-Insect-Numbers-Threaten-Collapse-Of-Nature.

309. Dunne, Daisy. Wildlife 6 February 2020 19:00 Climate Change Driving 'Rapid And Widespread' Decline Of Bumblebees. Carbon Brief. [Online] Carbon Brief, 06 02, 2020. [Cited: 05 16, 2021.] Https://Www.Carbonbrief.Org/Climate-Change-Driving-Rapid-And-Widespread-Decline-Of-Bumblebees?Utm_Campaign=Carbon%20brief%20daily%20briefing&Utm_Medium=Email&Utm_Source=Revue%20newsletter.

310. Jarvis, Brooke. The Insect Apocalypse Is Here: What Does It Mean For The Rest Of Life On Earth? [Online] New York Times, 11 27, 2018. [Cited: 12 20, 2019.] Https://Www.Nytimes.Com/2018/11/27/Magazine/Insect-Apocalypse.Html.

311. Carrington, Damian. Nsect Numbers Down 25% Since 1990, Global Study Finds. The Guardian. [Online] The Guardian, 04 23, 2020. [Cited: 05 16, 2021.] Https://Www.Theguardian.Com/Environment/2020/Apr/23/Insect-Numbers-Down-25-Since-1990-Global-Study-Finds.

312. Hunt, Katie. Fireflies Are Facing Extinction Due To Habitat Loss, Pesticides And Artificial Light. Cnn World. [Online] Cnn World, 02 03, 2020. [Cited: 05 16, 2021.] Https://Edition.Cnn.Com/2020/02/03/World/Fireflies-Extinction-Risk-Scn/Index.Html?

313. United Nations, Department Of Economic And Social Affairs, Population Division. The Speed Of Urbanization Around The World. [Online] 12 2018. [Cited: 01 07, 2020.] Https://Population.Un.Org/Wup/Publications/Files/Wup2018-Popfacts_2018-1.Pdf.

314. Carrington, Damian. Light Pollution Is Key 'Bringer Of Insect Apocalypse'. [Online] 11 22, 2019. [Cited: 01 07, 2020.] Https://Www.Theguardian.Com/Environment/2019/Nov/22/Light-Pollution-Insect-Apocalypse.

315. Laurance, Bill. Climate Change Is Killing Off Earth's Insects. [Online] Stuff, 02 13, 2019. [Cited: 01 08, 2020.] Https://Www.Stuff.Co.Nz/Environment/Climate-News/110570237/Climate-Change-Is-Killing-Off-Earths-Insects.

316. Guarino, Ben. 'Hyperalarming' Study Shows Massive Insect Loss. [Online] Washington Post, 10 16, 2018. [Cited: 10 16, 2020.] Https://Www.Washingtonpost.Com/Science/2018/10/15/Hyperalarming-Study-Shows-Massive-Insect-Loss/.

317. Tropical Forests Were The Primary Sources Of New. H. K. Gibbsa, A. S. Rueschb Et Al. 38, 2010, Pnas, Vol. 107, Pp. 16732–16737.

318. Ritchie, Hannah. Half Of The World's Habitable Land Is Used For Agriculture. [Online] Our World In Data, 11 11, 2019. [Cited: 01 08, 2020.] Https://Ourworldindata.Org/Global-Land-For-Agriculture.

319. Conis, Elena. Beyond Silent Spring: An Alternate History Of Ddt. [Online] Science History, 02 14, 2017. [Cited: 01 09, 2020.] Https://Www.Sciencehistory.Org/Distillations/Beyond-Silent-Spring-An-Alternate-History-Of-Ddt.

320. Pesticide Action Network. The Ddt Story. [Online] Pesticide Action Network. [Cited: 01 09, 2020.] Https://Www.Panna.Org/Resources/Ddt-Story.

321. Ddt - A Brief History And Status. [Online] Epa, 08 11, 2017. [Cited: 01 09, 2020.] Https://Www.Epa.Gov/Ingredients-Used-Pesticide-Products/Ddt-Brief-History-And-Status.

322. Johnson, Jake. Alarming Decline Of Insect Population Linked To Toxic Pesticides In U.S. Agriculture. [Online] Ecowatch, 08 07, 2019. [Cited: 01 10, 2020.] Https://Www.Ecowatch.Com/Insect-Population-Pesticides-Agriculture-2639705712.Html?Rebelltitem=3#Rebelltitem3.

323. An Assessment Of Acute Insecticide Toxicity Loading (Aitl) Of Chemical Pesticides Used On Agricultural Land In The United States . Michael Dibartolomeis, Susan Kegley ,Pierre Mineau,Rosemarie Radford And Kendra Klein Et Al. 8, S.L. : Plos One, 02 14, 2019, Vol. 14. Https://Doi.Org/10.1371/Journal.Pone.0220029.

324. Fao. Global Plans Of Action Endorsed To Halt The Escalating Degradation Of Soils. [Online] Fao, 07 24, 2014. [Cited: 01 12, 2020.] Http://Www.Fao.Org/News/Story/En/Item/239341/Icode/.

325. Campaign To Protect Rural England. Back To The Land: Rethinking Our Approach To Soil. [Online] Campaign To Protect Rural England, 12 2018. [Cited: 01 12, 2020.] Https://Static1.Squarespace.Com/Static/58cff61c414fb598d9e947ca/T/5c1b6cedb8a0450f181fb1dd/1545301245853/Cpre_F%26f3_Soil_26nov_Web.Pdf.

326. Environment Agency. The State Of The Environment: Soil. [Online] Environment Agency, 06 2019. [Cited: 01 12, 2020.] Https://Assets.Publishing.Service.Gov.Uk/Government/Uploads/System/Uploads/Attachment_Data/File/805926/State_Of_Th e_Environment_Soil_Report.Pdf.

327. Cho, Renee. Can Soil Help Combat Climate Change? [Online] State Of The Planet: Earth Institute, Columbia University, 02 21, 2018. [Cited: 01 12, 2020.] Https://Blogs.Ei.Columbia.Edu/2018/02/21/Can-Soil-Help-Combat-Climate-Change/.

328. The Role Of Soil Carbon In Natural Climate Solutions. D. A. Bossio, S. C. Cook-Patton, P. W. Ellis Etal. S.L. : Nature, 2020, Nature Sustainability, Vol. 3, Pp. 391-398.

329. Rao, Sailesh. Animal Agriculture Is The Leading Cause Of Climate Change - A Position Paper. Pune : Journal Of Ecological Society, 2021. Https://Climatehealers.Org/Wp-Content/Uploads/2021/04/Jes-Rao.Pdf.

330. Cosier, Susan. The World Needs Topsoil To Grow 95% Of Its Food – But It's Rapidly Disappearing. [Online] The Guardian, 05 30, 2019. [Cited: 01 12, 2020.] Https://Www.Theguardian.Com/Us-News/2019/May/30/Topsoil-Farming-Agriculture-Food-Toxic-America.

331. Harrabin, Roger. Climate Change Being Fuelled By Soil Damage - Report. [Online] Bbc, 04 29, 2019. [Cited: 01 12, 2020.] Https://Www.Bbc.Com/News/Science-Environment-48043134.

332. Fao. International Year Of Soil Conference. [Online] Fao, 07 06, 2015. [Cited: 01 11, 2020.] Http://Www.Fao.Org/Soils-2015/Events/Detail/En/C/338738/.

333. World Wildlife Fund. Soil Erosion And Degradation. [Online] World Wildlife Fund. [Cited: 01 11, 2020.] Https://Www.Worldwildlife.Org/Threats/Soil-Erosion-And-Degradation.

334. Club Of Mozambique. 40,000 Hectares Lost Because Of Pesticide Abuse. [Online] Club Of Mozambique, 07 18, 2019. [Cited: 01 11, 202.] Https://Clubofmozambique.Com/News/40000-Hectares-Lost-Because-Of-Pesticide-Abuse-137071/?Fbclid=Iwar3frorj82mtjcs8hxgmorh7knbom_Ftsv1pvrbwhctqunpxt-Foqs3l_Ku.

335. Iucn. To Feed Future Generations, Countries Must Invest In Living Soils. [Online] Iucn, 06 15, 2018. [Cited: 01 11, 2020.] Https://Www.Iucn.Org/News/Secretariat/201806/Feed-Future-Generations-Countries-Must-Invest-Living-Soils.

336. Gray, Richard. Why Soil Is Disappearing From Farms. [Online] Bbc. [Cited: 01 11, 2020.] Http://Www.Bbc.Com/Future/Bespoke/Follow-The-Food/Why-Soil-Is-Disappearing-From-Farms/.

337. The Parliamentary Office Of Science And Technology. Uk Soil Degradation. [Online] The Parliamentary Office Of Science And Technology, 07 2006. [Cited: 01 11, 2020.] Https://Www.Parliament.Uk/Documents/Post/Postpn265.Pdf.

338. Unah, Linus. Erosion Crisis Swallows Homes And Livelihoods In Nigeria. [Online] Erosion Crisis Swallows Homes And Livelihoods In Nigeria, 01 20, 2020. [Cited: 01 22, 2020.] Https://Www.Climatechangenews.Com/2020/01/20/Erosion-Crisis-Swallows-Homes-Livelihoods-Nigeria/?Utm_Campaign=Carbon%20brief%20daily%20briefing&Utm_Medium=Email&Utm_Source=Revue%20newsletter.

339. Howe, Marvine. Ddt's Use Backed By Nobel Winner. [Online] The New York Times, 11 09, 1971. [Cited: 01 14, 2020.] Https://Www.Nytimes.Com/1971/11/09/Archives/Ddts-Use-Backed-By-Nobel-Winner-Borlaug-Denounces-Efforts-To-Ban.Html.

340. Pesticide Action Network. Myths & Facts. Pesticide Action Network. [Online] Pesticide Action Network. [Cited: 05 16, 2021.] Https://Www.Panna.Org/Pesticides-Big-Picture/Myths-Facts.

341. Philpott, Tom. A Brief History Of Our Deadly Addiction To Nitrogen Fertilizer. [Online] Mother Jones, 04 19, 2013. [Cited: 01 14, 2020.] Https://Www.Motherjones.Com/Food/2013/04/History-Nitrogen-Fertilizer-Ammonium-Nitrate/.

342. Fernández, Lucía. Global Fertilizer Consumption By Nutrient 1965-2019 . Statista. [Online] Fao, 11 29, 2021. [Cited: 01 30, 2022.] Https://Www.Statista.Com/Statistics/438967/Fertilizer-Consumption-Globally-By-Nutrient/.

343. University Of Southern Denmark. Pesticides Make The Life Of Earthworms Miserable. [Online] Sciencedaily, 03 25, 2014. [Cited: 01 15, 2020.] Https://Www.Sciencedaily.Com/Releases/2014/03/140325113232.Htm.

344. Giordano, Chiara. British Farmland Is Losing Its Worms Due To Intensive Ploughing With Some Fields Having None At All, Say Researchers. [Online] The Scottish Sun, 02 24, 2019. [Cited: 01 15, 2020.] Https://Www.Thescottishsun.Co.Uk/News/3921205/British-Farmland-Is-Losing-Its-Worms-Due-To-Intensive-Ploughing-With-Some-Fields-Having-None-At-All-Say-Researchers/.

345. Critical Decline Of Earthworms From Organic Origins. Blakemore, Robert J. 33, S.L. : Soil Systems, 2018, Soilsystems, Vol. 2, Pp. 2-28. Https://Doi.Org/10.3390/Soilsystems2020033.

346. Pesticides And Soil Invertebrates: A Hazard Assessment. Gunstone T, Cornelisse T, Klein K, Dubey A And Donley N. Lausanne : Front. Environ. Sci., 2021. Https://Doi.Org/10.3389/Fenvs.2021.643847.

347. Cassella, Carly. 2 °C Of Warming Could Open The Floodgates For 230 Billion Tons Of Carbon To Escape. Science Alert. [Online] Science Alert, 11 07, 2020. [Cited: 05 16, 2021.] Https://Www.Sciencealert.Com/2-C-Of-Warming-Could-Open-The-Floodgates-For-Billions-Of-Tons-Of-Soil-Carbon?Utm_Campaign=Applenews&Utm_Medium=Referral&Utm_Source=Applenews.

348. Wikipedia. Medicinal Plants. [Online] Wikipedia. [Cited: 01 16, 2020.] Https://En.Wikipedia.Org/Wiki/Medicinal_Plants.

349. Heal, Geoffrey. Climate Change Is Killing Millions Of Plants And Animals — And Untold Scientific Discoveries In The Process. [Online] The Hill, 07 11, 2019. [Cited: 01 20, 2020.] Https://Thehill.Com/Opinion/Energy-Environment/452617-Climate-Change-Is-Killing-Millions-Of-Plants-And-Animals-And.

350. Cooke, Justin. 9 Famous Examples Of Drugs That Came From Plants. [Online] Thesunlightexperiment, 06 22, 2018. [Cited: 01 20, 2020.] Https://Thesunlightexperiment.Com/Blog/2018/6/7/9-Famous-Examples-Of-Drugs-That-Came-From-Plants.

351. Fao. Ten Things You May Not Know About Forests. [Online] Fao, 09 13, 2017. [Cited: 01 24, 2020.] Http://Www.Fao.Org/Zhc/Detail-Events/En/C/1033884/.

352. Marshall, Michael. First Land Plants Plunged Earth Into Ice Age. [Online] New Scientist, 02 01, 2012. [Cited: 01 16, 2020.] Https://Www.Newscientist.Com/Article/Dn21417-First-Land-Plants-Plunged-Earth-Into-Ice-Age/.

353. Mongabay. Rare Plant Species Are Especially Vulnerable To Climate Change, And Rarity Is More Common Than Previously Understood. *Mongabay*. [Online] Mongabay, 01 09, 2020. [Cited: 05 16, 2021.] Https://News.Mongabay.Com/2020/01/Rare-Plant-Species-Are-Especially-Vulnerable-To-Climate-Change-And-Rarity-Is-More-Common-Than-Previously-Understood/.

354. Carrington, Damian. 'Frightening' Number Of Plant Extinctions Found In Global Survey. [Online] The Guardian, 06 10, 2019. [Cited: 01 16, 2020.] Https://Www.Theguardian.Com/Environment/2019/Jun/10/Frightening-Number-Of-Plant-Extinctions-Found-In-Global-Survey.

355. Bressan, David. Rare Plant Species Face Extinction Due To World Population Growth And Climate Change. [Online] Forbes, 11 28, 2019. [Cited: 01 21, 2020.] Https://Www.Forbes.Com/Sites/Davidbressan/2019/11/28/Rare-Plant-Species-Face-Extinction-Due-To-Population-Growth-And-Climate-Change/#41e85f12d8cf.

356. International Union For Conservation Of Nature. Over Half Of Europe's Endemic Trees Face Extinction. [Online] International Union For Conservation Of Nature (Iucn), 09 27, 2019. [Cited: 02 03, 2020.] Https://Www.Iucn.Org/News/Species/201909/Over-Half-Europes-Endemic-Trees-Face-Extinction.

357. The Irish Times. More Than Two-Fifths Of Europe's Native Trees At Risk Of Extinction – Assessment. *The Irish Times.* [Online] The Irish Times, 09 27, 2019. [Cited: 05 16, 2021.] Https://Www.Irishtimes.Com/News/World/Europe/More-Than-Two-Fifths-Of-Europe-S-Native-Trees-At-Risk-Of-Extinction-Assessment-1.4032410.

358. Gurib-Fakim, Ameenah. Climate Change Is Wiping Out The Baobab, Africa's 'Tree Of Life'. [Online] The Guardian, 06 13, 2018. [Cited: 02 03, 2020.] Https://Www.Theguardian.Com/Commentisfree/2018/Jun/13/Climate-Change-Baobab-Africa-Tree-Of-Life.

359. Specktor, Brandon. Joshua Trees Will Be All-But-Extinct By 2070 Without Climate Action, Study Warns. [Online] Live Science, 07 17, 2019. [Cited: 02 03, 2020.] Https://Www.Livescience.Com/65953-Climate-Change-Destroying-Joshua-Trees.Html.

360. Lampcov, Patrick Greenfield And Mette. Beetles And Fire Kill Dozens Of 'Indestructible' Giant Sequoia Trees. [Online] The Guardian, 01 18, 2020. [Cited: 02 04, 2020.] Https://Www.Theguardian.Com/Environment/2020/Jan/18/Beetles-And-Fire-Kill-Dozens-Of-California-Indestructible-Giant-Sequoia-Trees-Aoe.

361. Kaufman, Mark. The Hard Truth About Being A 21st Century Tree In California. [Online] Mashable, 07 03, 2019. [Cited: 02 04, 2020.] Https://Mashable.Com/Article/California-Tree-Die-Off-Climate-Change/.

362. Milman, Oliver. Tree-Damaging Pests Pose 'Devastating' Threat To 40% Of Us Forests. [Online] The Guardian, 08 12, 2019. [Cited: 02 04, 2020.] Https://Www.Theguardian.Com/Environment/2019/Aug/12/Us-Forests-Pests-Risk-Climate-Crisis-Resource.

363. Martin-Luther-Universität Halle-Wittenberg. How Climate Change Alters Plant Growth. [Online] Phys Org, 01 15, 2018. [Cited: 01 21, 2020.] Https://Phys.Org/News/2018-01-Climate-Growth.Html.

364. Kolbert, Elizabeth. *The Sixth Extinction.* New York : Henry Holt, 2014.

365. Sugiyama, Ayumi. Tree Line On Mt. Fuji Reaches New Heights; Global Warming Cited. *The Asahi Shimbun.* [Online] The Asahi Shimbun, 01 05, 2021. [Cited: 05 16, 2021.] Http://Www.Asahi.Com/Ajw/Articles/14035103?Utm_Campaign=Carbon%20brief%20daily%20briefing&Utm_Content=20210105&Utm_Medium=Email&Utm_Source=Revue%20daily.

366. Watts, Jonathan. World's Great Forests Could Lose Half Of All Wildlife As Planet Warms – Report. [Online] The Guardian, 03 14, 2018. [Cited: 01 21, 2020.] Https://Www.Theguardian.Com/Environment/2018/Mar/14/Worlds-Great-Forests-Could-Lose-Half-Of-All-Wildlife-As-Planet-Warms-Report.

367. Luscombe, Belinda. Do We Need $75,000 A Year To Be Happy? [Online] Time, 09 06, 2010. [Cited: 01 21, 2020.] Http://Content.Time.Com/Time/Magazine/Article/0,9171,2019628,00.Html.

368. Dunne, Daisy. Tropical Forests Losing Ability To Absorb Co2, Study Says. [Online] Carbonbrief, 01 27, 2020. [Cited: 01 29, 2020.] Https://Www.Carbonbrief.Org/Tropical-Forests-Losing-Ability-To-Absorb-Co2-Study-Says?

369. Nuccitelli, Dana. New Study Undercuts Favorite Climate Myth 'More Co2 Is Good For Plants'. [Online] The Guardian, 09 19, 2016. [Cited: 01 21, 2020.] Https://Www.Theguardian.Com/Environment/Climate-Consensus-97-Per-Cent/2016/Sep/19/New-Study-Undercuts-Favorite-Climate-Myth-More-Co2-Is-Good-For-Plants.

370. Fogarty, David. Crops Face Toxic Timebomb In Warmer World: Study. [Online] Reuters, 06 29, 2009. [Cited: 01 21, 2020.] Https://Www.Reuters.Com/Article/Us-Climate-Crops/Crops-Face-Toxic-Timebomb-In-Warmer-World-Study-Idustre55s2ky20090629.

371. Earth Institute At Columbia University. Wine Regions Could Shrink Dramatically With Climate Change Unless Growers Swap Varieties. [Online] Phys.Org, 01 27, 2020. [Cited: 01 30, 2020.] Https://Phys.Org/News/2020-01-Wine-Regions-Climate-Growers-Swap.Html.

372. Flinders University. Climate Change Risks 'Extinction Domino Effect'. [Online] Flinders University, 11 29, 2018. [Cited: 01 21, 2020.] Https://Www.Sciencedaily.Com/Releases/2018/11/181129122506.Htm.

373. Rocky Mountain Tree-Ring Research. Oldlist, A Database Of Old Trees. [Online] Rocky Mountain Tree-Ring Research, 2018. [Cited: 01 23, 2020.] Http://Www.Rmtrr.Org/Oldlist.Htm.

374. Pomeroy, Ross. Do Trees Die Of Old Age? [Online] Real Clear Science, 11 18, 2014. [Cited: 01 23, 2020.] Https://Www.Realclearscience.Com/Blog/2014/11/Do_Trees_Die_Of_Old_Age.Html.

375. Un Department Of Economic And Social Affairs. On International Day, Unece/Fao Forestry And Timber Section Releases 10 Facts To Fall In Love With Forests. *Un Department Of Economic And Social Affairs.* [Online] Un Department Of Economic And Social Affairs, 03 20, 2019. [Cited: 05 16, 2021.] Https://Www.Un.Org/Esa/Forests/News/2019/03/On-International-Day-Unece-Fao-Forestry-And-Timber-Section-Releases-10-Facts-To-Fall-In-Love-With-Forests/Index.Html.

376. Mongabay. 10 Facts About Forests For International Forest Day. [Online] Mongabay, 03 21, 2016. [Cited: 01 24, 2020.] Https://News.Mongabay.Com/2016/03/10-Facts-About-Forests-For-International-Forest-Day/.

377. Al, Damian Carrington Et. One Football Pitch Of Forest Lost Every Second In 2017, Data Reveals. [Online] The Guardian, 06 27, 2017. [Cited: 01 24, 2020.] Https://Www.Theguardian.Com/Environment/Ng-Interactive/2018/Jun/27/One-Football-Pitch-Of-Forest-Lost-Every-Second-In-2017-Data-Reveals.

378. Neslen, Arthur. Tokyo Olympics Venues 'Built With Wood From Threatened Rainforests'. [Online] The Guardian, 29 11 2018. [Cited: 08 02 2020.] Https://Www.Theguardian.Com/Environment/2018/Nov/29/Tokyo-Olympics-Venues-Built-With-Wood-From-Threatened-Rainforests.

379. Wwf. Living Forests Report Saving Forests At Risk. S.L. : Wwf, 2015. Http://Awsassets.Panda.Org/Downloads/Living_Forests_Report_Chapter_5_1.Pdf.

380. Scheidler, Fabian. The End Of The Megamachine: A Brief History Of A Failing Civilization. Winchester : Zero Books, 2020. 978 1 78904 271 9 .

381. Michalak, Roman. The (Hi)Story About Europeans And Their Forests. [Online] Fao. [Cited: 01 29, 2020.] Https://Www.Unece.Org/Forests/News/The-History-About-Europeans-And-Their-Forests.Html.

382. Greenpeace. Over 71% Of Eu Farmland Dedicated To Meat And Dairy, New Research. [Online] Greenpeace, 02 12, 2019. [Cited: 02 20, 2020.] Https://Www.Greenpeace.Org/Eu-Unit/Issues/Nature-Food/1807/71-Eu-Farmland-Meat-Dairy/.

383. Usda. U.S. Forest Resource Facts And Historical Trends. [Online] Usda, 08 2014. [Cited: 01 29, 2020.] Https://Www.Fia.Fs.Fed.Us/Library/Brochures/Docs/2012/Forestfacts_1952-2012_English.Pdf.

384. Leatherby, Dave Merrill And Lauren. Here's How America Uses Its Land. [Online] Bloomberg, 07 31, 2018. [Cited: 02 20, 2020.] Https://Www.Bloomberg.Com/Graphics/2018-Us-Land-Use/.

385. Little Left To Lose: Deforestation And Forest Degradation In Australia Since European Colonization. Bradshaw, Corey J. A. 1, 2012, Journal Of Plant Ecology, Vol. 5, Pp. 109-120.

386. Pearce, Fred. Rivers In The Sky: How Deforestation Is Affecting Global Water Cycles. [Online] Yale School Of Forestry & Environmental Studies, 07 24, 2018. [Cited: 01 30, 2020.] Https://E360.Yale.Edu/Features/How-Deforestation-Affecting-Global-Water-Cycles-Climate-Change?

387. Watts, Jonathan. Amazon Deforestation Accelerating Towards Unrecoverable 'Tipping Point'. [Online] The Guardian, 07 25, 2019. [Cited: 01 31, 2020.] Https://Www.Theguardian.Com/World/2019/Jul/25/Amazonian-Rainforest-Near-Unrecoverable-Tipping-Point.

388. Reuters In Brasília. Brazil's Bolsonaro Unveils Bill To Allow Commercial Mining On Indigenous Land. The Guardian. [Online] The Guardian, 02 06, 2020. [Cited: 05 17, 2021.] Https://Www.Theguardian.Com/World/2020/Feb/06/Brazil-Bolsonaro-Commercial-Mining-Indigenous-Land-Bill

389. Cann, Oliver. Fifteen Years To Save The Amazon Rainforest From Becoming Savannah. [Online] World Economic Forum, 01 22, 2020. [Cited: 01 31, 2020.] Https://Www.Weforum.Org/Press/2020/01/Fifteen-Years-To-Save-The-Amazon-Rainforest-From-Becoming-Savannah/.

390. Nunez, Christina. Deforestation Explained. [Online] National Geographic, 02 07, 2019. [Cited: 01 31, 2020.] Https://Www.Nationalgeographic.Com/Environment/Global-Warming/Deforestation/.

391. Gouby, Melanie. World's Second Biggest Rainforest Will Soon Reopen To Large-Scale Logging . National Geographic. [Online] National Geographic, 09 30, 2021. [Cited: 0 31, 2022.] Https://Www.Nationalgeographic.Com/Environment/Article/Worlds-Second-Biggest-Rainforest-Will-Soon-Reopen-To-Large-Scale-Logging.

392. Kilvert, Nick. Australia Is The World's Third-Largest Exporter Of Co2 In Fossil Fuels, Report Finds. [Online] Abc News, 08 19, 2019. [Cited: 01 31, 2020.] Https://Www.Abc.Net.Au/News/Science/2019-08-19/Australia-Co2-Exports-Third-Highest-Worldwide/11420654.

393. Mazengarb, Michael. Australia To Become World's Biggest Dealer In Fossil Fuel Emissions. [Online] Renew Economy, 07 08, 2019. [Cited: 01 31, 2020.] Https://Reneweconomy.Com.Au/Australia-To-Become-Worlds-Biggest-Dealer-In-Fossil-Fuel-Emissions-71881/.

394. Chow, Denise. Australia Wildfires Unleash Millions Of Tons Of Carbon Dioxide. [Online] Nbc News, 01 23, 2020. [Cited: 01 31, 2020.] Https://Www.Nbcnews.Com/Science/Environment/Australia-Wildfires-Unleash-Millions-Tons-Carbon-Dioxide-N1120186.

395. Humans And Forests In Pre-Colonial Southeast Asia. Reid, Anthony. 1, 1995, Environment And History, Vol. 1, Pp. 93-110.

396. Ives, Mike. In War-Scarred Landscape, Vietnam Replants Its Forests. [Online] Yale School Of Forestry & Environmental Studies, 11 04, 2010. [Cited: 02 05, 2020.] Https://E360.Yale.Edu/Features/In_War-Scarred_Landscape_Vietnam_Replants_Its_Forests.

397. Drollette, Daniel. A Plague Of Deforestation Sweeps Across Southeast Asia. [Online] Yale School Of Forestry & Environmental Studies, 05 20, 2013. [Cited: 02 04, 2020.] Https://E360.Yale.Edu/Features/A_Plague_Of_Deforestation_Sweeps_Across_Southeast_Asia.

398. Sodhi, Navot S. Southeast Asian Biodiversity: An Impending Disaster. [Online] Trends In Ecology & Evolution, 01 2004. [Cited: 02 05, 2020.] Https://Www.Researchgate.Net/Publication/259043691_Southeast_Asian_Biodiversity_An_Impending_Disaster.

399. Union Of Concerned Scientists. What's Driving Deforestation? [Online] Union Of Concerned Scientists, 02 08, 2016. [Cited: 02 05, 2020.] Https://Www.Ucsusa.Org/Resources/Whats-Driving-Deforestation.

400. Wwf. What Are The Biggest Drivers Of Tropical Deforestation? [Online] Wwf, 2018. [Cited: 02 05, 2020.] Https://Www.Worldwildlife.Org/Magazine/Issues/Summer-2018/Articles/What-Are-The-Biggest-Drivers-Of-Tropical-Deforestation.

401. The Livestock, Environment And Development (Lead) Initiative. Livestock's Long Shadow. [Online] The Livestock, Environment And Development (Lead) Initiative, 2006. [Cited: 02 07, 2020.] Http://Www.Fao.Org/3/A-a0701e.Pdf.

402. Mendes, Karla. Deforestation Drops In Brazil's Atlantic Forest, But Risks Remain: Experts. [Online] Mongabay, 07 13, 2019. [Cited: 02 07, 2020.] Https://News.Mongabay.Com/2019/07/Deforestation-Drops-In-Brazils-Atlantic-Forest-But-Risks-Remain-Experts/.

403. Amigo, Ignacio. In Brazil's Atlantic Forest, Conservation Efforts Drown In A Sea Of Eucalyptus. [Online] Mongabay, 04 05, 2017. [Cited: 02 07, 2020.] Https://News.Mongabay.Com/2017/04/In-Brazils-Atlantic-Forest-Conservation-Efforts-Drown-In-A-Sea-Of-Eucalyptus/.

404. Carvalho, Paul Jepson And Sergio Henrique Collaco De. Brazil's Cerrado Forests Won't Be Saved By Corporate Pledges On Deforestation. [Online] The Conversation, 12 08, 2017. [Cited: 02 07, 2020.] Https://Theconversation.Com/Brazils-Cerrado-Forests-Wont-Be-Saved-By-Corporate-Pledges-On-Deforestation-87130.

405. Asher, Claire. Saving The Amazon Has Come At The Cost Of Cerrado Deforestation: Study. [Online] Mongabay, 11 15, 2018. [Cited: 02 07, 2020.] Https://News.Mongabay.Com/2018/11/Saving-The-Amazon-Has-Come-At-The-Cost-Of-Cerrado-Deforestation-Study/.

406. Wwf. Choco+Darien=Biodiversity. [Online] Wwf. [Cited: 02 07, 2020.] Https://Wwf.Panda.Org/Knowledge_Hub/Where_We_Work/Choco_Darien/.

407. N., Luis Fernando Gómez. Landscape Management In Chocó-Darién Priority Watersheds. S.L. : Wwf-Colombia, 2013. Https://Wwfint.Awsassets.Panda.Org/Downloads/Choco_Darien_32.Pdf.

408. Wwf. Gran Chaco. [Online] Wwf. [Cited: 02 07, 2020.] Https://Www.Worldwildlife.Org/Places/Gran-Chaco.

409. Chisleanschi, Rodolfo. Gran Chaco: South America's Second-Largest Forest At Risk Of Collapsing. [Online] Mongabay, 09 17, 2019. [Cited: 02 07, 2020.] Https://News.Mongabay.Com/2019/09/Gran-Chaco-South-Americas-Second-Largest-Forest-At-Risk-Of-Collapsing/.

410. Goñi, Uki. Soy Destruction In Argentina Leads Straight To Our Dinner Plates. [Online] The Guardian, 10 26, 2018. [Cited: 02 07, 2020.] Https://Www.Theguardian.Com/Environment/2018/Oct/26/Soy-Destruction-Deforestation-In-Argentina-Leads-Straight-To-Our-Dinner-Plates.

411. African Sisters Education Collaborative. 10 Interesting Facts About Forests And Trees In Africa. [Online] African Sisters Education Collaborative, 03 21, 2017. [Cited: 02 19, 2020.] Http://Asec-Sldi.Org/News/Current/Interesting-Facts-Forests-Trees-Africa/.

412. Wwf. Top 10 Facts About The Congo Basin. [Online] Wwf. [Cited: 02 19, 2020.] Https://Wwf.Panda.Org/?8825/Top-10facts-About-The-Congo.

413. —. Congo Basin Facts. [Online] Wwf. [Cited: 02 19, 2020.] Https://Www.Worldwildlife.Org/Places/Congo-Basin.

414. Butler, Rhett A. The Congo Rainforest. [Online] Mongabay, 04 01, 2019. [Cited: 02 19, 2020.] Https://Rainforests.Mongabay.Com/Congo/.

415. Fao. Chapter 16. East Africa. [Online] Fao. [Cited: 02 19, 2020.] Http://Www.Fao.Org/3/y1997e/y1997e0l.Htm.

416. Wwf. Eastern And Southern African Countries Announce Groundbreaking Move To Curb Regional Illegal Timber Trade. Wwf Panda. [Online] Wwf, 09 09, 2015. [Cited: 06 04, 2022.] Https://Wwf.Panda.Org/Wwf_News/?252213/Eastern-And-Southern-African-Countries-Announce-Groundbreaking-Move-To-Curb-Regional-Illegal-Timber-Trade.

417. Gokkon, Basten. Borneo, Ravaged By Deforestation, Loses Nearly 150,000 Orangutans In 16 Years, Study Finds. [Online] Mongabay, 02 15, 2018. [Cited: 02 09, 2020.] Https://News.Mongabay.Com/2018/02/Borneo-Ravaged-By-Deforestation-Loses-Nearly-150000-Orangutans-In-16-Years-Study-Finds/.

418. Gaveau, David. Is Deforestation In Borneo Slowing Down? [Online] Center For International Forestry Research (Cifor), 01 15, 2019. [Cited: 02 09, 2020.] Https://Forestsnews.Cifor.Org/59378/Has-Borneos-Deforestation-Slowed-Down?Fnl=En.

419. Murphy, Denis J. Palm Oil: Scourge Of The Earth, Or Wonder Crop? [Online] The Conversation, 07 01, 2015. [Cited: 02 09, 2020.] Http://Theconversation.Com/Palm-Oil-Scourge-Of-The-Earth-Or-Wonder-Crop-42165.

420. Wwf. Found Only In Borneo. [Online] Wwf. [Cited: 02 09, 2020.] Https://Wwf.Panda.Org/Knowledge_Hub/Where_We_Work/Borneo_Forests/About_Borneo_Forests/Borneo_Animals/.

421. Un Environment Programme. Deforestation In Borneo Is Slowing, But Regulation Remains Key. [Online] Un Environment Programme, 02 18, 2019. [Cited: 02 09, 2020.] Https://Www.Unenvironment.Org/News-And-Stories/Story/Deforestation-Borneo-Slowing-Regulation-Remains-Key.

422. Sekiguchi, Kei. Traders Expect More Cuts In Malaysian Log Exports. [Online] Nikkei Asian Review, 04 03, 2018. [Cited: 02 0, 2020.] Https://Asia.Nikkei.Com/Business/Business-Trends/Traders-Expect-More-Cuts-In-Malaysian-Log-Exports.

423. Neslen, Arthur. Tokyo Olympics Venues 'Built With Wood From Threatened Rainforests'. [Online] The Guardian, 11 2, 2018. [Cited: 02 09, 2020.] Https://Www.Theguardian.Com/Environment/2018/Nov/29/Tokyo-Olympics-Venues-Built-With-Wood-From-Threatened-Rainforests.

424. Mongabay. Deforestation Statistics For Papua New Guinea. [Online] Mongabay. [Cited: 02 0, 2020.] Https://Rainforests.Mongabay.Com/Deforestation/Archive/Papua_New_Guinea.Htm.

425. Adam, David. Satellite Images Show Papua New Guinea Deforestation At Critical Level. [Online] The Guardian, 06 02, 2008. [Cited: 02 09, 2020.] Https://Www.Theguardian.Com/Environment/2008/Jun/02/Forests.Conservation.

426. Gabbatiss, Josh. Alarming Photos Reveal Devastating Scale Of Rainforest Destruction In Papua New Guinea. [Online] The Independent, 03 21, 2018. [Cited: 02 09, 2020.] Https://Www.Independent.Co.Uk/Environment/Papua-New-Guinea-Rainforest-Destruction-Photos-Deforestation-Global-Witness-Illegal-Logging-A8265451.Html.

427. Wwf. Sumatra. [Online] Wwf. [Cited: 02 09, 2020.] Https://Wwf.Panda.Org/Knowledge_Hub/Where_We_Work/Sumatra/.

428. The Environmental Impacts Of Palm Oil In Context. Erik Meijaard, Thomas M. Brooks And Kimberly M. Carlson Et Al. S.L. : Nature Plants, 2020, Vol. 6. Https://Www.Nature.Com/Articles/S41477-020-00813-

492 - REFERENCES

W?Utm_Campaign=Carbon%20brief%20daily%20briefing&Utm_Content=20201208&Utm_Medium=Email&Utm_Source=Revu
e%20daily.
429. Vidal, John. This Article Is More Than 9 Years Old 'The Sumatran Rainforest Will Mostly Disappear Within 20 Years'. The
Guardian. [Online] The Guardian, 05 26, 2013. [Cited: 06 04, 2022.]
Https://Www.Theguardian.Com/World/2013/May/26/Sumatra-Borneo-Deforestation-Tigers-Palm-Oil.
430. Wwf. Greater Mekong. [Online] Wwf. [Cited: 02 18, 2020.] Https://Www.Worldwildlife.Org/Places/Greater-Mekong.
431. Vidal, John. Greater Mekong Countries 'Lost One-Third Of Forest Cover In 40 Years'. [Online] The Guardian, 05 02, 2013.
[Cited: 02 18, 2020.] Https://Www.Theguardian.Com/Environment/2013/May/02/Greater-Mekong-Forest-Cover.
432. Wilderness Society. 10 Facts About Deforestation In Australia. [Online] Wilderness Society, 01 18, 2018. [Cited: 02 19,
2020.] Https://Www.Wilderness.Org.Au/News-Events/10-Facts-About-Deforestation-In-Australia.
433. Wwf. Wwf Living Forests Report: Chapter 5 Saving Forests At Risk. Gland, Switzerland : Wwf, 2015.
434. Laidlaw, M.J. A Scientific Review Of The Impacts Of Land Clearing On Threatened Species In Queensland. Brisbane : The State Of
Queensland , 2017.
435. Food And Agriculture Organization, Electronic Files And Web Site. Forest Area (% Of Land Area) - Country Ranking.
[Online] Index Mundi, 06 30, 2016. [Cited: 01 30, 2020.]
Https://Www.Indexmundi.Com/Facts/Indicators/Ag.Lnd.Frst.Zs/Rankings.
436. Gakpo, Joseph Opoku. Agricultural Technology Key To Protecting Nature And Preventing Pandemics . Cornell: Alliance
For Science. [Online] Cornell: Alliance For Science, 07 24, 2020. [Cited: 05 15, 2021.]
Https://Allianceforscience.Cornell.Edu/Blog/2020/07/Agricultural-Technology-Key-To-Protecting-Nature-And-Preventing-
Pandemics/.
437. Patton, Dominique. China's Multi-Story Hog Hotels Elevate Industrial Farms To New Levels. Reuters. [Online] Reuters, 05
11, 2018. [Cited: 05 15, 2021.] Https://Www.Reuters.Com/Article/Us-China-Pigs-Hotels-Insight-Iduskbn1ib362.
438. Watson, Robert. Biodiversity Touches Every Aspect Of Our Lives – So Why Has Its Loss Been Ignored? [Online] The
Guardian, 09 19, 2019. [Cited: 02 21, 2020.] Https://Www.Theguardian.Com/Environment/2019/Sep/19/Biodiversity-Touches-
Every-Aspect-Of-Our-Lives-So-Why-Has-Its-Loss-Been-Ignored.
439. Aguilera, Jasmine. 'The Numbers Are Just Horrendous.' Almost 30,000 Species Face Extinction Because Of Human
Activity. [Online] Time, 07 18, 2009. [Cited: 02 21, 2020.] Https://Time.Com/5629548/Almost-30000-Species-Face-Extinction-
New-Report/?Fbclid=Iwar1k8uvv5aiyjaanw1_Vexaqb6m_Vkc0l_Duq0xt4mrau78qzqvnxrp81j0.
440. Wwf. How Many Species Are We Losing? [Online] Wwf. [Cited: 02 26, 2020.]
Https://Wwf.Panda.Org/Discover/Our_Focus/Biodiversity/Biodiversity/#:~:Text=But%20if%20the%20upper%20estimate,Are%2
0becoming%20extinct%20each%20year..
441. Zaraska, Marta. Meathooked: The History And Science Of Our 2.5-Million-Year Obsession With Meat. New York : Basic Books,
2016. 978-0465036622.
442. Maslin, Simon L. Lewis And Mark A. How Disease And Conquest Carved A New Planetary Landscape. [Online] The
Atlantic, 08 24, 2018. [Cited: 02 26, 2020.] Https://Www.Theatlantic.Com/Science/Archive/2018/08/Human-Planet-Migration-
Columbian-Exchange/568423/.
443. Churchill, Ward. History Not Taught Is History Forgot: Columbus' Legacy Of Genocide. Justice Initiative International.
[Online] Justice Initiative International, 1994. [Cited: 06 05, 2022.]
Https://Justiceinitiativeinternational.Wordpress.Com/2016/11/23/History-Not-Taught-Is-History-Forgot-Columbus-Legacy-
Of-Genocide/.
444. Gresham College. The Rise And Fall Of European Empires From The 16th To The 20th Century. [Online] Gresham
College. [Cited: 02 26, 2020.] Https://Www.Gresham.Ac.Uk/Series/The-Rise-And-Fall-Of-European-Empires-From-The-16th-
To-The-20th-Century/.
445. Thompsell, Angela. Real Men/Savage Nature: The Rise Of African Big Game Hunting, 1870–1914. Britain And The World.
London : Palgrave Macmillan, 2015, Pp. 12-41.
446. Murphy, Asia. Conservation's Biggest Challenge? The Legacy Of Colonialism. [Online] Livescience, 05 20, 2019. [Cited: 02
26, 2020.] Https://Www.Livescience.Com/65507-Conservation-Colonialism-Legacy.Html.
447. India Today. Killing Animals As Trophy: Shocking Facts About Trophy Hunting. [Online] India Today, 10 25, 2018.
[Cited: 02 26, 2020.] Https://Www.Indiatoday.In/Education-Today/Gk-Current-Affairs/Story/Killing-Animals-As-Trophy-
Shocking-Facts-About-Trophy-Hunting-1375145-2018-10-25.
448. U.S. Department Of The Interior. Share The Conservation Legacy Of Theodore Roosevelt. U.S. Department Of The
Interior. [Online] U.S. Department Of The Interior, 02 14, 2020. [Cited: 05 17, 2021.]
Https://Www.Doi.Gov/Blog/Conservation-Legacy-Theodore-Roosevelt.
449. Scully, Matthew. Dominion: The Power Of Man, The Suffering Of Animals, And The Call To Mercy. New York : St. Martin's
Griffin, 2003. 0-312-26147-0.
450. Wikipedia. Smithsonian–Roosevelt African Expedition. Wikipedia. [Online] Wikipedia. [Cited: 05 17, 2021.]
Https://En.Wikipedia.Org/Wiki/Smithsonian%E2%80%93roosevelt_African_Expedition.
451. Disappearing Elephants. Overview. [Online] Disappearing Elephants. [Cited: 02 27, 2020.]
Https://Disappearingelephants.Com/Overview/.
452. Roser, Max. Two Centuries Of Rapid Global Population Growth Will Come To An End. [Online] Our World In Data, 06
18, 2019. [Cited: 02 26, 2020.] Https://Ourworldindata.Org/World-Population-Growth-Past-Future.
453. Hoare, Philip. The Animal Victims Of The First World War Are A Stain On Our Conscience. [Online] The Guardian, 11
07, 2018. [Cited: 02 27, 2020.] Https://Www.Theguardian.Com/Commentisfree/2018/Nov/07/Animal-Victims-First-World-
War.

454. Keller, Tait. Destruction Of The Ecosystem. [Online] International Encyclopedia Of The First World War, 10 08, 2014. [Cited: 02 27, 2020.] Https://Encyclopedia.1914-1918-Online.Net/Article/Destruction_Of_The_Ecosystem.
455. The Impact Of W The Impact Of World War One On The F Ar One On The Forests And Soils Of E Ests And Soils Of Europe. Heiderscheidt, Drew. 3, 2018, The Undergraduate Research Journal At The Univesity Of Northern Colorado, Vol. 7, Pp. 1-17.
456. Feeney-Hart, Alison. The Little-Told Story Of The Massive Wwii Pet Cull. [Online] Bbc, 10 12, 2013. [Cited: 02 27, 2020.] Https://Www.Bbc.Com/News/Magazine-24478532.
457. White, Conan. When Ww2 Ended Where Did All The 100's Of Millions Of Weapons Go? [Online] War History Online, 11 21, 2018. [Cited: 02 27, 2020.] Https://Www.Warhistoryonline.Com/Instant-Articles/Wwii-Where-Weapons-Wind-Up.Html.
458. Lymbery, Philip. Dead Zone: Where The Wild Things Were. London : Bloomsbury, 2017. 978-1-4088-6826-3.
459. National Chicken Council. U.S. Chicken Industry History. [Online] National Chicken Council. [Cited: 04 08, 2020.] Https://Www.Nationalchickencouncil.Org/About-The-Industry/History/.
460. Lusk, Jayson. What Caused The Rise Of "Factory Farms"? [Online] Jayson Lusk: Food And Agriculture Economist, 10 29, 2013. [Cited: 04 08, 2020.] Http://Jaysonlusk.Com/Blog/2013/10/29/What-Caused-The-Rise-Of-Factory-Farms.
461. Merriam Webster. Mderriam Webster Online Dictionary. [Online] 2020. [Cited: 06 02, 2020.] Https://Www.Merriam-Webster.Com/Dictionary/Factory%20farm.
462. Livestock, Livelihoods And The Environment: Understanding The Trade-Offs. Mario Herrero, Philip K Thornton, Pierre Gerber And Robin S Reid. 2009, Environmental Sustainability, Vol. 1, Pp. 1-10.
463. Wills, Kendra. Where Do All These Soybeans Go? [Online] Michigan State University, 10 08, 2013. [Cited: 04 08, 2020.] Https://Www.Canr.Msu.Edu/News/Where_Do_All_These_Soybeans_Go.
464. Cox, Lisa. Beef Industry Linked To 94% Of Land Clearing In Great Barrier Reef Catchments. [Online] The Guardian, 08 07, 2019. [Cited: 04 08, 2020.] Https://Www.Theguardian.Com/Australia-News/2019/Aug/08/Beef-Industry-Linked-To-94-Of-Land-Clearing-In-Great-Barrier-Reef-Catchments.
465. Hoekstra, A. Y. The Hidden Water Resource Use Behind Meat And Dairy. Enschede : University Of Twente, 2012. Https://Doi.Org/10.2527/Af.2012-0038.
466. Wwf. Water Scarcity. Wwf. [Online] Wwf. [Cited: 06 08, 2020.] Https://Www.Worldwildlife.Org/Threats/Water-Scarcity.
467. Roy, Eleanor Ainge. 'Under Serious Threat': New Zealand Vows To Clean Up Its Polluted Waterways. The Guardian. [Online] 09 05, 2019. Https://Www.Theguardian.Com/World/2019/Sep/05/Under-Serious-Threat-New-Zealand-Vows-To-Clean-Up-Its-Polluted-Waterways.
468. Monbiot, George. Sheepwrecked. [Online] 2013. [Cited: 06 02, 2020.] Https://Www.Monbiot.Com/2013/05/30/Sheepwrecked/.
469. Total Global Agricultural Land Footprint Associated With Uk Food Supply 1986–2011. Ruiter, Henri De. 2017, Global Environmental Change, Vol. 43, Pp. 72-81.
470. Hayek, Helen Harwatt And Matthew N. Eating Away At Climate Change With Negative Emissions. Cambridge : Harvard Law School, 2019.
471. Roser, Hannah Ritchie And Max. Meat And Dairy Production. [Online] 2019. [Cited: 06 03, 2020.] Https://Ourworldindata.Org/Meat-Production.
472. Fao. Meat & Meat Products. [Online] 2019. [Cited: 06 02, 2020.] Http://Www.Fao.Org/Ag/Againfo/Themes/En/Meat/Home.Html#:~:Text=While%20meat%20consumption%20has%20been,Since%201980%20in%20developing%20countries.&Text=World%20meat%20production%20is%20projected,Is%20expected%20in%20developing%20countries..
473. Morell, Virginia. Meat-Eaters May Speed Worldwide Species Extinction, Study Warns. [Online] 2015. [Cited: 06 03, 2020.] Https://Www.Sciencemag.Org/News/2015/08/Meat-Eaters-May-Speed-Worldwide-Species-Extinction-Study-Warns.
474. Rebecca Wright, Ivan Watson, Tom Booth And Masrur Jamaluddin,. Borneo Is Burning. Cnn. [Online] Cnn, 11 19, 2019. [Cited: 06 04, 2020.] Https://Edition.Cnn.Com/Interactive/2019/11/Asia/Borneo-Climate-Bomb-Intl-Hnk/.
475. Potenza, Alessandra. When Humans Kill Each Other In War, Wildlife Dies Too. [Online] 2018. [Cited: 06 03, 2020.] Https://Www.Theverge.Com/2018/1/10/16871120/War-Wildlife-Population-Decline-Africa-Conservation.
476. The Humane Society. Trophy Hunting By The Numbers. Humane Society International. [Online] Humane Society International, 02 2016. [Cited: 06 04, 2020.] Https://Www.Hsi.Org/Wp-Content/Uploads/Assets/Pdfs/Report_Trophy_Hunting_By_The.Pdf.
477. Wildlife Rescue Organisation. Canned Hunts. [Online] 2016. [Cited: 06 03, 2020.] Https://Wildlife-Rescue.Org/Services/Advocacy/Canned-Hunts/#:~:Text=Most%20states%20allow%20canned%20hunting,Hunting%20preserves%2c%20or%20canned%20hunts..
478. Powers, Ronald. Humane Society: Zoo Animals Easy Targets At 'Canned Hunts'. [Online] 1994. [Cited: 06 03, 2020.] Https://Apnews.Com/9466b1d748bd6e9e38d50cd0192b40db.
479. Cross, Daniel T. Traditional Chinese Medicine Is A Scourge On Exotic Wildlife. [Online] Sustainability Times, 08 05, 2019. [Cited: 06 05, 2020.] Https://Www.Sustainability-Times.Com/Environmental-Protection/Traditional-Medicine-Is-A-Scourge-On-Exotic-Wildlife/.
480. Mccann, Gregory. China Is Decimating Southeast Asian Wildlife. The Diplomat. [Online] The Diplomat, 01 24, 2018. [Cited: 06 05, 2020.] Https://Thediplomat.Com/2018/01/China-Is-Decimating-Southeast-Asian-Wildlife/.
481. Lindmeier, Christian. Stop Using Antibiotics In Healthy Animals To Prevent The Spread Of Antibiotic Resistance. World Health Organisation. [Online] World Health Organisation, 11 07, 2017. [Cited: 05 17, 2021.] Https://Www.Who.Int/News/Item/07-11-2017-Stop-Using-Antibiotics-In-Healthy-Animals-To-Prevent-The-Spread-Of-Antibiotic-Resistance.

482. Woodward, Aylin. 18 Signs We're In The Middle Of A 6th Mass Extinction. *Business Insider.* [Online] Business Insider, 07 19, 2019. [Cited: 06 08, 2020.] Https://Www.Businessinsider.Com/Signs-Of-6th-Mass-Extinction-2019-3#The-Loss-Of-Even-One-Species-Could-Also-Cause-An-Extinction-Domino-Effect-To-Ripple-Through-An-Ecosystem-Causing-The-Entire-Community-To-Collapse-7.

483. Goldburg, Rebecca. Scientists Find That 30% Of Global Fish Catch Is Unreported. *Pew Trusts.* [Online] Pew Trusts, 01 19, 2016. [Cited: 06 12, 2020.] Https://Www.Pewtrusts.Org/En/Research-And-Analysis/Articles/2016/01/19/Scientists-Find-That-30-Percent-Of-Global-Fish-Catch-Is-Unreported.

484. The World Bank. *Fish To 2030 Prospects For Fisheries And Aquaculture.* Washington Dc : The World Bank, 2013. 83177-Glb.

485. Stanford News. Science Study Predicts Collapse Of All Seafood Fisheries By 2050. *Stanford News.* [Online] Stanford News, 11 02, 2006. [Cited: 06 09, 2020.] Https://News.Stanford.Edu/News/2006/November8/Ocean-110806.Html.

486. Morsink, Kalila. With Every Breath You Take, Thank The Ocean. *Smithsonian: Ocean.* [Online] Smithsonian: Ocean, 07 2017. [Cited: 06 10, 2020.] Https://Ocean.Si.Edu/Ocean-Life/Plankton/Every-Breath-You-Take-Thank-Ocean

487. Science History Institute. Science Matters: The Case Of Plastics. *Science History Institute.* [Online] Science History Institute. [Cited: 06 10, 2020.] Https://Www.Sciencehistory.Org/The-History-And-Future-Of-Plastics.

488. Tiseo, Ian. Global Plastic Production 1950-2020. *Statista.* [Online] Statista, 01 12, 2022. [Cited: 02 07, 2022.] Https://Www.Statista.Com/Statistics/282732/Global-Production-Of-Plastics-Since-1950/.

489. Knapton, Sarah. Plastic Weighing Equivalent Of One Billion Elephants Has Been Made Since 1950s And Most Is Now Landfill. *The Telegraph.* [Online] The Telegraph, 07 19, 2017. [Cited: 06 10, 2020.] Https://Www.Telegraph.Co.Uk/Science/2017/07/19/Plastic-Weighing-Equivalent-One-Billion-Elephants-Has-Made-Since/.

490. Taylor, Sandra Laville And Matthew. A Million Bottles A Minute: World's Plastic Binge 'As Dangerous As Climate Change'. *The Guardian.* [Online] The Guardian, 06 28, 2017. [Cited: 06 11, 2020.] Https://Www.Theguardian.Com/Environment/2017/Jun/28/A-Million-A-Minute-Worlds-Plastic-Bottle-Binge-As-Dangerous-As-Climate-Change.

491. Alberts, Elizabeth Claire. These Creepy 'Ghost Nets' Are Killing Thousands Of Animals Every Year. *The Dodo.* [Online] The Dodo, 03 19, 2018. [Cited: 06 12, 2020.] Https://Www.Thedodo.Com/In-The-Wild/Ocean-Animals-Dying-In-Lost-Fishing-Gear.

492. Parker, Laura. The Great Pacific Garbage Patch Isn't What You Think It Is. *National Geographic.* [Online] National Geographic, 03 2018. [Cited: 06 10, 2020.] Https://Www.Nationalgeographic.Com/News/2018/03/Great-Pacific-Garbage-Patch-Plastics-Environment/#:~:Text=The%20study%20also%20found%20that,From%20othe%202011%20japanese%20tsunami..

493. Mcveigh, Karen. New Rules To Tackle 'Wild West' Of Plastic Waste Dumped On Poorer Countries. *The Guardian.* [Online] The Guardian, 12 29, 2020. [Cited: 05 17, 2021.] Https://Www.Theguardian.Com/Environment/2020/Dec/29/New-Rules-To-Tackle-Wild-West-Of-Plastic-Waste-Dumped-On-Poorer-Countries.

494. Shukman, David. Just 20 Firms Behind More Than Half Of Single-Use Plastic Waste - Study. *Yahoo News.* [Online] Bbc News, 05 18, 2021. [Cited: 05 19, 2021.] Https://Www.Bbc.Com/News/Science-Environment-57149741.

495. National Geographic. 7 Things You Didn't Know About Plastic. *National Geographic.* [Online] National Geographic, 04 04, 2018. [Cited: 06 11, 2020.] Https://Blog.Nationalgeographic.Org/2018/04/04/7-Things-You-Didnt-Know-About-Plastic-And-Recycling/

496. History Of Fishing. The History Of Fishing. *The History Of Fishing.* [Online] The History Of Fishing. [Cited: 06 12, 2020.] Http://Www.Historyoffishing.Com/.

497. Fao. The State Of World Fisheries And Aquaculture 2020. *Fao.* [Online] Fao, 2020. [Cited: 06 12, 2020.] Https://www.fao.org/documents/card/en/c/ca9229en/

498. Shiffman, David. Predatory Fish Have Declined By Two Thirds In The 20th Century . *Sientific American.* [Online] Scientific American, 10 20, 2014. [Cited: 06 15, 2020.] Https://Www.Scientificamerican.Com/Article/Predatory-Fish-Have-Declined-By-Two-Thirds-In-The-20th-Century/.

499. Hays, Jeffrey. Bluefin Tuna Fish Farming . *Facts And Details.* [Online] Facts And Details, 01 2012. [Cited: 06 15, 2020.] Http://Factsanddetails.Com/World/Cat53/Sub340/Item2188.Html.

500. My News Desk. African Penguins Face Extinction Within 15 Years, Warn Campaigners On World Penguin Day. *My News Desk.* [Online] My News Desk, 04 25, 2018. [Cited: 06 15, 2020.] Http://Www.Mynewsdesk.Com/Uk/Minerva-Communications/Pressreleases/African-Penguins-Face-Extinction-Within-15-Years-Warn-Campaigners-On-World-Penguin-Day-2485462.

501. Taylor, Matthew. Decline In Krill Threatens Antarctic Wildlife, From Whales To Penguins. *The Guardian.* [Online] The Guardian, 02 14, 2018. [Cited: 06 15, 2020.] Https://Www.Theguardian.Com/Environment/2018/Feb/14/Decline-In-Krill-Threatens-Antarctic-Wildlife-From-Whales-To-Penguins.

502. Cecco, Leyland. Saving Canada's Wild Salmon: Rescuers Pin Hopes On Fish Ladder And Salmon Cannon. *The Guardian.* [Online] The Guardian, 06 15, 2020. [Cited: 06 16, 2020.] Https://Www.Theguardian.Com/Environment/2020/Jun/15/Canada-Wild-Salmon-Fish-Ladder-Salmon-Cannon-British-Columbia-Aoe.

503. Wobble, The Big. Fish All Gone! Gulf Of Alaska Fishery To Close For The First Time Ever: No More Cod: Salmon All But Gone: Millions Of Small Sea Birds Died Since 2015. *The Big Wobble.* [Online] The Big Wobble, 12 08, 2019. [Cited: 06 18, 2020.] Http://www.thebigwobble.org/2019/12/fish-all-gone-gulf-of-alaska-fishery-to.html

504. Mcveigh, Karen. Revealed: 97% Of Uk Marine Protected Areas Subject To Bottom Trawling. *The Guardian.* [Online] The Guardian, 10 09, 2020. [Cited: 05 17, 2021.] Https://Www.Theguardian.Com/Environment/2020/Oct/09/Revealed-97-Of-Uk-Offshore-Marine-Parks-Subject-To-Destructive-Fishing.

505. The Rt Hon Lord Goldsmith, And The Rt Hon Theresa Villiers Mp. Uk Creates Global Alliance To Help Protect The World's Ocean. *Department For Environment, Food & Rural Affairs,*. [Online] Gov.Uk, 09 24, 2019. [Cited: 05 17, 2021.] Https://Www.Gov.Uk/Government/News/Uk-Creates-Global-Alliance-To-Help-Protect-The-Worlds-Ocean.

506. Cecco, Leyland. Canada Ignored Warnings Of Virus Infecting Farmed And Wild Salmon . *The Guardian.* [Online] The Guardian, 04 14, 2022. [Cited: 04 15, 2022.] Https://Www.Theguardian.Com/World/2022/Apr/14/Infected-Farmed-Wild-Salmon-Canada-Virus-Report.

507. Neslen, Arthur. Global Fish Production Approaching Sustainable Limit, Un Warns. *The Guardian.* [Online] The Guardian, 07 07, 2016. [Cited: 06 16, 2020.] Https://Www.Theguardian.Com/Environment/2016/Jul/07/Global-Fish-Production-Approaching-Sustainable-Limit-Un-Warns.

508. Mcveigh, Karen. Banned Pesticide Blamed For Killing Bees May Be Approved For Fish Farms. *The Guardian.* [Online] The Guardian, 05 27, 2021. [Cited: 06 17, 2021.] Https://Www.Theguardian.Com/Environment/2021/May/27/Novichok-For-Insects-May-Be-Approved-For-Scottish-Fish-Farms.

509. Stanford Encyclopedia Of Philosophy. The Moral Status Of Animals. *Stanford Encyclopedia Of Philosophy.* [Online] Stanford Encyclopedia Of Philosophy, 08 23, 2017. [Cited: 05 17, 2021.] Https://Plato.Stanford.Edu/Entries/Moral-Animal/.

510. Jabr, Ferris. It's Official: Fish Feel Pain. *Smithsonianmag.* [Online] Smithsonianmag, 01 08, 2018. [Cited: 06 16, 2020.] Https://Www.Smithsonianmag.Com/Science-Nature/Fish-Feel-Pain-180967764/.

511. The Editors Of Encyclopaedia Britannica. Whale Oil. *Britannica.* [Online] Encyclopædia Britannica, Inc., 06 02, 2011. [Cited: 06 18, 2020.] Https://Www.Britannica.Com/Technology/Whale-Oil.

512. Cressey, Daniel. World's Whaling Slaughter Tallied At 3 Million. *Scientific American.* [Online] Nature Magazine, 03 12, 2015. [Cited: 06 18, 2020.] Https://Www.Scientificamerican.Com/Article/World-S-Whaling-Slaughter-Tallied-At-3-Million/.

513. Kyodo. Japan Back In Whaling Business, But Quotas And Cost Make Meat A Hard Sell. *The Japan Times.* [Online] The Japan Times, 04 21, 2020. [Cited: 06 18, 2020.] Https://Www.Japantimes.Co.Jp/News/2020/04/21/National/Japan-Commercial-Whaling-Meat-Hard-Sell/#.Xuqkp2ozzqi.

514. Lee, Jane J. Faroe Island Whaling, A 1,000-Year Tradition, Comes Under Renewed Fire. *National Geographic.* [Online] National Geographic, 09 12, 2014. [Cited: 06 18, 2020.] Https://Www.Nationalgeographic.Com/News/2014/9/140911-Faroe-Island-Pilot-Whale-Hunt-Animals-Ocean-Science/.

515. Press, Hanna. Norway Now Kills More Whales Than Japan. *Save Dolphins.* [Online] International Marine Mammal Project, 07 29, 2019. [Cited: 06 18, 2020.] Http://Savedolphins.Eii.Org/News/Entry/Norway-Now-Kills-More-Whales-Than-Japan.

516. Mcgrath, Matt. Whale Killing: Dna Shows Iceland Whale Was Rare Hybrid . *Bbc.* [Online] Bbc, 07 20, 2018. [Cited: 06 18, 2020.] Https://Www.Bbc.Com/Science-Environment-44809115.

517. Whale And Dolphin Conservation. How Intelligent Are Whales And Dolphins? *Whale And Dolphin Conservation.* [Online] Whale And Dolphin Conservation. [Cited: 06 23, 2020.] Https://Uk.Whales.Org/Whales-Dolphins/How-Intelligent-Are-Whales-And-Dolphins/.

518. Kropshofer, Katharina. Whales And Dolphins Lead 'Human-Like Lives' Thanks To Big Brains, Says Study. *The Guardian.* [Online] The Guardian, 10 16, 2017. [Cited: 06 23, 2020.] Https://Www.Theguardian.Com/Science/2017/Oct/16/Whales-And-Dolphins-Human-Like-Societies-Thanks-To-Their-Big-Brains.

519. Cbc News. Killer Whale Lets Her Dead Newborn Go After Carrying Body For 17 Days. *Cbc News.* [Online] Cbc News, 08 12, 2018. [Cited: 06 24, 2020.] Https://Www.Cbc.Ca/News/Canada/British-Columbia/Killer-Whale-Dead-Calf-1.4782542#:-:Text=A%20female%20killer%20whale%20known%20as%20j%2d35%20has%20stopped,Of%20victoria%20on%20saturday%20afternoon..

520. National Geographic. Blue Whales And Communication . *National Geographic.* [Online] National Geographic, 03 26, 2011. [Cited: 06 23, 2020.] Https://Www.Nationalgeographic.Com.Au/Science/Blue-Whales-And-Communication.Aspx#:-:Text=Sound%20is%20the%20most%20effective,With%20sound%20across%20the%20water..

521. Pbs. A Whale Of A Business. *Pbs Frontline.* [Online] Pbvs, 2014. [Cited: 06 24, 2020.] Https://Www.Pbs.Org/Wgbh/Pages/Frontline/Shows/Whales/Etc/Cron.Html.

522. Change For Animals Foundation. Cfaf's Work For Whales And Dolphins In Captivity. *Change For Animals Foundation.* [Online] Change For Animals Foundation. [Cited: 06 24, 2020.] Https://Www.Changeforanimals.Org/Whales-And-Dolphins-In-Captivity.

523. Harvey, Chelsea. Double Whammy Of Warming, Overfishing Could Spell Disaster For Antarctic Krill. *Scientific American.* [Online] Scientific American, 10 27, 2020. [Cited: 05 18, 2021.] Https://Www.Scientificamerican.Com/Article/Double-Whammy-Warming-Overfishing-Could-Spell-Disaster-For-Antarctic-Krill/.

524. Byrne, Jane. Https://Www.Feednavigator.Com/Article/2018/03/27/How-Sustainable-Is-The-Krill-Meal-Supply-Chain. *Feed Navigator.* [Online] Feed Navigator, 03 27, 2018. [Cited: 05 18, 2021.] Https://Www.Feednavigator.Com/Article/2018/03/27/How-Sustainable-Is-The-Krill-Meal-Supply-Chain.

525. Dunne, Daisy. Southern Right Whales Could Face New Threat From Ocean Heating, Study Warns. *Independent.* [Online] Independent, 12 22, 2020. [Cited: 05 18, 2021.] Https://Www.Independent.Co.Uk/Climate-Change/News/Southern-Right-Whales-Climate-Change-B1777222.Html.

526. Fao. 5. Products. *Fao.* [Online] Fao. [Cited: 05 18, 2021.] Http://Www.Fao.Org/3/w5911e/W5911e08.Htm.

527. Wwf. Warmer Ocean Temperatures And Melting Sea Ice In The Polar Regions May Jeopardise The Ecology Of The Arctic And Antarctic Feeding Grounds Of Many Large Whales. *Wwf.* [Online] Wwf. [Cited: 05 18, 2021.] Https://Wwf.Panda.Org/Discover/Knowledge_Hub/Endangered_Species/Cetaceans/Threats/Climate_Change/?

528. Stone, Madelein. How Much Is A Whale Worth? . *National Geographic.* [Online] National Geographic, 09 24, 2019. [Cited: 06 24, 2020.] Https://Www.Nationalgeographic.Com/Environment/2019/09/How-Much-Is-A-Whale-Worth/

496 ~ REFERENCES

529. *Global Catches, Exploitation Rates, And Rebuilding Options For Sharks.* Boris Worm, Brendal Davis, Lisa Kettemer, Christine A. Ward-Paige, Demian Chapman, Michael R. Heithaus, Steven T. Kessel, Samuel H. Gruber. S.L. : Marine Policy, 2013, Vol. 40. 0308-597x.
530. The Florida Museum. Yearly Worldwide Shark Attack Summary. *Florida Museum.* [Online] The Florida Museum, 01 22, 2020. [Cited: 07 15, 2020.] Https://Www.Floridamuseum.Ufl.Edu/Shark-Attacks/Yearly-Worldwide-Summary/.
531. Pederson, Mario. Shark Attacks At A Low For 2020, But Deaths Double 2019 Numbers. *Orlando Sentinel.* [Online] Orlando Sentinel, 08 05, 2020. [Cited: 07 14, 2021.] Https://Www.Orlandosentinel.Com/News/Florida/Os-Ne-Shark-Updates-Florida-America-Deaths-20200805-Daf5tjjm3ncktp20umfydxasm4-Story.Html.
532. Drury, Colin. Man Becomes Second Victim In 10 Days To Be Killed By Cows In Northern England. *Independent.* [Online] Independent, 09 26, 2020. [Cited: 05 18, 2021.] Https://Www.Independent.Co.Uk/News/Uk/Home-News/Killed-Cows-Walker-Carlisle-Pennine-Way-B620768.Html.
533. Dowling, David. How The Creator Of 'Jaws' Became The Shark's Greatest Defender. *Narratively.* [Online] Narratively, 08 14, 2014. [Cited: 09 29, 2020.] Https://Narratively.Com/How-The-Creator-Of-Jaws-Became-The-Sharks-Greatest-Defender/.
534. Bbc Magazine. How Jaws Misrepresented The Great White. *Bbc Magazine.* [Online] Bbc, 06 09, 2015. [Cited: 09 29, 2020.] Https://Www.Bbc.Com/News/Magazine-33049099.
535. Keegan, Matthew. Shark Finning: Why The Ocean's Most Barbaric Practice Continues To Boom. *The Guardian.* [Online] The Guardian, 07 06, 2020. [Cited: 10 02, 2020.] Https://Www.Theguardian.Com/Environment/2020/Jul/06/Shark-Finning-Why-The-Oceans-Most-Barbaric-Practice-Continues-To-Boom.
536. Denyer, Simon. Home Share Worldviews Even As China Turns Away From Shark Fin Soup, The Prestige Dish Is Gaining Popularity Elsewhere In Asia. *The Washington Post.* [Online] The Washington Post, 02 15, 2018. [Cited: 05 18, 2021.] Https://Www.Washingtonpost.Com/News/Worldviews/Wp/2018/02/14/Even-As-China-Turns-Away-From-Shark-Fin-Soup-The-Prestige-Dish-Is-Gaining-Popularity-Elsewhere-In-Asia/.
537. The Japanese Shark Fishing Industry. *Marine Bio.* [Online] Marine Bio, 07 18, 2010. [Cited: 10 02, 2020.] Https://Marinebio.Org/The-Japanese-Shark-Fishing-Industry/.
538. Hembery, Sandra. Dozens Of Dead Sharks Have Washed Up On A Welsh Beach. *Wales Online.* [Online] Wales Online, 05 15, 2019. [Cited: 10 02, 2020.] Https://Www.Walesonline.Co.Uk/News/Dozens-Dead-Sharks-Washed-Up-16279344.
539. *Trends In Shark Bycatch Research: Current Status And Research Needs.* Cooke, Juan M. Molina And Steven J. S.L. : Rev Fish Biol Fisheries, 2012, Vol. 22.
540. Wwf. Saving Sharks. *Panda.* [Online] Wwf, 05 11, 2006. [Cited: 10 05, 2020.] Https://Wwf.Panda.Org/Wwf_News/?68540/Saving-Sharks-With-Magnets.
541. Bettis, Stephanie M. *Shark Bycatch In Commercial Fisheries: A Global Perspective.* Davie, Florida : Nova Southeastern University, 2017.
542. Johnson, Ellen. Why We Need Sharks For Healthy Oceans—And A Healthy Planet. *Mystic Aquarium.* [Online] Mystic Aquarium, 07 26, 2017. [Cited: 10 05, 2020.] Https://Www.Mysticaquarium.Org/2017/07/26/Need-Sharks-Healthy-Oceans-Healthy-Planet/#:-:Text=Sharks%20keep%20ocean%20ecosystems%20in%20balance&Text=Because%20sharks%20directly%20or%20indirectly,Their%20prey%20species%20through%20fear..
543. Zahid, Aisha. Coronavirus: Half A Million Sharks 'Could Be Killed For Vaccine', Experts Warn. *Sky News.* [Online] Sky News, 09 28, 2020. [Cited: 10 14, 2020.] Https://News.Sky.Com/Story/Coronavirus-Half-A-Million-Sharks-Could-Be-Killed-For-Vaccine-Experts-Warn-12083167?Fbclid=Iwar1_Mph6gisq-Lc4nsn1rkm1zqfzypra1r49mn1afocr255hnd1k-3de6as.
544. Carrington, Damian. Global Warming Of Oceans Equivalent To An Atomic Bomb Per Second. *The Guardian.* [Online] The Guardian, 01 07, 2019. [Cited: 05 18, 2021.] Https://Www.Theguardian.Com/Environment/2019/Jan/07/Global-Warming-Of-Oceans-Equivalent-To-An-Atomic-Bomb-Per-Second.
545. Freeman., Devon Godek And Andrew M. Physiology, Diving Reflex. *Ncbi.* [Online] 08 24, 2020. [Cited: 10 07, 2020.] Https://Www.Ncbi.Nlm.Nih.Gov/Books/Nbk538245/#:-:Text=The%20diving%20reflex%20commonly%20referred,In%20response%20to%20water%20submersion..
546. Jones, Nicola. As Waters Warm, Ocean Heatwaves Are Growing More Severe . *Yale Environment 360.* [Online] Yale School Of The Environment, 10 13, 2020. [Cited: 10 14, 2020.] Https://E360.Yale.Edu/Features/As-Waters-Warm-Ocean-Heatwaves-Are-Growing-More-Severe?
547. Monnier, Jen. Marine Heat Waves Are Becoming More Common And Intense. What Can We Do To Minimize Harm? *Ensia.* [Online] Ensia, 08 11, 2020. [Cited: 10 13, 2020.] Https://Ensia.Com/Features/Ocean-Heat-Waves-Marine-Ecosystems/.
548. Gibbens, Sarah. Ocean Heat Waves Are Killing Underwater Life, Threatening Biodiversity. *National Geographic.* [Online] National Geographic, 03 04, 2019. [Cited: 10 15, 2020.] Https://Www.Nationalgeographic.Com/Environment/2019/03/Ocean-Heat-Waves-Threaten-Sea-Life-Biodiversity/.
549. Ewing-Chow, Daphne. Marine Heat Waves Are Putting Caribbean Fisheries In Hot Water . *Forbes.* [Online] Forbes, 09 30, 2020. [Cited: 10 15, 2020.] Https://Www.Forbes.Com/Sites/Daphneewingchow/2020/09/30/Marine-Heat-Waves-Are-Putting-Caribbean-Fisheries-In-Hot-Water/#63ce43c54691.
550. Quartz India. *"There Is No Fish In The Ocean": The Future Has Arrived In This Indian Fishing Village.* [Online] Quartz India, 02 19, 2010. [Cited: 10 16, 2020.] Https://Qz.Com/India/1804562/For-Indias-Karnataka-Fishermen-Climate-Changes-Just-Barren-Sea/
551. Holden, Emily. Temperatures Of Deepest Ocean Rising Quicker Than Previously Thought. *The Guardian.* [Online] The Guardian, 10 14, 2020. [Cited: 10 15, 2020.] Https://Www.Theguardian.Com/Environment/2020/Oct/14/Enormous-Amount-Of-Heat-Even-Deepest-Ocean-Is-Warming-Study.

552. National Ocean Service. What Is A Red Tide? *National Ocean Service.* [Online] National Ocean Service. [Cited: 05 18, 2021.] Https://Oceanservice.Noaa.Gov/Facts/Redtide.
553. Corbett, Jessica. Scientists Behind New Study Warn Increasingly Stable Oceans Are 'Very Bad News'. *Eco Watch.* [Online] Eco Watch, 09 29, 2020. [Cited: 06 12, 2022.] Https://Www.Ecowatch.Com/Stable-Oceans-Climate-Crisis-2647855196.Html.
554. Adam, David. Explainer: How Does Climate Change Affect The Ocean? *China Dialogue Ocean.* [Online] China Dialogue, 09 25, 2020. [Cited: 10 17, 2020.] Https://Chinadialogueocean.Net/15101-How-Does-Climate-Change-Affect-The-Ocean/.
555. Upton, John. Bleak Picture Of Great Barrier Reef's Changing Chemistry. *Climate Central.* [Online] Climate Central, 02 24, 2016. [Cited: 06 12, 2022.] Https://Www.Climatecentral.Org/News/Bleak-Picture-Of-Great-Barrier-Reefs-Future-20060.
556. *Calcium Carbonate Production, Coral Cover And Diversity Along A Distance Gradient From Stone Town: A Case Study From Zanzibar, Tanzania.* Al, Natalia Herran Et. Lausanne : Frontiers In Marine Science, 2017.
557. Davidson, Jordan. Mysterious Disease Ravages Caribbean Reefs. *Ecowatch.* [Online] Ecowatch, 06 11, 2019. [Cited: 10 17, 2020.] Https://Www.Ecowatch.Com/Mysterious-Disease-Caribbean-Reefs-2639164702.Html?Rebelltitem=2#Rebelltitem2.
558. Visontay, Elias. Great Barrier Reef Corals Have More Than Halved In Past 25 Years, Study Shows. *The Guardian.* [Online] The Guardian, 10 14, 2020. [Cited: 10 17, 2020.] Https://Www.Theguardian.Com/Environment/2020/Oct/14/Great-Barrier-Reef-Corals-Have-More-Than-Halved-In-Past-25-Years-Study-Shows
559. Chinn, Hannah. 'Dire Outlook': Scientists Say Florida Reefs Have Lost Nearly 98% Of Coral This Article Is More Than 5 Mo. *The Guardian.* [Online] The Guardian, 11 18, 2020. [Cited: 05 18, 2021.] Https://Www.Theguardian.Com/Environment/2020/Nov/18/Coral-Reefs-Florida-Dire-Outlook
560. Holthaus, Eric. Heartbroken Scientists Lament The Likely Loss Of 'Most Of The World's Coral Reefs' . *Grist.* [Online] Grist, 01 05, 2018. [Cited: 10 17, 2020.] Https://Grist.Org/Article/Heartbroken-Scientists-Lament-The-Likely-Loss-Of-Most-Of-The-Worlds-Coral-Reefs/
561. American Geophysical Union. Warming, Acidic Oceans May Nearly Eliminate Coral Reef Habitats By 2100. *Advancing Earth And Space Science.* [Online] American Geophysical Union, 02 17, 2020. [Cited: 10 17, 2020.] Https://News.Agu.Org/Press-Release/Warming-Acidic-Oceans-May-Nearly-Eliminate-Coral-Reef-Habitats-By-2100/.
562. National Ocean Service. How Much Oxygen Comes From The Ocean? *National Ocean Service.* [Online] National Ocean Service. [Cited: 05 18, 2021.] Https://Oceanservice.Noaa.Gov/Facts/Ocean-Oxygen.Html#:~:Text=At%20least%20half%20of%20earth's%20oxygen%20comes%20from%20the%20ocean.&Text=Scientists%20estimate%20that%2050%2d80,Some%20bacteria%20that%20can%20photosynthesize..
563. Reuters Graphics. Carbon Recyclers: How Ocean Ecosystems Help Fight Climate Change. *Reuters Graphics.* [Online] Reuters Graphics. [Cited: 05 18, 2021.] Https://Graphics.Reuters.Com/Climate-Change/Seagrass/Xklvyyeqjvg/.
564. Pancia, Anthony. Blue Whale, World's Largest Animal, Caught On Camera Having A Poo. *Abc News.* [Online] Abc News, 11 16, 2019. [Cited: 06 12, 2022.] Https://Www.Abc.Net.Au/News/2019-11-16/Blue-Whale-Worlds-Largest-Animal-Caught-On-Camera-Having-A-Poo/11708368.
565. Angus, Ian. Triple Crisis In The Anthropocene Ocean. Part Two: Running Low On Oxygen. *Resilence.* [Online] Resilience, 09 22, 2020. [Cited: 10 16, 2020.] Https://Www.Resilience.Org/Stories/2020-09-22/Triple-Crisis-In-The-Anthropocene-Ocean-Part-Two-Running-Low-On-Oxygen/.
566. Erickson-Davis, Morgan. New Study Finds Mangroves May Store Way More Carbon Than We Thought. *Mongabay.* [Online] Mongabay, 05 02, 2018. [Cited: 05 19, 2021.] Https://News.Mongabay.Com/2018/05/New-Study-Finds-Mangroves-May-Store-Way-More-Carbon-Than-We-Thought/.
567. The Economist. Prawns And Shrimps Can Be Four Times Worse For The Environment Than Beef — But Artificial Alternatives Could Help. *I News.* [Online] I News, 07 10, 2020. [Cited: 05 19, 2021.] Https://Inews.Co.Uk/Inews-Lifestyle/Food-And-Drink/Shrimps-Prawns-Climate-Change-Mangroves-397502.
568. Medianet. Queensland Prawns Suffering Horrific Eye Mutilations . *Medianet.* [Online] Medianet, 09 06, 2017. [Cited: 05 20, 2021.] Https://Www.Medianet.Com.Au/Releases/142686/.
569. Bassem, Samah M. Water Pollution And Aquatic Biodiversity. *Water Pollution And Aquatic Biodiversity.* Http://Medcraveonline.Com/Bij/Water-Pollution-And-Aquatic-Biodiversity.Html, 2020, Vol. 4, 1.
570. Tickner D, Opperman Jj, Abell R, Et Al. Bending The Curve Of Global Freshwater Biodiversity Loss: An Emergency Recovery Plan. *Bioscience.* 2020, Vol. 70, 4.
571. Shams M. Galib, A.B.M. Mohsin And Md. Taskin Parvez Et Al. Municipal Wastewater Can Result In A Dramatic Decline In Freshwater Fishes: A Lesson From A Developing Country. *Knowledge & Management Of Aquatic Ecosystems.* Https://Www.Kmae-Journal.Org/Articles/Kmae/Full_Html/2018/01/Kmae180036/Kmae180036.Html, 2018, Vol. 37, 419.
572. Mapes, Lynda V. Drugs Found In Puget Sound Salmon From Tainted Wastewater. *The Seattle Times.* [Online] The Seattle Times, 02 23, 2016. [Cited: 10 27, 2020.] Https://Www.Seattletimes.Com/Seattle-News/Environment/Drugs-Flooding-Into-Puget-Sound-And-Its-Salmon/.
573. Gómez-Upegui, Salomé. Fish On Drugs: Cocktail Of Medications Is 'Contaminating Ocean Food Chain' . *The Guardian.* [Online] The Guardian, 04 29, 2022. [Cited: 04 30, 2022.] Https://Www.Theguardian.Com/Environment/2022/Apr/29/Drugs-Medications-Contaminate-Ocean-Food-Chains-Fish-Florida-Bonefish.
574. Laville, Sandra. Shocking State Of English Rivers Revealed As All Of Them Fail Pollution Tests. *The Guardian.* [Online] The Guardian, 09 17, 2020. [Cited: 10 27, 2020.] Https://Www.Theguardian.Com/Environment/2020/Sep/17/Rivers-In-England-Fail-Pollution-Tests-Due-To-Sewage-And-Chemicals.
575. —. Water Firms Discharged Raw Sewage Into English Waters 400,000 Times Last Year. *The Guardian.* [Online] The Guardian, 03 21, 2021. [Cited: 06 18, 2021.] Https://Www.Theguardian.Com/Environment/2021/Mar/31/Water-Firms-Discharged-Raw-Sewage-Into-English-Waters-400000-Times-Last-Year.

576. Carrington, Damian. Fishery Collapse 'Confirms Silent Spring Pesticide Prophecy'. *The Guardian.* [Online] The Guardian, 10 31, 2019. [Cited: 10 27, 2020.] Https://Www.Theguardian.Com/Environment/2019/Oct/31/Fishery-Collapse-Confirms-Silent-Spring-Pesticide-Prophecy.

577. Milman, Oliver. Pollution From Car Tires Is Killing Off Salmon On Us West Coast, Study Finds. *The Guardian.* [Online] The Guardian, 12 03, 2020. [Cited: 05 19, 2021.] Https://Www.Theguardian.Com/Environment/2020/Dec/03/Coho-Salmon-Pollution-Car-Tires-Die-Off

578. Renner, Rebecca. More Than 430 Manatees Have Perished In 2021. Why Are They Dying? *National Geographic.* [Online] National Geographic, 03 18, 2021. [Cited: 04 14, 2022.] Https://Www.Nationalgeographic.Com/Animals/Article/430-Florida-Manatees-Have-Died-In-2021.

579. Florida Fish And Wildlife Conservation Commission Marine Mammal Pathobiology Laboratory. *Preliminary 2021 Manatee Mortality Table By County.* S.L. : Florida Fish And Wildlife Conservation Commission Marine Mammal Pathobiology Laboratory, 2021. Https://Myfwc.Com/Media/25429/Yeartodate.Pdf.

580. Webber, Kathleen. How Fast Fashion Is Killing Rivers Worldwide. *Ecowatch.* [Online] Ecowatch, 03 22, 2017. [Cited: 10 27, 2020.] Https://Www.Ecowatch.Com/Fast-Fashion-Riverblue-2318389169.Html.

581. Loisel, Angela Gallego-Sala And Julie. Guest Post: How Human Activity Threatens The World's Carbon-Rich Peatlands. *Carbonbrief.* [Online] Carbonbrief, 12 21, 2020. [Cited: 05 19, 2021.] Https://Www.Carbonbrief.Org/Guest-Post-How-Human-Activity-Threatens-The-Worlds-Carbon-Rich-Peatlands?

582. International Union For Conservation Of Nature (Iucn). Peatland Ecosystems. *International Union For Conservation Of Nature (Iucn).* [Online] International Union For Conservation Of Nature (Iucn). [Cited: 05 19, 2021.] Https://Www.Iucn.Org/Commissions/Commission-Ecosystem-Management/Our-Work/Cems-Specialist-Groups/Peatland-Ecosystems.

583. Harris, Nancy. Destruction Of Tropical Peatland Is An Overlooked Source Of Emissions. *World Resources Institute.* [Online] World Resources Institute, 04 21, 2016. [Cited: 05 19, 2021.] Https://Www.Wri.Org/Insights/Destruction-Tropical-Peatland-Overlooked-Source-Emissions.

584. Dalton, Jane. Ban On Burning Grouse Moor Peatlands 'Not Enough To Tackle Climate Crisis'. *Independent.* [Online] Independent, 01 30, 2021. [Cited: 05 19, 2021.] Https://Www.Independent.Co.Uk/Climate-Change/News/Ban-Peat-Land-Grouse-Moor-Bog-Burn-Fire-B1794947.Html.

585. Carrington, Damian. Sales Of Peat Compost To Gardeners To Be Banned From 2024. *The Guardian.* [Online] The Guardian, 05 18, 2021. [Cited: 05 19, 2021.] Https://Www.Theguardian.Com/Environment/2021/May/18/Sales-Of-Peat-Compost-To-Gardeners-To-Be-Banned-From-2024

586. Goodd, *A Global Dataset Of More Than 38,000 Georeferenced Dams.* Mark Mulligan, Arnoput Van Soesbergen And Leonardo Saenz. 31, S.L. : Springer Nature Limited, 2020, Vol. 7.

587. Lovgren, Stevan. Two-Thirds Of The Longest Rivers No Longer Flow Freely—And It's Harming Us. *National Geographic.* [Online] National Geographic, 05 08, 2019. [Cited: 11 24, 2020.] Https://Www.Nationalgeographic.Com/Environment/2019/05/Worlds-Free-Flowing-Rivers-Mapped-Hydropower/.

588. Main, Douglas. The Chinese Paddlefish, One Of World's Largest Fish, Has Gone Extinct . *National Geographic.* [Online] National Geographic, 01 08, 2020. [Cited: 11 24, 2020.] Https://Www.Nationalgeographic.Com/Animals/2020/01/Chinese-Paddlefish-One-Of-Largest-Fish-Extinct/.

589. Aguilera, Jasmine. 'The Numbers Are Just Horrendous.' Almost 30,000 Species Face Extinction Because Of Human Activity. *Time.* [Online] Time, 07 18, 2019. [Cited: 11 24, 2020.] Https://Time.Com/5629548/Almost-30000-Species-Face-Extinction-New-Report/?Fbclid=Iwar1k8uvv5aiyjaanw1_Vexaqb6m_Vkcol_Duqoxt4mrau78qzqvnxrp81jo.

590. Diehn, Sonya Angelica. Five Ways Mega-Dams Harm The Environment. *Dw .* [Online] Dw, 06 25, 2020. [Cited: 11 24, 2020.] Https://Www.Dw.Com/En/Five-Ways-Mega-Dams-Harm-The-Environment/A-53916579.

591. Afp. Cambodia's Giant Life-Giving Tonle Sap Lake In Peril. *Bangkok Post.* [Online] Bangkok Post, 12 21, 2020. [Cited: 05 19, 2021.] Https://Www.Bangkokpost.Com/World/2038855/Cambodias-Giant-Life-Giving-Tonle-Sap-Lake-In-Peril?Fbclid=Iwar1ysxromqnmloltvqzz8ywqt-Jqctll3kiogxiiwrlmf_3wtilo9s8rciw.

592. Zvomuya, Fidelis. The Zambezi River, Drained Bone Dry . *International Rivers.* [Online] International Rivers, 11 30, 2017. [Cited: 11 25, 2020.] Https://Www.Internationalrivers.Org/News/Blog-The-Zambezi-River-Drained-Bone-Dry/.

593. Reuters Staff. China Eyes 60 Gw Of Hydropower On Tibet's Brahmaputra River - State Media. *Independent.* [Online] Independent, 11 30, 2020. [Cited: 05 19, 2021.] China Eyes 60 Gw Of Hydropower On Tibet's Brahmaputra River - State Media.

594. Withnall, Adam. India Dam Bursts Amid Monsoon Deluge, Flooding Seven Villages And Sweeping Away Homes. *Independent.* [Online] Independent, 07 03, 2019. [Cited: 05 19, 2021.] Https://Www.Independent.Co.Uk/News/World/Asia/India-Dam-Burst-Flooding-Deaths-Monsoon-Rain-Tiware-Maharashtra-Mumbai-A8986036.Html.

595. Muktarsingh, Victoria Elms And Natasha. Uttarakhand Dam Disaster: What Caused India's Deadly Flood? *Sky News.* [Online] Sky News, 02 16, 2021. [Cited: 05 19, 2021.] Https://News.Sky.Com/Story/Uttarakhand-Dam-Disaster-What-Caused-Indias-Deadly-Flood

596. Borunda, Brian Clark Howard & Alejandra. 8 Mighty Rivers Run Dry From Overuse . *National Geographic.* [Online] National Geographic. [Cited: 11 24, 2020.] Https://Www.Nationalgeographic.Com/Environment/Photos/Rivers-Run-Dry/.

597. Langford, David Lewis & John. Why Has The Darling Dried Up? . *Inside Story.* [Online] Inside Story, 05 08, 2019. [Cited: 11 24, 2020.] Https://Insidestory.Org.Au/How-Come-The-Darlings-Dried-Up/.

598. Un Department Of Economic And Social Affairs. Water Scarcity. *Un.* [Online] Un, 11 24, 2014. [Cited: 11 25, 2020.] Https://Www.Un.Org/Waterforlifedecade/Scarcity.Shtml#:~:Text=Did%20you%20know%3f,Living%20under%20water%20stressed%20conditions..

599. Axelsen, Winnie. Worldwide Water Shortage By 2040 . *Science Daily.* [Online] Science Daily, 07 29, 2014. [Cited: 05 19, 2021.] Https://Www.Sciencedaily.Com/Releases/2014/07/14

600. Carrington, Damian. Avoiding Meat And Dairy Is 'Single Biggest Way' To Reduce Your Impact On Earth. *The Guardian.* [Online] The Guardian, 05 31, 2018. [Cited: 11 25, 2020.] Https://Www.Theguardian.Com/Environment/2018/May/31/Avoiding-Meat-And-Dairy-Is-Single-Biggest-Way-To-Reduce-Your-Impact-On-Earth.

601. Borunda, Alejandra. How Beef Eaters In Cities Are Draining Rivers In The American West . *National Geographic.* [Online] National Geographic, 03 02, 2020. [Cited: 11 25, 2020.] Https://Www.Nationalgeographic.Com/Science/2020/03/Burger-Water-Shortages-Colorado-River-Western-Us/.

602. Oppenlander, Richard. Freshwater Depletion: Realities Of Choice. *Comfortably Unaware.* [Online] Comfortably Unaware, 11 26, 2014. [Cited: 11 25, 2020.] Http://Comfortablyunaware.Com/Blog/Freshwater-Depletion-Realities-Of-Choice/.

603. Farah, Troy. Us Rivers And Lakes Are Shrinking For A Surprising Reason: Cows. *The Guardian.* [Online] The Guardian, 07 02, 2020. [Cited: 11 25, 2020.] Https://Www.Theguardian.Com/Environment/2020/Jul/02/Agriculture-Cattle-Us-Water-Shortages-Colorado-River.

604. One Green Planet. How Fracking And Animal Agriculture Is Draining All The Water In Drought-Ridden California. *One Green Planet.* [Online] One Green Planet, 2014. [Cited: 11 26, 2020.] Https://Www.Onegreenplanet.Org/Environment/Fracking-Animal-Agriculture-Drought-California-Water/.

605. Food & Agriculture Organisation. Livestock's Long Shadow: Livestock's Role In Water Depletion And Pollution. *Food & Agriculture Organisation.* [Online] Food & Agriculture Organisation, 2006. [Cited: 11 25, 2020.] Http://Www.Fao.Org/3/a0701e/A0701e04.Pdf.

606. New Zealand: Gdp Share Of Agriculture . *Thye Global Economy.* [Online] The Global Economy, 2018. [Cited: 11 26, 2020.] Https://Www.Theglobaleconomy.Com/New-Zealand/Share_Of_Agriculture/.

607. Infoshare Stats. Agricultural And Horticultural Land Use. *Stats Nz.* [Online] Infoshare Stats Nz, 2016. [Cited: 11 26, 2020.] Https://Www.Stats.Govt.Nz/Topics/Agriculture.

608. Ropere Consulting. Dairy Farms Using Same Amount Of Water As 60 Million People . *Scoop Business.* [Online] Scoop Business, 09 18, 2017. [Cited: 11 26, 2020.] Https://Www.Scoop.Co.Nz/Stories/Bu1709/S00523/Dairy-Farms-Using-Same-Amount-Of-Water-As-60-Million-People.Htm.

609. Roy, Eleanor Ainge. 'Their Birthright Is Being Lost': New Zealanders Fret Over Polluted Rivers. *The Guardian.* [Online] The Guardian, 03 04, 2019. [Cited: 11 26, 2020.] Https://Www.Theguardian.Com/Environment/2019/Mar/04/Their-Birthright-Is-Being-Lost-New-Zealanders-Fret-Over-Polluted-Rivers.

610. Menon, Praveen. New Zealand's Deteriorating Water Raises A Stink. *Rnz.* [Online] Rnz, 01 21, 2019. [Cited: 11 26, 2020.] Https://Www.Rnz.Co.Nz/News/National/380544/New-Zealand-S-Deteriorating-Water-Raises-A-Stink#:~:Text=Coli%20bacteria%2c%20an%20indicator%20of,Past%20for%20contaminating%20these%20waters.&Text=Only%20a bout%2015%20percent%20of,Through%20dairy%20farms%2c%20ohe.

611. Tomazin, Farrah. Public Health Fears Rise As Cows Lay Waste To Rivers. *The Sydney Morning Herald.* [Online] The Sydney Morning Herald, 09 25, 2011. [Cited: 11 26, 2020.] Https://Www.Smh.Com.Au/Environment/Sustainability/Public-Health-Fears-Rise-As-Cows-Lay-Waste-To-Rivers-20110924-1kqth.Html.

612. Colley, Claire. River Pollution Leads To Welsh Demand For Halt To Intensive Chicken Farms . *The Guardian.* [Online] The Guardian, 10 05, 2020. [Cited: 11 26, 2020.] Https://Www.Theguardian.Com/Environment/2020/Oct/05/River-Pollution-Leads-To-Welsh-Demand-For-Halt-To-Intensive-Poultry-Units.

613. Yeoman, Barry. Here Are The Rural Residents Who Sued The World's Largest Hog Producer Over Waste And Odors – And Won. . *The Fern.* [Online] The Fern, 12 20, 2019. [Cited: 11 26, 2020.] Https://Thefern.Org/2019/12/Rural-North-Carolinians-Won-Multimillion-Dollar-Judgments-Against-The-Worlds-Largest-Hog-Producer-Will-Those-Cases-Now-Be-Overturned/.

614. Wilson, Sacoby. Rural Americans' Struggles Against Factory Farm Pollution Find Traction In Court. *Greenbiz.* [Online] Greenbiz, 08 06, 2018. [Cited: 05 19, 2021.] Https://Www.Greenbiz.Com/Article/Rural-Americans-Struggles-Against-Factory-Farm-Pollution-Find-Traction-Court.

615. Duke Health News. N.C. Residents Living Near Large Hog Farms Have Elevated Disease, Death Risks. *Duke Department Of Surgery.* [Online] Duke Surgery: Duke University School Of Medicine, 09 19, 2018. [Cited: 05 19, 2021.] Https://Surgery.Duke.Edu/News/Nc-Residents-Living-Near-Large-Hog-Farms-Have-Elevated-Disease-Death-Risks.

616. Marx, David Jackson And Gary. Spills Of Pig Waste Kill Hundreds Of Thousands Of Fish In Illinois . *Chicago Tribune.* [Online] Chicago Tribune, 08 05, 2016. [Cited: 11 26, 2020.] Https://Www.Chicagotribune.Com/Investigations/Ct-Pig-Farms-Pollution-Met-20160802-Story.Html.

617. Foodprint. How Industrial Agriculture Affects Our Water. *Foodprint.* [Online] Grace Communications Foundation . [Cited: 11 26, 2020.] Https://Foodprint.Org/Issues/How-Industrial-Agriculture-Affects-Our-Water/.

618. Vaughan, Adam. Meat And Dairy Production Emit More Nitrogen Than Earth Can Cope With Read More: Https://Www.Newscientist.Com/Article/2248000-Meat-And-Dairy-Production-Emit-More-Nitrogen-Than-Earth-Can-Cope-With/#Ixzz6es8x6yck. *New Scientist.* [Online] New Scientist, 07 06, 2020. [Cited: 11 26, 2020.] Https://Www.Newscientist.Com/Article/2248000-Meat-And-Dairy-Production-Emit-More-Nitrogen-Than-Earth-Can-Cope-With/?Fbclid=Iwar2bosd2bsuagrjwmynsc5xblsgxisaxhwmlljıdeuzbn2fopzs1uvhbjja#Ixzz6s2axucxe.

619. Gustin, Georgina. As The Livestock Industry Touts Manure-To-Energy Projects, Environmentalists Cry 'Greenwashing'. *Inside Climate News.* [Online] Inside Climate News, 12 07, 2020. [Cited: 05 19, 2021.] Https://Insideclimatenews.Org/News/07122020/Livestock-Industry-Manure-Energy-Natural-Gas/.

620. Hendry, Lisa. Why Are Birds The Only Surviving Dinosaurs? *Natural History Museum*. [Online] Natural History Museum. [Cited: 12 17, 2020.] Https://Www.Nhm.Ac.Uk/Discover/Why-Are-Birds-The-Only-Surviving-Dinosaurs.Html#:~:Text=The%20beginning%20of%20birds,About%20150%20million%20years%20old..

621. Ferrer, Benjamin. Rising Uk Welfare Poultry Supply Pushes Out Intensive Factory Farming, Says Bcc . *Foodingredientsfirst*. [Online] Foodingredientsfirst, 06 02, 2020. [Cited: 12 19, 2020.] Https://Www.Foodingredientsfirst.Com/News/Rising-Uk-Welfare-Poultry-Supply-Pushes-Out-Intensive-Factory-Farming-Says-Bcc.Html.

622. Matsuo, Yohei. Japan's Eco-Friendly Chicken And Eggs Receive New Label. *Nikkei Asia*. [Online] Nikkei Asia, 01 14, 2020. [Cited: 12 19, 2020.] Https://Asia.Nikkei.Com/Business/Agriculture/Japan-S-Eco-Friendly-Chicken-And-Eggs-Receive-New-Label.

623. Briggs, Helen. 'Planet Of The Chickens': How The Bird Took Over The World . *Bbc*. [Online] Bbc, 12 12, 2018. [Cited: 12 21, 2020.] Https://Www.Bbc.Com/News/Science-Environment-46506184.

624. Gandhi, Maneka. Here's Why Beak Trimming Of Chicks In Poultry Farms Is A Senseless Act Of Cruelty . *Firstpost*. [Online] Firstpost, 08 22, 2016. [Cited: 12 17, 2020.] Https://Www.Firstpost.Com/Living/Heres-Why-Beak-Trimming-Of-Chicks-In-Poultry-Farms-Is-A-Senseless-Act-Of-Cruelty-2969954.Html.

625. Lawrence, Felicity. If Consumers Knew How Farmed Chickens Were Raised, They Might Never Eat Their Meat Again . *The Guardian*. [Online] The Guardian, 04 24, 2016. [Cited: 12 17, 2020.] Https://Www.Theguardian.Com/Environment/2016/Apr/24/Real-Cost-Of-Roast-Chicken-Animal-Welfare-Farms.

626. Peta. Chickens Used For Food. *Peta*. [Online] Peta. [Cited: 12 17, 2020.] Https://Www.Peta.Org/Issues/Animals-Used-For-Food/Factory-Farming/Chickens/.

627. Rspca. Laying Hens - Farming (Egg Production). *Rspca*. [Online] Rspca. [Cited: 12 19, 2020.] Https://Www.Rspca.Org.Uk/Adviceandwelfare/Farm/Layinghens/Farming#:~:Text=Around%2038%20million%20commercial%20egg,Eggs%20produced%20in%20battery%20cages.

628. 2014 Livestock-Related Academic Research Commission Survey Report Research On Chicken Egg Sales Trends Centered On Flat-Reared Eggs Possibility Of Supporting Animal Welfare. [Online] [Cited: 19 12 2020.] Https://Www.Alic.Go.Jp/Content/000120154.Pdf?Fbclid=Iwar0c5scazdkouhjcuajn70230cmp8jd68j2ztohencyrsm459nv-Migs9ra.

629. Mendez, Samara. *Us Egg Production Data Set*. Rockville, Md : The Humane League Labs, 2020. E00801.

630. Thompson, Julier. Here's What Farms Do To Hens Who Are Too Old To Lay Eggs. *Huffington Post*. [Online] Huffington Post, 05 14, 2018. [Cited: 12 17, 2020.] Https://Www.Huffpost.Com/Entry/Egg-Laying-Hens

631. Capps, Ashley. Eggs: What Are You Really Eating? . *Free From Harm*. [Online] Free From Harm, 02 12, 2014. [Cited: 12 17, 2020.] Https://Freefromharm.Org/Eggs-What-Are-You-Really-Eating

632. United Poultry Concerns. Forced Molting. *United Poultry Concerns*. [Online] United Poultry Concerns, 02 03, 2018. [Cited: 12 17, 2020.] Https://www.upc-online.org/molting/

633. World Animal Production. 10 Facts You Should Know About Factory-Farmed Chickens. *World Animal Production*. [Online] World Animal Production, 11 29, 2016. [Cited: 12 17, 2020.] Https://Www.Worldanimalprotection.Org/News/10-Facts-You-Should-Know-About-Factory-Farmed-Chickens.

634. Stop Force Feeding. Foie Gras Facts. *Stop Force Feeding*. [Online] Stop Force Feeding. [Cited: 12 17, 2020.] Https://Www.Stopforcefeeding.Com/Facts#:~:Text=Over%20a%20three%20to%20four,Lbs)%20of%20grain%20every%20year..

635. Gandhi, Maneka. A Gruesome Fact On 'Pate Foie Gras'. *The Statesmen*. [Online] The Statesmen, 05 12, 2019. [Cited: 12 17, 2020.] Https://Www.Thestatesman.Com/Supplements/8thday/Gruesome-Fact-Pate-Foie-Gras-1502754358.Html.

636. Dalton, Jane. Chicken From Uk Supermarkets And Fast-Food Chains 'Fuelling Mass Forest Loss In South America'. *Independent*. [Online] Independent, 01 01, 2020. [Cited: 12 17, 2020.] Https://Www.Independent.Co.Uk/Home-News/Chicken-Forests-Greenpeace-Supermarket-Fast-Food-Wildlife-Trees-Climate-Deforestation-A9292576.Html.

637. Reidy, Susan. Japan Expected To Import More Soybeans. *World-Grain*. [Online] World-Grain, 09 04, 2020. [Cited: 12 17, 2020.] Https://Www.World-Grain.Com/Articles/13533-Japan-Expected-To-Import-More-Soybeans.

638. Ackerman, Jennifer. Describing Someone As "Birdbrained" Is Misguided, Unless You're Talking About Emus. *Smithsonian Magazine*. [Online] Smithsonian Magazine, 05 03, 2016. [Cited: 12 19, 2020.] Https://Www.Smithsonianmag.Com/Science-Nature/Describing-Someone-Birdbrained-Misguided-Unless-Youre-Talking-About-Emus-180958981/.

639. Hance, Jeremy. Birds Are More Like 'Feathered Apes' Than 'Bird Brains'. *The Guardian*. [Online] The Guardian, 11 05, 2016. [Cited: 12 19, 2020.] Https://Www.Theguardian.Com/Environment/Radical-Conservation/2016/Nov/05/Birds-Intelligence-Tools-Crows-Parrots-Conservation-Ethics-Chickens.

640. Barras, Colin. Despite What You Might Think, Chickens Are Not Stupid. *Bbc Earth*. [Online] Bbc, 01 11, 2017. [Cited: 12 19, 2020.] Http://Www.Bbc.Com/Earth/Story/20170110-Despite-What-You-Think-Chickens-Are-Not-Stupid#:~:Text=There%20is%20something%20odd%20about,Vertebrate%20species%20on%20the%20planet.&Text=But%20chicken s%20are%2c%20in%20fact%2c%20anything%20but%20dumb.

641. Alexander C. Lees, Lucy Haskell, Tris Allinson, Simeon B. Bezeng, Ian J. Burfield, Luis Miguel Renjifo, Kenneth V. Rosenberg, Ashwin Viswanathan, And Stuart H.M. Butchart. *Annual Review Of Environment And Resources State Of The World's Birds*. S.L. : Annual Reviews, 2022. Https://Doi.Org/10.1146/Annurev-Environ-112420-014642.

642. Weston, Phoebe. The Guardian. *Atlas Reveals Birds Pushed Further North Amid Climate Crisis*. [Online] The Guardian, 12 03, 2020. [Cited: 12 21, 2020.] Https://Www.Theguardian.Com/Environment/2020/Dec/03/Atlas-Reveals-Birds-Pushed-Further-North-Amid-Climate-Crisis-

643. Paul, Kari. Dying Birds And The Fires: Scientists Work To Unravel A Great Mystery. *The Guardian*. [Online] The Guardian, 10 18, 2020. [Cited: 12 22, 2020.] Https://Www.Theguardian.Com/Environment/2020/Oct/18/Dying-Birds-And-The-Fires-Scientists-Work-To-Unravel-A-Great-Mystery.

644. Abc7 News. Hundreds Of Thousands Of Migratory Birds Found Dead In New Mexico, Mystifying Scientistshttps://Abc7.Com/Dead-Birds-New-Mexico-Migratory-Hundreds/6426322/. *Abc7 News.* [Online] Abc7 News, 09 17, 2020. [Cited: 12 22, 2020.] Https://Abc7.Com/Dead-Birds-New-Mexico-Migratory-Hundreds/6426322/.

645. Fish All Gone! Gulf Of Alaska Fishery To Close For The First Time Ever: No More Cod: Salmon All But Gone: Millions Of Small Sea Birds Died Since 2015. *The Big Wobble.* [Online] The Big Wobble, 12 08, 2019. [Cited: 12 21, 2020.] Http://Www.Thebigwobble.Org/2019/12/Fish-All-Gone-Gulf-Of-Alaska-Fishery-To

646. Bbc News. Pacific 'Blob' Heatwave Feared To Have Killed A Million Birds . *Bbc News.* [Online] The Bbc. [Cited: 12 21, 2020.] Https://Www.Bbc.Com/News/Science-Environment-51140869?Intlink_From_Url=&Link_Location=Live-Reporting-Story&Utm_Campaign=Carbon%20brief%20daily%20briefing&Utm_Medium=Email&Utm_Source=Revue%20newsletter.

647. *Long-Term Declines In Common Breeding Seabirds In Japan.* Masayuki Senzaki, Akira Terui , Naoki Tomita Et Al. 3, Cambridge : Bird Conservation International, 2020, Vol. 30.

648. Zabarenko, Deborah. One-Third Of World Fish Catch Used For Animal Feed. *Reuters.* [Online] Reuters, 10 29, 2008. [Cited: 12 21, 2020.] Https://Www.Reuters.Com/Article/Us-Fish-Food-Idustre49soxh20081029.

649. Bowen, Devon. African Penguins Will Be Extinct In Just A Few Years - Unless We Do Something Right Now! *Two Oceans Aquarium.* [Online] Two Oceans Aquarium, 04 25, 2018. [Cited: 12 21, 2020.] Https://Www.Aquarium.Co.Za/Blog/Entry/Prevent-African-Penguin-Extinction-In-8-Years.

650. Taylor, Matthew. Antarctica's King Penguins 'Could Disappear' By The End Of The Century. *The Guardian.* [Online] The Guardian, 02 26, 2018. [Cited: 12 21, 2020.] Https://Www.Theguardian.Com/Environment/2018/Feb/26/Antarcticas-King-Penguins-Could-Disappear-By-The-End-Of-The-Century.

651. Burri, Lena. Antarctic Krill - A Sustainable Protein Source For Fish And Shrimp. *Qrill.* [Online] Qrill, 20. Https://Www.Qrillaqua.Com/Blog-And-News/Krill-A-Sustainable-Protein-Source-For-Shrimp-And-Fish-Feed#:~:Text=Krill%20is%20a%20feed%20attractant%20and%20shrimp%20and%20fish%20growth%20accelerator.&Text=In%20shri mp%20farming%2c%20krill%20is,Quality%20and%20o.

652. Hance, Jeremy. After 60 Million Years Of Extreme Living, Seabirds Are Crashing. *The Guardian.* [Online] The Guardian, 09 22, 2015. [Cited: 12 21, 2020.] Https://Www.Theguardian.Com/Environment/Radical-Conservation/2015/Sep/22/After-60-Million-Years-Of-Extreme-Living-Seabirds-Are-Crashing.

653. Mcveigh, Karen. Industrial Fishing Ushers The Albatross Closer To Extinction, Say Researchers . *The Guardian.* [Online] The Guardian, 01 31, 2019. [Cited: 12 21, 2020.] Https://Www.Theguardian.Com/Environment/2019/Jan/31/Industrial-Fishing-Ushers-Albatross-Closer-To-Extinction-Say-Researchers.

654. Orea R. J. Anderson, Cleo J. Small, John P. Croxall Et Al. Global Seabird Bycatch In Longline Fisheries. *Inter-Research Science Publisher.* [Online] 06 08, 2011. [Cited: 12 21, 2020.] Https://Www.Int-Res.Com/Abstracts/Esr/V14/N2/P91-106/.

655. Davis, Josh. Fishermen Are Cutting Off The Beaks Of Endangered Albatrosses. *Natural History Museum.* [Online] Natural History Museum, 11 17, 2020. [Cited: 12 22, 2020.] Https://Www.Nhm.Ac.Uk/Discover/News/2020/November/Fishermen-Are-Cutting-Off-The-Beaks-Of-Endangered-Albatrosses.Html.

656. Birdlife International. Why We Need Birds (Far More Than They Need Us) . *Birdlife International.* [Online] Birdlife International, 01 04, 2019. [Cited: 12 21, 2020.] Https://Www.Birdlife.Org/Worldwide/News/Why-We-Need-Birds-Far-More-They-Need-Us.

657. Damian Carrngton, Niko Kommenda, Pablo Guierrez And Cath Levett. One Football Pitch Of Forest Lost Every Second In 2017, Data Reveals. *The Guardian.* [Online] The Guardian, 06 27, 2018. [Cited: 12 22, 2020.] Https://Www.Theguardian.Com/Environment/Ng-Interactive/2018/Jun/27/One-Football-Pitch-Of-Forest-Lost-Every-Second-In-2017-Data-Reveals.

658. Woodward, Aylin. North America Has Lost Nearly 3 Billion Birds In The Last 50 Years — Another Sign That We're In The Middle Of A 6th Mass Extinction. *Business Insider.* [Online] Business Insider, 09 20, 2019. [Cited: 12 22, 2020.] Https://Www.Businessinsider.Com/3-Billion-Birds-Disappeared-Across-Us-Canada-Since-1970-2019-9.

659. Geffroy, Laurianne. Where Have All The Farmland Birds Gone? *Cnrs News.* [Online] Cnrs News, 03 21, 2018. [Cited: 12 22, 2020.] Https://News.Cnrs.Fr/Articles/Where-Have-All-The-Farmland-Birds-Gone.

660. Press Association . Farmland Bird Decline Prompts Renewed Calls For Agriculture Overhaul. *The Guardian.* [Online] The Guardian, 11 23, 2017. [Cited: 12 22, 2020.] Https://Www.Theguardian.Com/Environment/2017/Nov/23/Farmland-Bird-Decline-Prompts-Renewed-Calls-For-Agriculture-Overhaul.

661. *The Recent Declines Of Farmland Bird Populations In Britain: An Appraisal Of Causal Factors And Conservation Actions.* Newton, Ian. Peterborough : British Ornithologists Union, 2004, Vol. 146.

662. Youth, Howard. Insect Freefall: What Does It Mean For Birds? *American Bird Conservancy.* [Online] American Bird Conservancy, 08 29, 2019. [Cited: 12 22, 2020.] Https://Abcbirds.Org/Blog/Insect-Freefall/

663. Springer. Birds Eat 400 To 500 Million Tons Of Insects Annually . *Science Daily.* [Online] Springer, 07 09, 2018. [Cited: 12 22, 2020.] Https://www.springer.com/gp/about-springer/media/research-news/all-english-research-news/birds-eat-400-to-500-million-tonnes-of-insects-annually-/15910278

664. Barkham, Patrick. Hooded Vultures 'On Brink Of Extinction' In Africa After Mass Poisoning. *The Guardian.* [Online] The Guardian, 03 06, 2020. [Cited: 12 22, 2020.] Https://Www.Theguardian.Com/World/2020/Mar/06/Hooded-Vultures-Extinction-Africa-Mass-Poisoning.

665. Gray, Dr. Thomas. Empty Forest Syndrome – A Hauntingly Quiet Crisis . *Wildlife Alliance.* [Online] Wildlife Alliance, 04 20, 2017. [Cited: 12 22, 2020.] Https://Www.Wildlifealliance.Org/Empty-Forest-Syndrome/?Fbclid=Iwar1lmrpe61bncp1zskg1o5kqger5-Fsaebgwllef5itkpyd4hitm8y5h_Nc.

666. Bbc. Parrots Found Stuffed In Plastic Bottles In Indonesia. *Bbc.* [Online] Bbc, 11 20, 2020. [Cited: 12 22, 2020.] Https://Www.Bbc.Com/News/World-Asia-55016513?Fbclid=Iwar1qyumbgk6yo8zry9ri8fdlbtfwd3wiassbfedjgic3ehaqcmzpp3jkico.

667. —. Exotic Indonesian Birds Smuggled In Drain Pipes. *Bbc.* [Online] Bbc, 11 17, 2017. [Cited: 12 22, 2020.] Https://Www.Bbc.Com/News/World-Asia-42021915.

668. Bergman, Charles. Wildlife Trafficking . *Smithsonian Magazine.* [Online] Smithsonian Magazine, 12 2009. [Cited: 12 22, 2020.] Https://Www.Smithsonianmag.Com/Travel/Wildlife-Trafficking-149079896/#:~:Text=Birds%20are%20the%20most%20common,Traded%20illegally%20worldwide%20every%20year..

669. Wwf. Draining The Amazon Of Wildlife. *Wwf.* [Online] Wwf. [Cited: 12 22, 2020.] Https://Wwf.Panda.Org/Discover/Knowledge_Hub/Where_We_Work/Amazon/Amazon_Threats/Other_Threats/Illegal_Wildlife_Trade_Amazon/?.

670. Laville, Sandra. Human-Made Materials Now Outweigh Earth's Entire Biomass – Study . *The Guardian.* [Online] The Guardian, 12 09, 2020. [Cited: 12 22, 2020.] Https://Www.Theguardian.Com/Environment/2020/Dec/09/Human-Made-Materials-Now-Outweigh-Earths-Entire-Biomass-Study

671. End Of Neanderthals Linked To Flip Of Earth's Magnetic Poles, Study Suggests. *The Guardian.* [Online] The Guardian, 02 18, 2021. [Cited: 02 19, 2021.] Https://Www.Theguardian.Com/Science/2021/Feb/18/End-Of-Neanderthals-Linked-To-Flip-Of-Earths-Magnetic-Poles-Study-Suggests.

672. *Fermi And Lotka: The Long Odds Of Survival In A Dangerous Universe.* Peacock, Kent A. 3, London : Journal Of The British Interplanetary Society, 2018, Vol. 71.

673. Mann, Michael. *The Sources Of Social Power: A History Of Power From The Beginning To A.D. 1760.* New York : Cambridge University Press, 1986. 978-0-521-30851-9.

674. Jackson, Henry. New Data Reveal The Hidden Mechanisms Of The Collapse Of The Roman Empire. *Resilience.* [Online] Resilience, 05 29, 2018. [Cited: 02 11, 2021.] Https://Www.Resilience.Org/Stories/2018-05-29/New-Data-Reveal-The-Hidden-Mechanisms-Of-The-Collapse-Of-The-Roman-Empire/.

675. Neale, Jonathan. Lithium, Batteries And Climate Change. *Climate & Capitalism.* [Online] Climate & Capitalism, 02 11, 2021. [Cited: 02 15, 2021.] Https://Climateandcapitalism.Com/2021/02/11/Lithium-Batteries-And-Climate-Change/.

676. Telesur English. Elon Musk Confesses To Lithium Coup In Bolivia. *Telesur English.* [Online] Telesur English, 07 25, 2020. [Cited: 02 15, 2021.] Https://Www.Telesurenglish.Net/News/Elon-Musk-Confesses-To-Lithium-Coup-In-Bolivia-20200725-0010.Html.

677. Pew Research Center. 3. How Religious Commitment Varies By Country Among People Of All Ages . *Pew Research Center.* [Online] Pew Research Center, 06 13, 2018. [Cited: 02 17, 2021.] Https://Www.Pewforum.Org/2018/06/13/How-Religious-Commitment-Varies-By-Country-Among-People-Of-All-Ages/.

678. The Harris Poll. Americans' Belief In God, Miracles And Heaven Declines. *Pr Newswire.* [Online] The Harris Poll, 12 16, 2013. [Cited: 06 15, 2021.] Https://Www.Prnewswire.Com/News-Releases/Americans-Belief-In-God-Miracles-And-Heaven-Declines-236051651.Html.

679. Almond, Philip C. Five Aspects Of Pentecostalism That Shed Light On Scott Morrison's Politics. *The Conversation.* [Online] The Conversation, 05 23, 2019. [Cited: 03 22, 2021.] Https://Theconversation.Com/Five-Aspects-Of-Pentecostalism-That-Shed-Light-On-Scott-Morrisons-Politics-117511.

680. Green, Emma. A Christian Insurrection . *The Atlantic.* [Online] The Atlantic, 01 08, 2021. [Cited: 02 17, 2021.] Https://Www.Theatlantic.Com/Politics/Archive/2021/01/Evangelicals-Catholics-Jericho-March-Capitol/617591/.

681. Hedges, Chris. *War Is A Force That Gives Us Meaning.* New York : Publicaffairs, 2002. 1-58648-049-9 .

682. Karkabi, Barbara. Deepak Chopra Says Christ's Teachings Reach Beyond The Christian Church. *Chron.* [Online] Chron, 03 08, 2008. [Cited: 03 18, 2021.] Https://Www.Chron.Com/Life/Houston-Belief/Article/Deepak-Chopra-Says-Christ-S-Teachings-Reach-1675413.Php.

683. Bryant, Miranda. Catholic Order Pledges $100m In Reparations To Descendants Of Enslaved People. *The Guardian.* [Online] The Guardian, 03 16, 2021. [Cited: 03 17, 2021.] Https://Www.Theguardian.Com/World/2021/Mar/16/Jesuit-Conference-Canada-Us-Catholic-Reparations-Descendants-Enslaved-People.

684. Guardian Staff And Agencies. Racist Extremists Pose Most Deadly Terrorist Threat To Us, Intelligence Report Warns. *The Guardian.* [Online] The Guardian, 03 17, 2021. [Cited: 03 18, 2021.] Https://Www.Theguardian.Com/Us-News/2021/Mar/17/Racist-Extremists-Us-Domestic-Terrorism-Intelligence-Report.

685. Luo, Michael. American Christianity's White-Supremacy Problem. *The New Yorker.* [Online] The New Yorker, 09 02, 2020. [Cited: 02 15, 2021.] Https://Www.Newyorker.Com/Books/Under-Review/American-Christianitys-White-Supremacy-Problem.

686. Dawkins, Richard. *The God Delusion.* New York : Mariner Books, 2008. 978-0-618-68000-9.

687. Filipovic, Jill. A New Poll Shows What Really Interests 'Pro-Lifers': Controlling Women . *The Guardian.* [Online] The Guardian, 08 22, 2019. [Cited: 03 01, 2021.] Https://Www.Theguardian.Com/Commentisfree/2019/Aug/22/A-New-Poll-Shows-What-Really-Interests-Pro-Lifers-Controlling-Women.

688. Worldometers. Abortions Worldwide This Year:. *Worldometers.* [Online] Worldometers, 03 01, 2021. [Cited: 03 01, 2021.] Https://Www.Worldometers.Info/Abortions/.

689. Butt, Riazat. Pope Claims Condoms Could Make African Aids Crisis Worse. *The Guardian.* [Online] The Guardian, 03 17, 2009. [Cited: 03 01, 2021.] Https://Www.Theguardian.Com/World/2009/Mar/17/Pope-Africa-Condoms-Aids.

690. Stille, Alexander. What Pope Benedict Knew About Abuse In The Catholic Church . *The New Yorker.* [Online] The New Yorker, 01 14, 2016. [Cited: 03 01, 2021.] Https://Www.Newyorker.Com/News/News-Desk/What-Pope-Benedict-Knew-About-Abuse-In-The-Catholic-Church.

691. Hitchens, Christopher. *The Missionary Position: Mother Teresa In Theory And Practice*. London : Verso Books, 1995. 1-85984-054-X.

692. Ariely, Dan. #30 - 92y With Dan Ariely. *Yuval Noah Harari: Getting Interviewed*. 22:00~, Singapore : Spotify, 03 02, 2021.

693. Richard, Matthieu. *A Plea For The Animals*. Boulder : Shambhala Publications, Inc., 2014. 9781611803051.

694. Konisky, David. Science Daily. *Survey Results Show Christians Becoming Less Concerned About The Environment*. [Online] Science Daily, 01 23, 2018. [Cited: 03 06, 2021.] Https://Www.Sciencedaily.Com/Releases/2018/01/180123113020.Htm.

695. *The Impact Of Religious Faith On Attitudes To Environmental Issues And Carbon Capture And Storage (Ccs) Technologies: A Mixed Methods Study*. R.Jones, Aimie L.B.Hope And Christopher. S.L. : Technology In Society, 2014, Vol. 38.

696. Matthews, Dylan. 9 Reasons Christopher Columbus Was A Murderer, Tyrant, And Scoundrel. *Vox*. [Online] Vox, 10 12, 2015. [Cited: 03 18, 2021.] Https://Www.Vox.Com/2014/10/13/6957875/Christopher-Columbus-Murderer-Tyrant-Scoundrel.

697. Szarek, Jessica. Bolivia's Massive Supply Of Lithium, And The Implications For The Local Community Of Potosi. *The Upstream Journal*. [Online] The Upstream Journal, 11 11, 2019. [Cited: 03 16, 2021.] Https://Www.Upstreamjournal.Org/Bolivia-Lithium/

698. Francia, Luis H. *A History Of The Philippines: From Indios Bravos To Filipinos*. New York : Peter Maya Publishers, 2010. 978-1-59020-285-2.

699. Pilger, John. John Pilger Q&A: 'Us Missiles Are Pointed At China'. *Aljazeera*. [Online] Aljazeera, 12 06, 2017. [Cited: 03 24, 2021.] Https://Www.Aljazeera.Com/Features/2017/12/6/John-Pilger-Qa-Us-Missiles-Are-Pointed-At-China#:-:Text=The%20us%20%E2%80%9cpivot%20to%20asia,Since%20the%20second%20world%20war.&Text=%E2%80%9cameric an%20bases%20form%20a%20giant,And%20beyond%2c%E2%80%9d%20pil.

700. Library Of Congress. Religion And The Founding Of The American Republic. *Library Of Congress*. [Online] Library Of Congress. [Cited: 05 20, 2021.] Https://Www.Loc.Gov/Exhibits/Religion/Rel01.Html.

701. Plymouth Patuxet. Mayflower And Mayflower Compact. *Plymouth Patuxet*. [Online] Plymouth Patuxet. [Cited: 05 20, 2021.] Https://Www.Plimoth.Org/Learn/Just-Kids/Homework-Help/Mayflower-And-Mayflower-Compact#:-:Text=After%20more%20than%20months,Cod%20on%20november%2011%2c%201620..

702. Cliopatria. How Many People Are Descended From The Mayflower Passengers? *History News Network*. [Online] History News Network, 10 23, 2004. [Cited: 05 21, 2021.] Https://Historynewsnetwork.Org/Blog/7360.

703. Johnson, Perry Miller And Thomas H. Introduction: The Puritans. *English Department Brooklyn College*. [Online] English Department Brooklyn College, 01 23, 2005. [Cited: 05 19, 2021.] Http://Academic.Brooklyn.Cuny.Edu/English/Melani/English2/Puritans_Intro.Html.

704. History.Com Editors. Pequot Massacres Begin. *History*. [Online] A&E Television Networks, 03 01, 2010. [Cited: 05 19, 2021.] Https://Www.History.Com/This-Day-In-History/Pequot-Massacres-Begin#:-:Text=On%20may%2026%2c%201637%2c%20two,A%20handful%20of%20its%20inhabitants.&Text=On%20july%2028%2c%2 0a%20third,War%20came%20to%20an%20end..

705. Zinn, Howard. *A Peoples History Of The United States*. New York : Harper Perennial, 1980. 978-0060838652.

706. Fausz, J. Frederick. An "Abundance Of Blood Shed On Both Sides". *The Virginia Magazine*. 1990, Vol. 98, 1.

707. Hume, Ivor Noël. "We Are Starved". *Colonial Williamsburg*. [Online] The Colonial Williamsburg Foundation, 2007. [Cited: 05 20, 2021.] Https://Web.Archive.Org/Web/20180306161448/Http://Www.History.Org/Foundation/Journal/Winter07/Starving.Cfm.

708. European Colonizers Killed So Many Native Americans That It Changed The Global Climate, Researchers Say. *Cnn World*. [Online] Cnn World, 02 02, 2019. [Cited: 05 20, 2021.] Https://Edition.Cnn.Com/2019/02/01/World/European-Colonization-Climate-Change-Trnd/Index.Html.

709. *Puritanism, Evangelicalism And The Evangelical Protestant Tradition*. Coffey, John. Leicester : Paper For Ets Conference, 2005.

710. Rice, Doyle. Was One Of Your Ancestors On The Mayflower? You Can Find Out Now . *Usa Today* . [Online] Usa Today , 06 13, 2018. [Cited: 06 18, 2022.] Https://Www.Usatoday.Com/Story/News/Nation/2018/06/13/Mayflower-Ancestors-Pilgrim-Database/699277002/.

711. Tharoor, Shashi. *An Era Of Darkness: The British Empire In India*. Mumbai : Aleph Book Company, 2016. 978-93-83064-65-6.

712. Facing History And Ourselves. "Expansion Was Everything". *Facing History And Ourselves*. [Online] Facing History And Ourselves. [Cited: 03 24, 2021.] Https://Www.Facinghistory.Org/Holocaust-And-Human-Behavior/Chapter-2/Expansion-Was-Everything.

713. Bouamane, Geoffrey Jones And Loubna. *"Power From Sunshine": A Business History Of Solar Energy* . Cambridge : Harvard Business School, 2012. Https://Www.Hbs.Edu/Ris/Publication%20files/12-105.Pdf.

714. Dearden, Nick. Africa Is Not Poor, We Are Stealing Its Wealth. *Aljazeera*. [Online] Aljazeera, 05 24, 2017. [Cited: 12 30, 2020.] Https://Www.Aljazeera.Com/Opinions/2017/5/24/Africa-Is-Not-Poor-We-Are-Stealing-Its-Wealth.

715. Mckie, Robin. Child Labour, Toxic Leaks: The Price We Could Pay For A Greener Future. *The Guardian*. [Online] The Guardian, 01 03, 2021. [Cited: 01 04, 2021.] Https://Www.Theguardian.Com/Environment/2021/Jan/03/Child-Labour-Toxic-Leaks-The-Price-We-Could-Pay-For-A-Greener-Future.

716. Burgis, Tom. *The Looting Machine Warlords, Oligarchs, Corporations, Smugglers, And The Theft Of Africa's Wealth*. New York : Public Affairs, 2015. 978-1-61039-439-0.

717. Jason Hickel, Dyland Sullivan And Huzaifa Zoomkawala. Rich Countries Drained $152tn From The Global South Since 1960. *Aljazeera*. [Online] Aljazeera, 05 06, 2021. [Cited: 05 19, 2021.] Https://Www.Aljazeera.Com/Opinions/2021/5/6/Rich-Countries-Drained-152tn-From-The-Global-South-Since-1960.

718. Murphy, Richard. *Dirty Secrets: How Tax Havens Destroy The Economy*. London : Verso, 2017. 978-1-78663-167-1.

719. Wikipedia. List Of People Named In The Panama Papers. *Wikipedia*. [Online] Wikipedia. [Cited: 12 30, 2020.] Https://En.Wikipedia.Org/Wiki/List_Of_People_Named_In_The_Panama_Papers.

720. Paradise Papers Reporting Team. Paradise Papers: Tax Haven Secrets Of Ultra-Rich Exposed . *Bbc News*. [Online] Bbc Panorama, 11 06, 2017. [Cited: 12 30, 2020.] Https://Www.Bbc.Com/News/Uk-41876942.
721. Garside, Juliette. Fund Run By David Cameron's Father Avoided Paying Tax In Britain. *The Guardian*. [Online] The Guardian, 04 04, 2016. [Cited: 12 30, 2020.] Https://Www.Theguardian.Com/News/2016/Apr/04/Panama-Papers-David-Cameron-Father-Tax-Bahamas#:-:Text=David%20cameron's%20father%20ran%20an,Bishop%20%E2%80%93%20to%20sign%20its%20paperwork.&Text=The%20guardian%20has%20confirmed%20that,The%20uk%20on%20its%20.
722. Shaxson, Nicholas. Tackling Tax Havens. *International Monetary Fund*. [Online] International Monetary Fund, 09 2019. [Cited: 12 30, 2020.] Https://Www.Imf.Org/External/Pubs/Ft/Fandd/2019/09/Tackling-Global-Tax-Havens-Shaxon.Htm.
723. Rajeev Syal, Simon Bowers And Patrick Wintour. 'Big Four' Accountants 'Use Knowledge Of Treasury To Help Rich Avoid Tax'. *The Guardian*. [Online] The Guardian, 04 26, 2013. [Cited: 12 30, 2020.] Https://Www.Theguardian.Com/Business/2013/Apr/26/Accountancy-Firms-Knowledge-Treasury-Avoid-Tax.
724. Wright, Oliver. Top Accountancy Firms Accused Of Exploiting Tax Laws They Helped To Draft. *Independent*. [Online] Independent, 01 31, 2013. [Cited: 12 30, 2020.] Https://Www.Independent.Co.Uk/News/Uk/Home-News/Top-Accountancy-Firms-Accused-Exploiting-Tax-Laws-They-Helped-Draft-8476028.Html.
725. *Ending The Deepression Through Planned Obsolescence.* London, Bernard. New York : University Of Wisconsin, 1932. Https://Babel.Hathitrust.Org/Cgi/Pt?Id=Wu.89097035273&View=1up&Seq=6.
726. Kwitny, Jonathan. The Great Transportation Conspiracy. [Book Auth.] W.B. Thompson. *Controlling Technology: Contemporary Issues.* Buffalo : Prometheus Books, 1991.
727. Gershon, Livia. The Birth Of Planned Obsolescence. *Jstor Daily* . [Online] Jstor, 04 10, 2017. [Cited: 03 27, 2021.] Https://Daily.Jstor.Org/The-Birth-Of-Planned-Obsolescence/.
728. Leonard, Annie. Story Of Stuff, Referenced And Annotated Script. *Story Of Stuff.* [Online] Story Of Stuff. [Cited: 03 27, 2021.] Https://Www.Storyofstuff.Org/Wp-Content/Uploads/2020/01/Storyofstuff_Annotatedscript.Pdf.
729. Mirowski, Philip. *The Road From Mont Pelerin: The Making Of The Neoliberal Thought Collective.* Cambridge : Harvard University Press, 2015. 978-0-674-03318-4.
730. Monbiot, George. *Out Of The Wreckage.* London : Verso, 2017. 13-978-1-78663-288-3.
731. Klein, Naomi. *The Shock Doctrine.* New York : Metropolitan Books, 2007. 978-1-4299-1948-7 .
732. Amnesty International. Life Under Pinochet: "They Were Taking Turns To Electrocute Us One After The Other" . *Amnesty International.* [Online] Amnesty International, 09 11, 2013. [Cited: 03 28, 2021.] Https://Www.Amnesty.Org/En/Latest/News/2013/09/Life-Under-Pinochet-They-Were-Taking-Turns-Electrocute-Us-One-After-Other/.
733. Military Wiki. Augusto Pinochet. *Military Wiki.* [Online] Military Wiki. [Cited: 03 28, 2021.] Https://Military.Wikia.Org/Wiki/Augusto_Pinochet.
734. Salinas, Eva. Chile — From Neoliberal To Social Policy Experiment? *Open Canada.* [Online] Open Canada, 10 31, 2019. [Cited: 03 28, 2021.] Https://Opencanada.Org/Chile-From-Neoliberal-To-Social-Policy-Experiment/.
735. Jenkins, Simon. How Margaret Thatcher's Falklands Gamble Paid Off. *The Guardian.* [Online] The Guardian, 04 09, 2013. [Cited: 03 30, 2021.] Https://Www.Theguardian.Com/Politics/2013/Apr/09/Margaret-Thatcher-Falklands-Gamble.
736. Inman, Phillip. Poorer Countries Spend Five Times More On Debt Than Climate Crisis – Report. *The Guardian.* [Online] The Guardian, 10 27, 2021. [Cited: 11 08, 2021.] Https://Www.Theguardian.Com/Environment/2021/Oct/27/Poorer-Countries-Spend-Five-Times-More-On-Debt-Than-Climate-Crisis-Report#:-:Text=Lower%20income%20countries%20spend%20five,A%20leading%20anti%20poverty%20charity.&Text=However%2 c%20the%20241.4m%20ann.
737. Busby, Mattha. Four-Fifths Of Sudan's £861m Debt To Uk Is Interest. *The Guardian.* [Online] The Guardian, 03 28, 2021. [Cited: 04 04, 2021.] Https://Www.Theguardian.Com/World/2021/Mar/28/Four-Fifths-Of-Sudans-861m-Debt-To-Uk-Is-Interest
738. Searcy, Kim. Sudan In Crisis . *Origins: Current Events In Historical Perspective .* [Online] Origins: Current Events In Historical Perspective , 07 2019. [Cited: 04 04, 2021.] Https://Origins.Osu.Edu/Article/Sudan-Darfur-Al-Bashir-Colonial-Protest
739. Chestney, Nina. 100 Million Will Die By 2030 If World Fails To Act On Climate: Report. *Reuters.* [Online] Reuters, 09 26, 2012. [Cited: 04 12, 2021.] Https://Www.Reuters.Com/Article/Us-Climate-Inaction-Idusbre8801hg20120925.
740. Spratt, David. At 4°C Of Warming, Would A Billion People Survive? What Scientists Say. . *Climate Code Red.* [Online] Climate Code Red, 08 18, 2019. [Cited: 04 11, 2021.] Http://Www.Climatecodered.Org/2019/08/At-4c-Of-Warming-Would-Billion-People.
741. Rich, Nathaniel. Losing Earth: The Decade We Almost Stopped Climate Change. *New York Times.* [Online] New York Times, 01 08, 2018. [Cited: 04 13, 2021.] Https://Www.Nytimes.Com/Interactive/2018/08/01/Magazine/Climate-Change-Losing-Earth.Html?Fbclid=Iwar3kvq6fxwsqm5l6pefqux8r0-_Yxs7zvboiqb83xn7nnwd1czv5gprtzi8.
742. Pbs. Interviews James Hansen. *Pbs Frontline.* [Online] Pbs. [Cited: 04 12, 2021.] Https://Www.Pbs.Org/Wgbh/Pages/Frontline/Hotpolitics/Interviews/Hansen.Html.
743. Dockrill, Peter. The Coal Industry Was Well Aware Of Climate Change Predictions Over 50 Years Ago. *Science Alert.* [Online] Science Alert, 11 26, 2019. [Cited: 04 12, 2021.] Https://Www.Sciencealert.Com/Coal-Industry-Knew-About-Climate-Change-In-The-60s-Damning-Revelations-Show#:-:Text=Environment-,The%20coal%20industry%20was%20well%20aware%20of,Predictions%20over%2050%20years%20ago&Text=A%20rediscovered%20journal%20from%20th.

744. Bengtsson, Suzanne Goldenberg And Helena. Biggest Us Coal Company Funded Dozens Of Groups Questioning Climate Change. *The Guardian*. [Online] The Guardian, 06 13, 2016. [Cited: 04 12, 2021.] Https://Www.Theguardian.Com/Environment/2016/Jun/13/Peabody-Energy-Coal-Mining-Climate-Change-Denial-Funding.

745. Conway, Naomi Oreskes And Eric M. *The Merchants Of Doubt: How A Handful Of Scientists Obscured The Truth On Issues From Tobacco Smoke To Climate Change.* New York : Bloomsbury Press, 2011. 978-1608193943 .

746. Hall, Shannon. Exxon Knew About Climate Change Almost 40 Years Ago . *Scientific American*. [Online] Scientific American, 10 26, 2015. [Cited: 04 14, 2021.] Https://Www.Scientificamerican.Com/Article/Exxon-Knew-About-Climate-Change-40-Years-Ago/.

747. Climate Reality Project. The Climate Denial Machine: How The Fossil Fuel Industry Blocks Climate Action. . *Climate Reality Project*. [Online] Climate Reality Project, 09 15, 2019. [Cited: 04 14, 2021.] Https://Www.Climaterealityproject.Org/Blog/Climate-Denial-Machine-How-Fossil-Fuel-Industry-Blocks-Climate-Action.

748. Climate Files. 1988 Shell Confidential Report "The Greenhouse Effect". *Climate Files*. [Online] Climate Files. [Cited: 04 14, 2021.] Http://Www.Climatefiles.Com/Shell/1988-Shell-Report-Greenhouse/.

749. Etal, Scott Waldman And Benjamin Hulac. This Is When The Gop Turned Away From Climate Policy. *E&E News*. [Online] E&E News, 12 05, 2018. [Cited: 04 14, 2021.] Https://Www.Eenews.Net/Stories/1060108785/.

750. Curry, Rex. Exxon's Climate Denial History: A Timeline . *Greenpeace*. [Online] Greenpeace. [Cited: 04 14, 2021.] Https://Www.Greenpeace.Org/Usa/Ending-The-Climate-Crisis/Exxon-And-The-Oil-Industry-Knew-About-Climate-Change/Exxons-Climate-Denial-History-A-Timeline/.

751. Keane, Phoebe. How The Oil Industry Made Us Doubt Climate Change . *Bbc News*. [Online] Bbc News, 09 20, 2020. [Cited: 04 14, 2021.] Https://Www.Bbc.Com/News/Stories-53640382.

752. Cook, John. A Brief History Of Fossil-Fuelled Climate Denial . *The Conservation*. [Online] The Conservation, 06 21, 2016. [Cited: 04 14, 2021.] Https://Theconversation.Com/A-Brief-History-Of-Fossil-Fuelled-Climate-Denial-61273.

753. Desmog. The Mont Pelerin Society (Mps). *Desmog*. [Online] Desmog. [Cited: 04 14, 2021.] Https://Www.Desmogblog.Com/Mont-Pelerin-Society/.

754. Snowden, Edward. *Permanent Record.* London : Macmillan , 2019. "978-1-5290-3567-4" Excerpt From: Edward Snowden. "Permanent Record." Ibooks. .

755. Cromwell, David Edwards And David. *Propaganda Blitz: How The Corporate Media Distort Reality.* London : Pluto Press, 2018. 978 1 7868 0330 6.

756. Goldenberg, John Vidal And Suzanne. Snowden Revelations Of Nsa Spying On Copenhagen Climate Talks Spark Anger. *The Guardian*. [Online] The Guardian, 01 30, 2014. [Cited: 02 03, 2021.] Https://Www.Theguardian.Com/Environment/2014/Jan/30/Snowden-Nsa-Spying-Copenhagen-Climate-Talks.

757. Frumin, Aliyah. How Much Does It Cost To Win A Seat In Congress? If You Have To Ask... Msnbc. [Online] Msnbc, 03 12, 2013. [Cited: 05 03, 2021.] Https://Www.Msnbc.Com/Hardball/How-Much-Does-It-Cost-Win-Seat-Congre-Msna19696.

758. Economic Times Of India. How Much Does It Cost To Become A Us President? Here's A Look. *Economic Times Of India*. [Online] Economic Times Of India, 09 14, 2020. [Cited: 05 03, 2021.] Https://Economictimes.Indiatimes.Com/News/International/World-News/How-Much-Does-It-Cost-To-Become-A-Us-President-Heres-A-Look/Increasing-Costs/Slideshow/78100825.Cms.

759. Toomey, Diane. How Big Money In Politics Blocked U.S. Action On Climate Change. *Yale Environment 360*. [Online] Yale Environment 360, 05 10, 2017. [Cited: 05 03, 2021.] Https://E360.Yale.Edu/Features/How-Big-Money-In-Politics-Blocked-U-S-Action-On-Climate-Change.

760. Studee. Mps And Their Degrees: Here's Where And What Our Uk Politicians Studied. *Studee*. [Online] Studee, 12 13, 2019. [Cited: 05 04, 2021.] Https://Studee.Com/Media/Mps-And-Their-Degrees-Media/.

761. Coughlan, Sean. Oxbridge 'Over-Recruits From Eight Schools'. *Bbc News*. [Online] Bbc News, 12 07, 2018. [Cited: 05 04, 2021.] Https://Www.Bbc.Co.Uk/News/Education-46470838.

762. Gov Uk. Elitism In Britain, 2019. *Gov Uk*. [Online] Gov Uk, 06 12, 2019. [Cited: 05 04, 2021.] Https://Www.Gov.Uk/Government/News/Elitism-In-Britain-2019.

763. *Private Education And Inequality In The Knowledge Economy.* Evelyn Huber, Jacob Gunderson & John D. Stephens. 2, London : Informa Uk Limited , 2019, Vol. 39. 10.1080/14494035.2019.1636603.

764. Dodd, Vikram. Tackle Poverty And Inequality To Reduce Crime, Says Police Chief. *Guardian*. [Online] Guardian, 04 18, 2021. [Cited: 05 06, 2021.] Https://Www.Theguardian.Com/Uk-News/2021/Apr/18/Tackle-Poverty-And-Inequality-To-Reduce-Says-Police-Chief

765. Hazell, Will. Climate Change: 70% Of Uk Teachers Say They Have Not Been Properly Trained To Teach About Crisis. *Inews*. [Online] Inews, 03 16, 2021. [Cited: 05 05, 2021.] Https://Inews.Co.Uk/News/Education/Climate-Change-Uk-Teachers-Training-Teach-Global-Warming-Schools-914301.

766. Mark Maslin, Alice Larkin, Shaun Fitzgerald Et Al. Universities Must Act Swiftly And Independently On Climate Change. *Times Higher Education*. [Online] Times Higher Education, 10 29, 2019. [Cited: 05 05, 2021.] Https://Www.Timeshighereducation.Com/Opinion/Universities-Must-Act-Swiftly-And-Independently-Climate-Change.

767. Gorvett, Zaria. How A Giant Space Umbrella Could Stop Global Warming. *Bbc Future*. [Online] Bbc, 04 27, 2016. [Cited: 05 05, 2021.] Https://Www.Bbc.Com/Future/Article/20160425-How-A-Giant-Space-Umbrella-Could-Stop-Global-Warming.

768. Nasa. Global Effects Of Mount Pinatubo. *Earth Observatory*. [Online] Nasa, 06 15, 2001. [Cited: 05 05, 2021.] Https://Earthobservatory.Nasa.Gov/Images/1510/Global-Effects-Of-Mount-Pinatubo

769. Radford, Tim. Geoengineering Could Trigger Disaster In Parts Of Africa . *Climate Central*. [Online] Climate Central, 04 07, 2013. [Cited: 05 05, 2021.] Https://Www.Climatecentral.Org/News/Geoengineering-Could-Trigger-Disaster-In-Parts-Of-Africa-15820.

770. Lutken, Thomas. India And Atmospheric Sulfate Injection: A Double-Edged Sword. *Center For Strategic & International Studies.* [Online] Center For Strategic & International Studies, 05 26, 2020. [Cited: 05 05, 2021.] Https://Www.Csis.Org/Blogs/New-Perspectives-Asia/India-And-Atmospheric-Sulfate-Injection-Double-Edged-Sword.

771. Ibbot, Samantha. Solar Geoengineering Not A 'Sensible Rescue Plan', Say Scientists. *Imperial College London.* [Online] Imperial College London, 02 15, 2021. [Cited: 05 05, 2021.] Https://Www.Imperial.Ac.Uk/News/213175/Solar-Geoengineering-Sensible-Rescue-Plan-Scientists/.

772. *Solar Geoengineering May Not Prevent Strong Warming From Direct Effects Of Co2 On Stratocumulus Cloud Cover.* Tapio Schneider, Colleen M. Kaul And Kyle G. Pressel. 48, Washington Dc : National Academy Of Sciences., 2020, Vol. 117. Https://Doi.Org/10.1073/Pnas.2003730117.

773. Ambrose, Julian. Record Metals Boom May Threaten Transition To Green Energy. *The Guardian.* [Online] The Guardian, 05 15, 2021. [Cited: 05 16, 2021.] Https://Www.Theguardian.Com/Business/2021/May/15/Record-Metals-Boom-May-Threaten-Transition-To-Green-Energy.

774. Nunez, Christina. How Green Are Those Solar Panels, Really? *National Geographic.* [Online] National Geographic, 11 12, 2014. [Cited: 05 05, 2021.] Https://Www.Nationalgeographic.Com/Science/Article/141111-Solar-Panel-Manufacturing-Sustainability-Ranking.

775. Vasagar, Jeevan. German Grids Restored To Public Ownership. *Financial Times.* [Online] Financial Times, 11 26, 2013. [Cited: 05 05, 2021.] Https://Www.Ft.Com/Content/2f3b0b1e-4dee-11e3-8fa5-00144feabdc0.

776. Wettengel, Julian. Citizens' Participation In The Energiewende. *Clean Energy Wireac.* [Online] Citizens' Participation In The Energiewende, 10 25, 2018. [Cited: 05 05, 2021.] Https://Www.Cleanenergywire.Org/Factsheets/Citizens-Participation-Energiewende.

777. Colthorpe, Andy. German Town Disconnects From Grid, Goes 100% Renewable - For An Hour. *Energy Storage News.* [Online] Energy Storage News, 12 05, 2019. [Cited: 05 05, 2021.] Https://Www.Energy-Storage.News/News/German-Town-Disconnects-From-Grid-Goes-100-Renewable-For-An-Hour.

778. World Health Organisation. Road Traffic Injuries. *World Health Organisation.* [Online] World Health Organisation, 02 07, 2020. [Cited: 05 05, 2021.] Https://Www.Who.Int/News-Room/Fact-Sheets/Detail/Road-Traffic-Injuries.

779. Oil Conglomerates Made Record Profits In 2021 . *Clean Technica.* [Online] Clean Technica, 03 29, 2021. [Cited: 04 26, 2022.] Https://Cleantechnica.Com/2022/03/29/Oil-Conglomerates-Made-Record-Profits-In-2021

780. Frangoul, Ammar. Coal Helped Drive Energy-Related Co2 Emissions To A Record High Last Year, Research Says. *Cnbc.* [Online] Cnbc, 03 09, 2022. [Cited: 04 26, 2022.] Https://Www.Cnbc.Com/2022/03/09/Energy-Related-Co2-Emissions-Hit-Highest-Ever-Level-In-2021-Iea.Html#:~:Text=The%20iea%20found%20energy%2drelated,Main%20driver%20behind%20the%20growth..

781. Milman, Oiiver. Largest Oil And Gas Producers Made Close To $100bn In First Quarter Of 2022. *The Guardian.* [Online] The Guardian, 05 13, 2022. [Cited: 05 15, 2022.] Https://Www.Theguardian.Com/Business/2022/May/13/Oil-Gas-Producers-First-Quarter-2022-Profits.

782. International Energy Agency. *World Energy Investment 2021.* Paris : International Energy Agency, 2021. Https://Iea.Blob.Core.Windows.Net/Assets/5e6b3821-Bb8f-4df4-a88b-e891cd8251e3/Worldenergyinvestment2021.Pdf.

783. Harvey, Fiona. No New Oil, Gas Or Coal Development If World Is To Reach Net Zero By 2050, Says World Energy Body . *The Guardian.* [Online] The Guardian, 05 18, 2021. [Cited: 04 29, 2022.] Https://Www.Theguardian.Com/Environment/2021/May/18/No-New-Investment-In-Fossil-Fuels-Demands-Top-Energy-Economist.

784. *Unextractable Fossil Fuels In A 1.5 °C World* . Dan Welsby, James Price, Steve Pye & Paul Ekins. S.L. : Nature, 2021, Vol. 597. Https://Www.Nature.Com/Articles/S41586-021-03821-8.

785. Mingle, Jonathan. How U.S. Gas Exports To Europe Could Lock In Future Emissions. *Yale Environment 360.* [Online] Yale Environment 360, 04 21, 2022. [Cited: 04 29, 2022.] Https://E360.Yale.Edu/Features/How-U.S.-Gas-Exports-To-Europe-Could-Lock-In-Future-Emissions

786. Pollin, Robert. Nationalize The U.S. Fossil Fuel Industry To Save The Planet . *The American Prospect.* [Online] The American Prospect, 04 08, 2022. [Cited: 04 29, 2022.] Https://Prospect.Org/Environment/Nationalize-Us-Fossil-Fuel-Industry-To-Save-The-Planet/.

787. Hanna, Thomas. A History Of Nationalization In The United States: 1917-2009. *Democracy Collaborative.* [Online] Democracy Collaborative, 11 04, 2019. [Cited: 04 29, 2022.] Https://Democracycollaborative.Org/Learn/Publication/History-Nationalization-United-States-1917-2009.

788. Consiglio, Dave. 10 Reasons Why Educated Professionals Tend To Have Fewer Kids. *Huffpost.* [Online] Huffpost, 10 06, 2017. [Cited: 05 12, 2021.] Https://Www.Huffpost.Com/Entry/10-Reasons-That-Educated-Professionals-Tend-To-Have_B_59d66a8de4bocf2548b3354c.

789. Wilson, Edward O. *Half-Earth: Our Planet's Fight For Life.* New York : Liveright Publishing Corporation, 2017. 9781631492525.

790. Reuters. Explainer: What To Expect From U.N. Conference On Biodiversity. *Reuters.* [Online] Reuters, 10 08, 2021. [Cited: 10 13, 2021.] Https://Www.Reuters.Com/Business/Environment/What-Expect-Un-Conference-Biodiversity-2021-10-08/.

791. Conservatory Land. Biggest Garden Capitals In The Uk. *Conservatory Land.* [Online] Conservatory Land, 05 2021. [Cited: 06 20, 2022.] Https://Www.Conservatoryland.Com/Blog/Biggest-Garden-Capitals-In-The-Uk/.

792. Iannotti, Marie. How Much To Plant Per Person In The Vegetable Garden. *The Spruce.* [Online] The Spruce, 02 10, 2019. [Cited: 05 05, 2021.] Https://Www.Thespruce.Com/How-Many-Vegetables-Per-Person-In-Garden-1403355.

793. Project Drawdown. Table Of Solutions. *Project Drawdown.* [Online] Project Drawdown. [Cited: 05 06, 2021.] Https://Drawdown.Org/Solutions/Table-Of-Solutions.

794. *Reducing Food's Environmental Impacts Through Producers And Consumers.* Nemecek, J. Poore And T. 6392, S.L. : Science, 2018, Vol. 360. Https://Www.Science.Org/Doi/10.1126/Science.Aaq0216.

795. Johnston, Ian. Carbon Dioxide Must Be Removed From The Atmosphere To Avoid Extreme Climate Change, Say Scientists. *Independent.* [Online] Independent, 07 18, 2017. [Cited: 05 06, 2021.] Https://Www.Independent.Co.Uk/Climate-Change/News/Carbon-Dioxide-Remove-Atmosphere-Climate-Change-Greenhouse-Gas-Scientists-Jim-Hansen-A7847426.Html.

796. *Global Priority Areas For Ecosystem Restoration.* Bernardo B. N. Strassburg, Alvaro Iribarrem, Hawthorne L. Beyer Etal. 14, S.L. : Nature, 2020, Vol. 10. Https://Www.Nature.Com/Articles/S41586-020-2784-9.

797. Rao, Sailesh. Animal Agriculture Is The Leading Cause Of Climate Change A Climate Healers Position Paper. *Climate Healers.* [Online] Climate Healers, 11 2019. [Cited: 09 08, 2021.] Https://Climatehealers.Org/The-Science/Animal-Agriculture-Position-Paper/.

798. Wwf: 60% Of Global Biodiversity Loss Due To Land Cleared For Meat-Based Diets. *Ecowatch.* [Online] Ecowatch, 10 05, 2017. [Cited: 05 06, 02021.] Https://www.ecowatch.com/biodiversity-meat-wwf-2493305671.html#:-:text=What%20is%20this%3F&text=In%20all%2C%20meat%2Dbased%20diets,Williamson%20explained%20to%20the%20Guardian.

799. Kolbert, Elizabeth. *Under A White Sky: The Nature Of The Future.* New York : Random House, 2021. 9780593238776.

800. Oxford Martin School. Plant-Based Diets Could Save Millions Of Lives And Dramatically Cut Greenhouse Gas Emissions. *Oxford Martin School.* [Online] University Of Oxford, 03 21, 2016. [Cited: 05 11, 2021.] Https://Www.Oxfordmartin.Ox.Ac.Uk/News/201603-Plant-Based-Diets/.

801. Carrington, Damian. Healthy Diet Means A Healthy Planet, Study Shows. *The Guardian.* [Online] The Guardian, 10 28, 2019. [Cited: 05 11, 2021.] Https://Www.Theguardian.Com/Environment/2019/Oct/28/Healthy-Diet-Means-A-Healthy-Planet-Study-Shows.

802. Chrisafis, Angelique. French Law Forbids Food Waste By Supermarkets. *Guardian.* [Online] Guardian, 02 04, 2016. [Cited: 05 06, 2021.] Https://Www.Theguardian.Com/World/2016/Feb/04/French-Law-Forbids-Food-Waste-By-Supermarkets.

803. Talkdeath. Environmental Impact Of Funerals Infographic. *Talkdeath.* [Online] Talkdeath, 2021. [Cited: 10 20, 2021.] Https://Www.Talkdeath.Com/Environmental-Impact-Funerals-Infographic/.

804. Hurst, Nathan. A California Startup Is Using Ashes To Protect Forests. *Smithsonian Magazine.* [Online] Smithsonian Magazine, 12 05, 2016. [Cited: 10 20, 2021.] Https://Www.Smithsonianmag.Com/Innovation/California-Startup-Using-Ashes-Protect-Forests-180961289/.

805. Seselja, Edwina. Forest Cemeteries Could Revive Koala Habitats, Protecting The Species By Charging $1,100 Per Burial. *Abc News.* [Online] Abc News, 09 24, 2020. [Cited: 10 20, 2021.] Https://Www.Abc.Net.Au/News/2020-09-24/Koala-Habitats-Could-Be-Revived-By-Forest-Cemeteries/12692006.

806. Bowles, Nellie. Could Trees Be The New Gravestones? *New York Times.* [Online] New York Times, 06 12, 2019. [Cited: 10 20, 2021.] Https://Www.Nytimes.Com/2019/06/12/Style/Forest-Burial-Death.Html.

807. University Of East Anglia. It's Official -- Spending Time Outside Is Good For You. *Science Daily.* [Online] Science Daily, 07 16, 2018. [Cited: 06 19, 2021.] Https://Www.Sciencedaily.Com/Releases/2018/07/180706102842.Htm.

808. Cockburn, Harry. The Fastest Way To Get Extinction Rebellion To Stop Is To Listen To Them . *The Independent.* [Online] The Independent, 04 13, 2022. [Cited: 04 13, 2022.] Https://Www.Independent.Co.Uk/Independentpremium/News-Analysis/Extinction-Rebellion-Oil-Energy-Ipcc-B2056438.Html?Fbclid=Iwar1bylfu9ovlh3ienjcxawpndk1wg6sawywpykroby-Cls2uytm5crpckjo.

809. Dyke, James. The Most Maddening Thing About Our Climate Change Situation Is That We Aren't Doomed . *Inews.* [Online] Inews, 04 08, 2022. [Cited: 04 21, 2022.] Https://Inews.Co.Uk/Opinion/Columnists/The-Most-Maddening-Thing-About-Our-Climate-Change-Situation-Is-That-We-Arent-Doomed-1561325?Fbclid=Iwar3ramgtnu5yo7blga8hddofuxtu3sjge5djzazww56esu_Cyfyojqho-Ju.

810. Grey, Carmody. Why We Need To Confront Denial Over The Climate Crisis . *The Tablet.* [Online] The Tablet, 04 15, 2022. [Cited: 04 21, 2022.] Https://Www.Thetablet.Co.Uk/Blogs/1/2041/Why-We-Need-To-Confront-Denial-Over-The-Climate-Crisis?

811. United Nations. Meetings Coverage And Press Releases. *United Nations.* [Online] United Nations, 04 04, 2022. [Cited: 04 21, 2022.] Https://Www.Un.Org/Press/En/2022/Sgsm21228.Doc.Htm.

812. Centre For Alternative Technology. Can We Reach Zero Carbon By 2025? *Centre For Alternative Technology.* [Online] Centre For Alternative Technology, 04 25, 2019. [Cited: 05 06, 2021.] Https://Cat.Org.Uk/Can-We-Reach-Zero-Carbon-By-2025/.

813. United Nations Environment Programme. *Emissions Gap Report 2019.* Nairobi : United Nations Environment Programme, 2019. Http://Www.Unenvironment.Org/Emissionsgap.

814. —. *The Heat Is On: A World Of Climate Promises Not Yet Delivered.* Nairobi : United Nations Environment Programme, 2021. Https://Www.Unep.Org/Emissions-Gap-Report-2021.

815. Klebnikov, Sergei. Stopping Global Warming Will Cost $50 Trillion: Morgan Stanley Report. *Forbes.* [Online] Forbes, 10 24, 2019. [Cited: 01 26, 2021.] Https://Www.Forbes.Com/Sites/Sergeiklebnikov/2019/10/24/Stopping-Global-Warming-Will-Cost-50-Trillion-Morgan-Stanley-Report/?Sh=569aba925102.

816. Stockholm International Peace Research Institute. *World Military Spending Rises To Almost $2 Trillion In 2020.* Solna : Stockholm International Peace Research Institute, 2021. Https://Sipri.Org/Media/Press-Release/2021/World-Military-Spending-Rises-Almost-2-Trillion-2020.

817. Oxfam International And Stockholm Environment Institute. *The Carbon Inequality Era.* Oxford : Oxfam International And Stockholm Environment Institute, 2020. Doi: 10.21201/2020.6492.

818. Oxford. *Taxing Extreme Wealth An Annual Tax On The World's Multi-Millionaires And Billionaires: What It Would Raise And What It Could Pay For.* Oxford : Oxfam, 2022. Https://Www.Fightinequality.Org/Sites/Default/Files/2022-01/Taxing-Extreme-Wealth-What-It-Would-Raise-What-It-Could-Pay-For.Pdf.

819. Carlin, David. A 5 Trillion Dollar Subsidy: How We All Pay For Fossil Fuels. *Forbes.* [Online] Forbes, 06 02, 2020. [Cited: 01 26, 2021.] Https://Www.Forbes.Com/Sites/Davidcarlin/2020/06/02/A-5-Trillion-Dollar-Subsidy-How-We-All-Pay-For-Fossil-Fuels/?Sh=22d56ace7ea1.

820. Samuel, Sigal. Everywhere Basic Income Has Been Tried, In One Map. *Vox.* [Online] Vox, 10 20, 2020. [Cited: 05 05, 2021.] Https://Www.Vox.Com/Future-Perfect/2020/2/19/21112570/Universal-Basic-Income-Ubi-Map.

821. Intergovernmental Panel On Climate Change. *Climate Change 2022 Mitigation Of Climate Change.* S.L. : Ipcc, 2022. Https://Report.Ipcc.Ch/Ar6wg3/Pdf/Ipcc_Ar6_Wgiii_Finaldraft_Fullreport.Pdf.

822. Kochhar, Rakesh. Are You In The Global Middle Class? Find Out With Our Income Calculator. *Pew Research Center.* [Online] Pew Research Center, 07 21, 2021. [Cited: 10 27, 2021.] Https://Www.Pewresearch.Org/Fact-Tank/2021/07/21/Are-You-In-The-Global-Middle-Class-Find-Out-With-Our-Income-Calculator/#:-:Text=The%20other%20four%20income%20groups,Income%20on%20more%20than%20%2450..

823. Hickel, Jason. Degrowth: A Response To Branko Milanovic. *Jason Hickel.* [Online] Jason Hickel, 10 27, 2020. [Cited: 10 27, 2021.] Https://Www.Jasonhickel.Org/Blog/Tag/Degrowth.

824. Vega, Malvina. 17+ Statistics About Jobs Lost To Automation And The Future Of Employment. *Tech Jury.* [Online] Tech Jury, 05 07, 2021. [Cited: 05 07, 2021.] Https://Techjury.Net/Blog/Jobs-Lost-To-Automation-Statistics/#:-:Text=What%20percentage%20of%20jobs%20will,Jobs%20lost%20to%20automation%20statistics..

825. Guest Authors. Guest Post: Refining The Remaining 1.5c 'Carbon Budget'. *Carbon Brief.* [Online] Carbon Brief, 01 19, 2021. [Cited: 05 09, 2021.] Https://Www.Carbonbrief.Org/Guest-Post-Refining-The-Remaining-1-5c-Carbon-Budget.

826. The World Bank. Total Greenhouse Gas Emissions (Kt Of Co2 Equivalent). *The World Bank.* [Online] World Resources Institute, 2020. [Cited: 06 21, 2022.] Https://Data.Worldbank.Org/Indicator/En.Atm.Ghgt.Kt.Ce.

827. *Trajectories Of The Earth System In The Anthropocene.* Will Steffen, Johan Rockström, Katherine Richardson Et Al. S.L. : National Academy Of Sciences, 2018. Https://Www.Scienceopen.Com/Document/Read?Vid=598b524f-729d-43fc-8faa-a9c6bf86of40.

828. *Use And Non-Use Value Of Nature And The Social Cost Of Carbon.* Moore, Bernardo A. Bastien-Olvera And Frances C. S.L. : Nature Sustainability, 2020, Vol. 4. Https://Doi.Org/10.1038/S41893-020-00615-0.

829. *Introducing A Global Carbon Allowance Trading System (G-Cats) As An Ecological Alternative To Neoliberalism.* Whalley, Simon. Boston : Science Open, 2022. 10.14293/S2199-1006.1.Sor-.Ppbmmhi.V1.

830. Freeland, Chrystia. *Plutocrats.* New York : Penguin Group, 2012. 978-1-59420-409-8.

831. Shaw, Matt. Billionaire Capitalists Are Designing Humanity's Future. Don't Let Them . *The Guardian.* [Online] The Guardian, 02 05, 2021. [Cited: 02 06, 2021.] Https://Www.Theguardian.Com/Commentisfree/2021/Feb/05/Jeff-Bezos-Elon-Musk-Spacex-Blue-Origin.

832. Survival Condo. Survival Condo. *Survival Condo.* [Online] Survival Condo, 2017. [Cited: 02 06, 2021.] Https://Survivalcondo.Com/.

833. Hannah Ritchie, Edouard Mathieu, Lucas Rodés-Guirao, Cameron Appel, Charlie Giattino, Esteban Ortiz-Ospina, Joe Hasell, Bobbie Macdonald, Diana Beltekian And Max Roser. Coronavirus Pandemic (Covid-19). *Our World In Data.* [Online] Our World In Data, 2020. [Cited: 03 22, 2022.] Https://Ourworldindata.Org/Covid-Vaccinations?Country=Owid_Wrl#Citation.

About the Author

Simon Whalley is a forty-four-year-old husband and father to an eleven-year-old son. He is originally from Wales but has spent the past twenty years in Asia where he has bought and sold Vespas, worked in the Vietnamese shipping industry, and taught English at every level from kindergarten to university in Vietnam, Taiwan and Japan. He has a master's degree in teaching English to Students of Other Languages (TESOL) and currently teaches English at an International Baccalaureate (IB) School on an island in the Seto Sea, Japan. Here, with his wife and son, he looks after a former mandarin farm which is being turned into an organic food forest. He was a co-founder of Extinction Rebellion Japan, creator of the website Foodfacts.jp and a contributing author to Commondreams.org, and the best-selling Carbon Almanac.